The **CROWN** and the **COSMOS**

The CROWN and the COSMOS

ASTROLOGY AND THE POLITICS OF MAXIMILIAN I

DARIN HAYTON

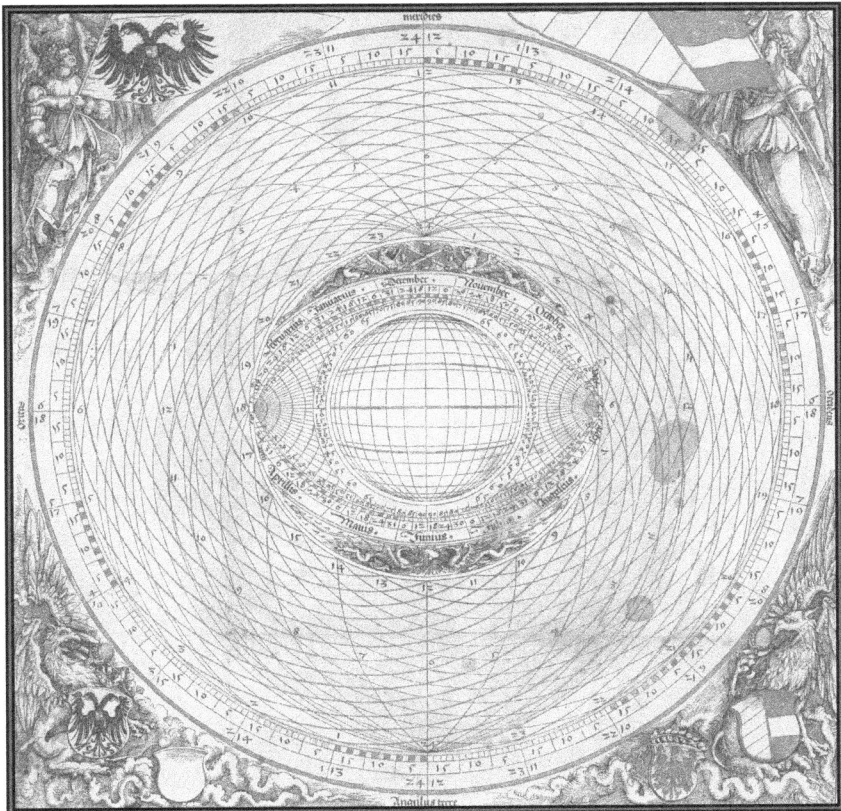

UNIVERSITY OF PITTSBURGH PRESS

Published by the University of Pittsburgh Press, Pittsburgh, Pa., 15260
This paperback edition, Copyright © 2025, University of Pittsburgh Press
Copyright © 2015, University of Pittsburgh Press
All rights reserved
Manufactured in the United States of America
10 9 8 7 6 5 4 3 2 1

Cataloging-in-Publication data is available from the Library of Congress

ISBN 13: 978-0-8229-6787-3
ISBN 10: 0-8229-6787-1

Publisher: University of Pittsburgh Press, 7500 Thomas Blvd., 4th floor, Pittsburgh, PA 15260, United States, www.upittpress.org

EU Authorized Representative: Easy Access System Europe, Mustamäe tee 50, 10621 Tallinn, Estonia, gpsr.requests@easproject.com

To my family

CONTENTS

ACKNOWLEDGMENTS

I alone did not write this book. Like all scholars, I have benefited from the input and assistance of other experts in my field and in closely related fields. I begin by acknowledging my intellectual debts. To Howard Louthan I owe a special thanks. He first brought the Habsburgs to my attention and encouraged me to spend a summer in Vienna exploring Maximilian's court. He guided my early faltering efforts to understand Maximilian's world and then read numerous half-baked versions of this project in the form of dissertation chapters. As my work matured, he remained an enthusiastic supporter. Other mentors and friends at the University of Notre Dame deserve mention. In the History and Philosophy of Science program there, Chris Hamlin, Cornelius O'Boyle, Phil Sloan, and the late Ernan McMullin, all encouraged me to think more critically about my work. I would not have survived without the support and irreverence of a small coterie of friends: Matt Dowd, Sofie Lachapelle, Bonnie Mak, Chris McClellan, and Steve Ruskin. Beyond the HPS program, Dave Jenkins offered invaluable observations and assistance.

Scholars at various archives and institutions across Europe were models of generosity and assistance. In Vienna I was fortunate to work with Helmuth Grössing, whose knowledge of the local archives saved me countless headaches. Karl Vocelka welcomed me into his circle of early modernists and listened to me as I mused about my research. Lonnie Johnson bridged the gap between social and scholarly life. Colleagues at the Seminar für Geistesgeschichte und Philosophie der Renaissance at Ludwig-Maximilians Universität in Munich offered not only helpful comments and feedback but also friendship: Professor Dr. Eckhart Kessler, Dr. Heinrich Kuhn, Zana Dodig, and Dr. Martin Schmeisser. At the Warburg Institute in London Charles Burnett, Isabelle Draelants, David Juste, Jill Kraye, and Alessandro Scaffi all listened to me think through my work, answered my questions, and saved me from making mistakes. While in London I came to know Patrick Curry and Steven Vanden Broecke, who prompted me to think about my work in more nuanced ways. Finally, during a postdoc at the Museum of the History of Science in Oxford I had the opportunity to work with Jim Bennett and Stephen Johnston, who reminded me that instruments and material objects play an important role in my work.

Archival work brings the added joy of meeting other scholars working on similar or related projects. In the long gestation of this book a number of scholars have given freely of their time and expertise. A few of these deserve special thanks here: Monica Azzolini, Peter Barker, Jean-Patrice Boudet, Tony Grafton, Rich Kremer, Darrel Rutkin, Mike Shank, and Bob Westman. More locally, colleagues at the Consortium for the History of Science, Technology and Medicine (formerly the Philadelphia Area Center for the History of Science) have listened to and offered excellent feedback on drafts of chapters: Babak Ashrafi, Jonathan Seitz, Amy Slaton, Elly Truit.

I want to thank the two reviewers who went above and beyond the call of duty in reading drafts of this book and providing detailed comments on every aspect. They prompted me to refine and clarify my argument and have saved me from countless mistakes. Any remaining errors are, of course, my own.

My research was possible only through the generous financial support of various institutions. The Nanovic Institute for European Studies awarded me a Summer Research Grant that allowed me to complete the preliminary research in Vienna. When the project I had proposed fell apart during the first morning in the archive, the Nanovic Institute encouraged me to use the funding to explore other possible research projects, thereby laying the foundation for this book. A Fulbright from the Austrian-American Educational Commission gave me the opportunity to spend a year in Vienna digging through the archives. A summer fellowship from the Herzog August Bibliothek in Wolfenbüttel allowed me to use the rich resources there and provided a base from which to explore other libraries in northern Germany. A Mellon Travel Fellowship from the Department of History at the University of Oklahoma and a generous invitation to give a talk there allowed me to use the university's rich holdings and to begin working through my ideas. The Deutscher Akademischer Austausch Dienst supported my research in Munich and archives in southern Germany. A Sophia Fellowship from the Warburg Institute and Sophia Trust provided the opportunity to spend four months writing at the Warburg. Under Don Howard's understanding and generous directorship, the History and Philosophy of Science Program at the University of Notre Dame supported both research and writing, as did the Edward M. and Ann Uhry Abrams Endowed Fellowship at Notre Dame. A research grant from the Renaissance Society of America allowed me to complete additional research. Finally, generous research funding from Haverford College has enabled me to return a number of times to different archives to check and recheck sources.

It is easy to forget the countless people behind desks and counters or in the stacks and workrooms who make research possible by doing the yeoman's labor of staffing archives, overseeing reading rooms, retrieving books and manuscripts, duplicating material, scanning and photographing sources, processing requests and payments, and other mundane and often invisible tasks. I want to thank the army of directors, archivists, and assistants who helped me locate, retrieve, and reproduce sources in libraries across Europe and the United States. To them I owe a special thanks. The staff at two libraries in particular deserve special mention. First, I had the good fortune to work in the beautiful Österreichische Nationalbibliothek in Vienna. I am grateful to Ernst Gamillscheg, Gertrud Oswald, Eike Zimmer, Hans Peter Zimmer, Mathias Böhm, Ingeborg Formann, Peter Prokop, Konstanze Mittendorfer, and the others whose names I have forgotten or failed to learn. Second, for a number of months the staff at the Bayerische Staatsbibliothek in Munich filled my interminable requests for sources and reproductions. Most recently, Sophie Schrader has been invaluable in helping me complete this book; I regret I cannot thank the others by name. Numerous other libraries allowed me to have access to their materials, often outside normal business hours, or provided copies of items in their collections, saving me the time and expense of traveling to see them. I want to thank: Dr. Brigitte Schürmann of the Staats- und Stadtbibliothek Augsburg; Dr. Paul Berthold Rupp at the Universitätsbibliothek, Augsburg; Renate Bauer at the Landesbibliothek Coburg; Pater Odo Lang, OSB, Maria Einsiedeln Stiftsbibliothek, Einsiedeln; Peter Zerlauth at the Universitäts- und Landesbibliothek Tirol, Innsbruck; Dr. Zdzislaw Pietrzyk, Deputy Director, Manuscripts and Rare Books, Jagiellonian University Library, Krakow; Dr. Rainer Schoch, Dr. Eberhard Slenczka, and Bianca Slowik at the Germanisches Nationalmuseum, Nürnberg; Dr. Adolf Hahnl at Erzabtei St. Peter, Salzburg; Fredi Hächler, Vadianische Sammlung, St. Gall; Dr. Michael Drucker of the Staatliche Bibliothek Regensburg; Dr. Ingrid Kastel and Olga Pohankova at the Albertina, Vienna; the staff at the Universitätsarchiv, Vienna; Dr. Leopold Cornaro at Universitätsbibliothek, Vienna; Henrietta Danker at the Herzog August Bibliothek, Wolfenbüttel; Joanna Madej, Department of Manuscripts, the University Library in Wrocław; Eric Pumroy Director of Special Collections at Bryn Mawr College; Beth Lander at the Historical Medical Library, the College of Physicians of Philadelphia; Kristen van der Veen, The Dibner Library, Smithsonian Institution; Jennifer B. Lee, Rare Book & Manuscript Library, Columbia University; Jennie Rathbun, Houghton Library, Harvard University; Eric Frazier, Boston Public Library; Stephen Greenberg, History of Medical Division, National Library of Medicine; Susan Bales, Rare Book Collection, UNC Chapel Hill;

Robin Rider, Curator of Special Collections at the University of Wisconsin-Madison.

I am indebted to the people at the University of Pittsburgh Press: Beth Davis, Peter Kracht, Sandy Crooms, and especially Abby Collier and Amberle Sherman. They shepherded my book through the various stages, from initial draft through readers' reports and to final revisions. Throughout they offered sage advice and timely encouragement.

Finally, I want to thank my family for putting up with my eccentricities, my frequent research trips, and my work habits. My parents encouraged me to pursue my interests and have offered unwavering support for my work, even if they did not understand why anybody would be interested in Renaissance astrology. Catherine was my constant companion on the journey that produced this book, a journey that lasted many years and took us across two continents. Her understanding and perseverance have reassured me, and her love has fortified me. She deserves as much credit for this book as I do. Pierce and Zoë have been a constant source of joy and much needed distraction. I cannot imagine writing this book without them.

ILLUSTRATIONS

The CROWN
and the COSMOS

Introduction

In 1509 Emperor Maximilian I was bogged down in a war with Venice. When his troops were unable to take the city by force, he appealed directly to its people for their support. On three separate occasions he instructed scholars to compose letters in Italian to be distributed to the Venetians, in which Maximilian praised them for their nobility and honor and reminded them of their former allegiance to his father, Emperor Frederick III, and to the House of Habsburg. Maximilian assured the Venetians that he and his army were there to free them from the tyrants controlling the city, just as he had freed other cities. He guaranteed to restore and protect their traditional laws and to let the Venetians choose their own rulers. He promised to grant the Venetians all the rights, privileges, and exemptions to trade in German cities that other cities in the empire enjoyed.[1] Maximilian had multiple copies of the letters printed, posted, and distributed to the people.[2] We do not know how persuasive the Venetians found his rhetoric, but Maximilian did succeed in bypassing the Venetian government and disseminating his message directly to the citizens. The contemporary Venetian chronicler Marino Sanuto recounts having seen a number of Maximilian's letters posted in various places throughout the city.[3] The emperor had crafted a message, used the latest technology to circumvent a foreign government, and broadcast his message directly to his enemy's citizenry. He offered them a particular image of the war and promised them a better future if they resisted their rulers and supported his cause.

Maximilian not only used what we would now call propaganda in his efforts to persuade enemies and advance his cause in the minds of foreigners; he also used it in his attempts to create an image of his rule and justify his actions in the minds of his own citizens. In 1499, after suffering a defeat by Swiss troops, Maximilian sought to boost morale among German princes and his own soldiers. When ambassadors from Milan offered him various astrological predictions about the successful outcome of his war with the Swiss and the growing tensions with the French, Maximilian apparently did not place much faith in them. The emperor did, however, encourage the ambassadors to continue bringing their predictions, as they were useful for reassuring German princes and his own court that victory over the Swiss was almost certain.[4] In Maximilian's use of these astrological predictions we can see a nuanced understanding of propaganda as a mechanism for shaping domestic opinion. He invoked independent experts who grounded their predictions in what was considered to be an authoritative body of knowledge, in astrology.

Astrology as Natural Knowledge

Propaganda as a category of rhetoric and representation is steeped in modern values and assumptions. We must use the term carefully to avoid treating it as a transhistorical category. But in considering Maximilian's communications as such, we begin to see how he tried to construct and convey authority. *The Crown and the Cosmos* focuses on one aspect of Maximilian's propaganda program: his use of astrology in his efforts to shape public opinion. Various systems of knowledge have long been used to assert and project authority. Well before Maximilian, princes and monarchs had invoked religion, prophetic knowledge and access to the divine, genealogies and historical analysis, and direct appeals to aristocracy to establish, justify, and bolster their rule. Maximilian made astrology an instrument of political power, an innovative use that points to an emerging role for natural knowledge in early modern political discourse. He took advantage of opportunities presented by the emerging print market to enlarge his audience and extend his base of political support, displaying an unprecedented concern with enlisting not merely the political elites, who already had power, but also popular audiences that extended into the lowest strata of society. In his program of political outreach, he enlisted astrology as a vehicle for communicating the Habsburg message to the broadest possible audience. As a traditional and academically respected body of knowledge that was embedded in popular and elite culture, astrology offered Maximilian a tool that used nature as evidence, guide, and justification for political actions. Controlling astrology and the experts who produced astrological knowledge played a key role in Maximilian's politics of representation.

While Ptolemy had distinguished between *astronomia* and *astrologia*, the early sixteenth-century actors in this book used these Latin terms in various and seemingly inconsistent ways. At times, the terms were interchangeable.[5] This flexibility reminds us that the two bodies of knowledge were complementary parts of a larger "science of the stars."[6] The flexibility also warns against reducing either term to its modern translation, astronomy or astrology. Similar terminological problems arise when trying to label the early sixteenth-century actors who engaged in these activities. Contemporaries identified themselves and others by a constellation of terms, such as *mathematicus, physicus, astrologus, astronomus*. Like the terms *astronomia* and *astrologia*, these markers of identification often varied. In this book I have opted to use the term *astrology* for the body of knowledge and *astrologer* for the person who produced that knowledge.[7]

Astrology derived its authority, on the one hand, from its empiricism, its grounding in purportedly objective natural phenomena that everybody could observe. On the other hand, such phenomena required interpretation by expert practitioners. Through them, astrology provided both explanatory and predictive knowledge. For the prince, this dual character of astrology, simultaneously visible and esoteric, paired with astrology's interpretive and predictive functions, made astrology an instrument of cultural persuasion and therefore a powerful political tool. I trace the different ways Maximilian used astrological expertise at various levels of his political program, from his own self-fashioning as both a skilled astrologer and an enthusiastic patron, to his patronage of astrologers who served the emperor's agenda throughout his reign and who communicated that agenda to different audiences that read and annotated their astrological divinations.

What distinguished Maximilian's reliance on astrology from that of his predecessors and contemporaries was his consistent and public use of astrologers and astrology to advance his political programs. While Maximilian's contemporaries relied on the advice of astrological counselors, they did not celebrate their use of such advisers. By contrast, Maximilian drew attention to his own astrological expertise and to his reliance on skilled astrologers. As with his artistic projects, Maximilian oversaw both the astrologers and their products.[8] In the process, Maximilian developed a broad political instrumentality for knowledge about the natural world and the purveyors of that knowledge. His astrologers moved between his court and the University of Vienna. They composed learned poetry and literature that highlighted Maximilian's status as the celestially chosen emperor, produced printed astrological instruments that were distributed to princes and elite courts as diplomatic gifts, strengthened and expanded the astrological curriculum at the university, and composed popular astrological pamphlets, both annual *practica* and prognostications for extraordinary events such as planetary conjunctions and comets. Within a century princes across Europe turned to scientific knowledge to construct their own image, shape public opinion, and advance their own political agendas.[9]

Propaganda, Print, and Early Modern Statecraft

Effective propaganda merges intelligible forms with credible content, plausible facts and evidence, and authoritative systems of knowledge. Considering any aspect of early modern statecraft as an example of propaganda risks distorting the past by viewing it through our modern categories;[10] however, the absence of the term in the early sixteenth century does not mean that early modern princes and audiences failed to recognize attempts at persuasion.[11] In early modern Europe art, ceremony, monuments, poetry, and literature as well as more overtly political rhetoric such as acts, laws, mandates and letters of patent all served as efforts to influence princes, aristocrats, and politically powerful subjects—to persuade them of the legitimacy of one's rule, that a course of action was justified, that one's authority was unassailable, and other such political aims for controlling and shaping one's image.[12] In his politics of representation, his attempt to project political values and shape opinion, Maximilian not only deployed these traditional rhetorical forms but also enlisted astrology and its practitioners, reflecting his understanding of astrology as an authoritative body of knowledge and an expectation that his audiences also considered it as such.

Early modern Europe experienced a profound shift in communication media with the spread of print.[13] Literacy rates were climbing and markets were emerging for printed texts and visual prints.[14] In response to an expanding consumer market, princes developed a broader and more nuanced politics of representation.[15] As they involved wider populations in the political process they put more effort into controlling the information that was transmitted to those audiences.[16] In this changing political space, propaganda became an indispensable tool of statecraft.

Historians of early modern England have detailed the Tudor monarchy's use of propaganda. Henry VII seemed to recognize the importance of representation when he adapted Emperor Maximilian I's coinage practices but lacked the channels available to his successors, who developed a coherent and widespread program.[17] Roy Strong has labeled Henry VIII's construction of a royal image through the patronage of artists and the portraiture of Hans Holbein the first propaganda campaign in English history.[18] Since then scholars have traced the many ways in which Tudor monarchs, especially Henry VIII and Elizabeth I, employed it to define the state and to promote allegiance to the monarch. Kevin Sharpe has recently argued that the best way to understand the Tudor monarchy is through its pervasive politics of representation, through which the Tudors established, sustained, and enhanced their reputations.[19] Controlling their image was particularly important during periods of uncertainty and crisis. While traditional forms of representation such as portraits and images formed the cornerstone of Tudor efforts,[20] scholars highlight the importance of the expanding audience for royal propaganda, increasingly through pamphlets and proclamations.[21] Sharpe grounds his work in the analysis of texts, images, and pageants, showing how the Tudors struggled to persuade their subjects of their right to rule and then exercised authority through communication with and appeal to those same subjects. What makes Sharpe's work so useful is his focus on the concrete mechanisms by which authority was constructed and legitimated. Because legitimation is a cultural process, Sharpe directs our attention to the cultural products that enacted that process—the histories, paintings, legends, and prophecies—whose goal was to make authority seem natural.[22] In France, King Louis XIV and his advisers made royal propaganda and representation omnipresent. Peter Burke's *The Fabrication of Louis XIV* provides a careful analysis of these efforts. Like scholars of early modern England, Burke focuses on various elite forms of propaganda such as pageantry, portraiture, medallions, architectural projects, and court-sponsored poetry and literature; that is, forms of expression commonly associated with aristocratic pastimes that together shaped public opinion, packaged the monarch, and constructed an ideology. His work elucidates the ways that art, magnificence, and charisma served power.[23]

Historians of early modern propaganda and the politics of representation owe a debt to R. J. W. Evans's study on Emperor Rudolf II.[24] Evans traced the ways that Rudolf shaped a coherent political program out of religion, humanist culture, the arts, and occult sciences. Although Evans avoided the term *propaganda*, Karl Vocelka made it the center of his detailed study of

Rudolf II's court. In addition to examining traditional elite forms, Vocelka drew attention to Rudolf's use of popular press.[25] Following Vocelka, Margit Altfahrt considered pamphlets an important vehicle for imperial propaganda at the court of Maximilian II.[26] For both Vocelka and Altfahrt, pamphlets function as important vehicles for princely propaganda during periods of crisis. Maximilian II and Rudolf II used ephemeral print because they recognized that pamphlets had been effective when used against them to foster sedition and unrest, especially among popular audiences.[27] They then sought to use this media to advance their own agenda in these popular audiences.

The Prince and the Public Image

The growing role of propaganda in early modern politics was not lost on sixteenth-century writers. In *The Prince* Machiavelli justified its use in early modern statecraft. He advised the prince to feign and dissimulate, to say whatever served his interests rather than be constrained by what he held to be true. Princes needed to appear to embody a set of virtues and characteristics whether or not they believed in them.[28] The prince's true nature was displaced by the prince's public image. A few years later Thomas Elyot wrote *The Book Named the Governor* in which he urged the monarch to display symbols of power in order to inspire his subjects' reverence and therefore obedience.[29] Both Machiavelli and Elyot point to the expanding role for representation in the exercise of rule as princes sought to gain support from traditional elite audiences and to secure compliance from their subjects in the lower registers of society.

By the mid-sixteenth century, Machiavelli and Elyot were justifying a set of practices that monarchs had already begun to adopt, albeit haltingly. We see in Maximilian an early understanding of the importance of public opinion and the prince's need to shape that opinion. Like other princes of his time, Maximilian considered his own image and representation an important piece of political rhetoric.[30] Compared to his predecessors and contemporaries, Maximilian employed a wider array of tools to construct his image and broader variety of channels to project it. He incorporated printed texts and visual arts, music, theater and ceremony, acts and mandates, and oral recitations into a coherent body of material that celebrated the emperor himself and the House of Habsburg.[31] Expanding what it meant to be a political actor, Maximilian broadcast his message to all registers of society. A letter from 1494 reveals his intended audience. Trying to generate support for a crusade

against the Turks, Maximilian cataloged the many levels of society he sought
to reach and to enlist in his program: "electors, spiritual and worldly, prelates,
counts, freemen, gentlemen, knights, servants, captains, magistrates, guard-
ians, administrators, officials, village mayors, lord mayors, judges, councilors,
citizens and parishioners, and otherwise all others of our and the empire's subjects and followers of whatever dignity, rank and occupation, who come forward or are shown this our royal letter or copy thereof to see or read, our every
grace and good."[32] Maximilian also projected his image through collections
of books, manuscripts, medals and coinage, and patronage of counselors.[33]
He disseminated his message through printed texts, letters of patent, proclamations, broadsheets, and pamphlets. Historians of print have cataloged the
emperor's tireless use of print as a political instrument.[34] Similarly, art historians have detailed Maximilian's efforts, especially the many woodcuts and
other visual representations that came from the emperor's coterie of artists.[35]

Maximilian understood these efforts and the expense associated with them
to be a necessary part of political praxis. In his autobiographical *Weisskunig*
he justified spending any amount of money and effort on crafting his image
by saying that those who failed to create their own memorials were destined
to be forgotten shortly after their death.[36] Maximilian also recognized that his
image was enhanced by being seen as a patron of learned men, and by their
service to him, two points he celebrated throughout his reign. Once again
Machiavelli seemed to confirm what Maximilian already understood: "The
first indications of the intelligence of a ruler are given by the quality of the
men around him. If they are capable and loyal, he should always be taken
to be shrewd, because he was able to recognize their ability and retain their
loyalty."[37] The prince's reputation was linked directly to the reputation and
expertise of the ministers, artists, and scholars he attracted to his court and
supported there.

Early Modern Propaganda and Courtly Science

Earlier scholarship has established that Maximilian exploited print as propaganda and has traced the many specific forms that propaganda has taken.
The emperor relied on dozens of printers to publish hundreds of mandates
and proclamations.[38] From the early 1480s, even before he was elected King of
the Romans, until his death in 1519, Maximilian orchestrated the production
and dissemination of printed materials intended to advance public opinion
of himself, the reputation of the House of Habsburg, and the authority of
the imperial office. Only recently have scholars begun to ask *how* the emper-

or's many projects constituted propaganda or *why* Maximilian's many edicts, pamphlets, broadsheets, and images might have been persuasive. Larry Silver's masterful study of the emperor's artistic program, *Marketing Maximilian*, explains how and why Maximilian expended so much energy on his genealogical projects and on portraying himself engaged in aristocratic pastimes.

The Crown and the Cosmos extends this scholarship both in what it considers propaganda and how that propaganda was intended for multiple and diverse audiences. In particular, I want to move our understanding of early modern propaganda beyond the traditional forms that historians typically understand to constitute a monarch's purview, such as poetry and literature, art and imagery, and the artists and scholars who produced them. In addition to these, Maximilian enlisted cheap ephemeral texts such as astrological pamphlets, wall calendars and timely broadsheets, and paper instruments. Along with these increasingly diverse forms, Maximilian also broadened the content of propaganda as it existed at the time. He recognized the strength of natural knowledge as a source of authority in persuading multiple audiences of his agenda. His use of astrology in light of his efforts to enlist all levels of society in his political program reveals an expanding role for scientific knowledge in politics and in shaping public opinion.

The present work contributes to a rich literature on Maximilian I. For more than a century political historians have assessed Maximilian's effectiveness as a political actor on the European stage and as an agent of political change. In these accounts the emperor is alternately viewed as regressive and an impediment to the formation of a German state, or as progressive and a stimulus for constitutional reform and the development of a multinational empire.[39] Along with these political histories, considerable scholarship has examined the emperor's efforts to construct and disseminate his image through literary and artistic works. Studies have shown how Maximilian used visual and literary arts to memorialize the emperor himself, to justify his claim to the imperial title, and to elevate the Habsburg dynasty.[40] Despite sustained interest in Maximilian's reign, little effort has been made, in any language, to investigate the scientific culture, specifically the astrological culture, at Maximilian's court.[41] No previous scholarship on courtly science has concentrated on Maximilian's court and his patronage practices.[42] Yet in order to understand Maximilian as a political actor we must take seriously his use of the science of astrology, which was highly innovative. The emperor imagined a much broader and more public role for astrology and astrologers in politics than any of his contemporaries.

This book draws on and contributes to scholarship on courtly science that elucidates how noble patronage shaped emerging attitudes about nature and enlisted natural knowledge to achieve commercial and material goals. With their carefully articulated codes of conduct and standards for authority, courts forged new ways of validating and using natural knowledge.[43] More recently, scholars have begun to study how purveyors of that knowledge established and maintained their places at court and the ends to which princes put their expertise.[44] Maximilian's use of astrology is an early example of a prince publicly invoking natural knowledge in the construction of his image, celebrating and rewarding the experts who produced that knowledge, and deploying it in early modern politics.

Despite some excellent early studies on astrology's importance for understanding European history, mainly by art historians and classicists, historians of science have only relatively recently come to terms with astrology's historical importance.[45] Studies have begun to illustrate astrology's central place in shaping how early modern Europeans understood the relationship between humans and the cosmos and how astrologers applied their science.[46] Perhaps the most thorough example is Robert Westman's magisterial *The Copernican Question*. Westman argues that a mixture of pragmatic concerns growing out of Copernicus's experience among Italian prognosticators—including how to go about making better and more accurate astrological predictions, as well as dealing with intellectual challenges, such as how to justify the ordering of the planets—stood at the center of his efforts to formulate a heliocentric model.[47]

Although some scholarship on the history of astrology in the Germanies confronts explicitly the relationship between astrology and politics—often in the context of the planetary conjunctions in 1524—other studies have only implicitly raised questions about the role of astrology in politics, concentrating instead on astrology's intellectual contexts.[48] Still, as Monica Azzolini pointed out in her book on astrology at the Sforza courts, *The Duke and the Stars*, considerable work remains to elucidate astrology's role in politics.[49] Two studies demonstrate how fruitful it is to consider astrology an integral facet of premodern politics. The first, Azzolini's careful analysis of the role of astrology at various Sforza courts in fifteenth-century Milan exposes the many different ways that Sforza dukes used astrology to understand and shape political situations.[50] Second, Michael Ryan's *A Kingdom of Stargazers* sheds light onto three fourteenth-century Aragonese courts, detailing how a strong monarch could use astrology to solidify one's authority and bolster one's rule. By contrast, a weak monarch's predilection for astrology was seen by contemporaries as evidence of an effete and inept ruler.[51] Both studies point to the value astrology could have as an instrument of propaganda. At the same time, they illustrate the very different ways princes who preceded Maximilian I used astrology. In contrast to Maximilian, other monarchs used astrology in an ad hoc way and did not generally draw attention to either their patronage of astrologers or their own knowledge of the subject.

This is the first book to link astrology to the Habsburg courts through the practices and products of a group of scholars active at those courts and at the University of Vienna, who benefited from Habsburg patronage. Such patronage sometimes assumed typical forms such as positions at court, stipends, or titles. Sometimes, however, it did not leave such easily identifiable tracks. And not all forms of patronage produced relationships in which a patron distributed favors to a client who, in turn, served the patron. Sometimes networks of patronage were grounded in shared obligations, mutual aid and benefit, and reciprocity. These more amorphous relationships are often difficult to demonstrate through surviving documents and have to be inferred by contextualizing authors, highlighting their various personal connections, and analyzing their immediate political and social arenas.

I have based my research on extensive manuscript and early printed materials from archives across northern and central Europe, including Austria, England, France, Germany, Poland, and Switzerland. Despite their contemporary importance, many of these sources have escaped scholarly consideration. Along with drawing our attention to these sources and their intellectual and cultural significance, I have tried to convey the sense and complexity of early modern astrology through nuanced translations of relevant and exemplary passages from the German and Latin originals. These translations reveal the contours of early modern astrology and give us an opportunity to watch historical actors teaching, learning, and using their science in some of its many contexts. Manuscript sources include letters and canons written explicitly for the emperor, pedagogical texts written for university students, students' lecture notes, and correspondence between individual astrologers. Printed materials range from technical treatises and university textbooks, at one end of the spectrum, through annual almanacs and ephemerides, to cheap ephemeral pamphlets, wall calendars, and broadsheets at the other. In addition to printed textual sources, I discuss paper astrological instruments and related visual material. Considering these sources as a coherent body of material allows us to reconstruct how the astrologers themselves understood these texts and images, and the relationships between them. In addition, it allows us to trace the lines of influence between the astrologers and the court, and to recover some of the concrete mechanisms that Maximilian used to disseminate his agenda through the various levels of society.

Chapter 1 argues that astrology was central to the emperor's efforts to fashion the ideal "modern" prince. Propaganda is inextricably linked to the image of the prince himself. Maximilian crafted his memorial in both words and visual representations, which were simultaneously idealized monuments shaping how contemporaries viewed him and normative portraits offering a model for his Habsburg successors. Throughout his autobiographical works, Maximilian underscored the importance of astrology.

The coherent system of predictive and explanatory knowledge proffered by astrologers became a cornerstone in Maximilian's courtly politics. Chapter 2 examines how the predictions of pro-Habsburg astrologers aligned with the emperor's goals, both promoting his agenda and advancing their own fortunes. In particular, I focus on the work of two astrologers in the 1490s, Sebastian Brant and Joseph Grünpeck. Brant produced a number of broadsheets supporting Maximilian's war against the French and his efforts to establish a centralized military, while Grünpeck used his astrological explanation of the spread of pox to argue for Maximilian's social reforms and produced pro-Habsburg astrological pamphlets timed to coincide with important moments or struggles throughout Maximilian's reign.

Chapter 3 illustrates how Maximilian relied on the University of Vienna both as a source from which to draw astrologers into his court and as a body of experts who could be tapped for advice and intellectual support in his political endeavors. This chapter details Maximilian's efforts to revitalize the university and to fund a series of institutional developments intended to reestablish the University of Vienna as an important center for teaching astrology and astronomy. It also shows how Maximilian developed patronage practices that extended beyond the court.

Chapter 4 focuses on the astrological instruments produced for Maximilian and important members of his court. In 1506 Andreas Stiborius developed an astrological instrument for use in and around Vienna that facilitated the calculation of propitious moments, which Maximilian used in concluding peace negotiations with the Hungarian forces that year. During the last decade of Maximilian's reign the imperial historian Johannes Stabius produced ornate printed astrological instruments along with his work on the imperial genealogies and Maximilian's *Ehrenpforte*. These instruments were functional devices used to calculate propitious times for various activities, but they were also distributed to important members of the emerging administrative class of lower nobility, imperial free knights, and upper bourgeoisie. These case studies illuminate how Maximilian used astrology to guide political decisions, to bolster his authority among the growing bureaucratic class, and to disseminate his image as a patron and student of astrology to rival courts.

The following chapters trace Maximilian's efforts to disseminate his political message through more popular texts. Capitalizing on the ubiquity of astrology, Maximilian exploited the power of print to publicize his political agenda to various audiences and all levels of society. Pamphlets, broadsheets, and advertisements were posted on notice boards, read out in town hall meetings, sold in the markets, becoming indispensable vehicles for communicating Habsburg and imperial interests. The emperor enlisted the astrologers at his court and the university in his propaganda campaign to promulgate a pro-Habsburg agenda to audiences beyond the narrow confines of elite society. Maximilian's coterie of astrologers used a variety of astrological genres to spread the emperor's message. Chapter 5 argues that astrological wall calendars and the annual *judicia* and *practica* that complemented them became important instruments in Habsburg politics. These texts, which drew on a visual vocabulary and used images along with words to convey their content, were wildly popular in the late fifteenth and early sixteenth centuries. For more than two decades they were produced by Georg Tannstetter, who was *Leibartz* and adviser to Maximilian and Archduke Ferdinand as well as a professor at the university. Analyzing his wall calendars and *practica* within the context of Tannstetter's activities at both the court and the university indicates how Tannstetter was able to publicize the Habsburg political and social programs to a popular audience. Chapter 6 focuses on Andreas Perlach, a master at the University of Vienna and an adviser to the Habsburg court in Vienna who for more than a decade produced astrological almanacs and ephemerides that supported the Habsburg political agenda.

Prognostications composed in response to particular celestial phenomena were another popular astrological genre. Unusual events such as the appearance of a comet or significant conjunctions of planets attracted widespread public attention and offered astrologers further opportunities to interpret the natural world. Chapter 7 details how astrologers at the Habsburg court seized on these prodigious events as evidence of Maximilian's preordained right to rule and justification for Habsburg authority within the empire. Three case studies stand at the center of this chapter: Johannes Stabius's *Pronosticon* on the planetary conjunction in 1503/1504, Tannstetter's *Libellus consolatorius* on the series of planetary conjunctions in 1524, and Perlach's *Des Cometen und ander Erscheinung in den Lüfften*, his tract on the comet in 1531.

Finally, I look beyond Maximilian's reign to his legacy for the Habsburg dynasty and, more broadly, the relationship between science and politics in early modern Europe. The importance Maximilian attached to scientific knowledge became a key component of Habsburg politics, which found its most mature expression a century later at the court of Emperor Rudolf II. Maximilian's efforts to establish patronage networks that linked individual experts as well as institutions to the court emerged as a central characteristic of early modern politics, especially in the Germanies where princes increasingly viewed local universities as corporate bodies of academic experts to be consulted in political affairs. Similarly, Maximilian's representation of himself as both a skilled practitioner and generous patron of the sciences prefigured the expanding roles for scientific knowledge in politics and at princely courts in later sixteenth- and seventeenth-century Europe.

Chapter One

Astrology and Maximilian's Autobiography

From the earliest years of his life Emperor Maximilian I was surrounded by astrologers and their interpretations of the natural world. Shortly after Maximilian's birth on March 22, 1459, Emperor Frederick III and Empress Eleanor summoned Regiomontanus, one of the most skilled astrologers in Europe, to the imperial residence in Wiener Neustadt to cast and interpret the future emperor's birth horoscope. At the time, Regiomontanus was a master of arts at the University of Vienna, where he collaborated with his teacher and friend Georg Peuerbach.[1] For nearly a decade Peuerbach had provided astrological advice to Frederick III's court in Wiener Neustadt as well as to Frederick's relatives, King Ladislaus Posthumus and Sigismund of Tyrol.[2] In 1452 Frederick had asked Peuerbach to cast a geniture for his future wife, Eleanor of Portugal. Peuerbach rushed to complete the geniture before Frederick departed for Rome, where he would be crowned emperor and marry Eleanor.[3] Throughout the following decade Frederick continued to cultivate his relationship with Regiomontanus and Peuerbach and turned to them for advice on political, social, and economic matters. He reportedly relied on astrologers to cast the horoscopes of contemporaries to learn their characters, their fates, and even when they would die.[4] By 1459 Peuerbach and Regiomontanus had established close ties to the imperial court. When Regiomontanus cast Maximilian's horoscope, he drew on his experience working with Peuer-

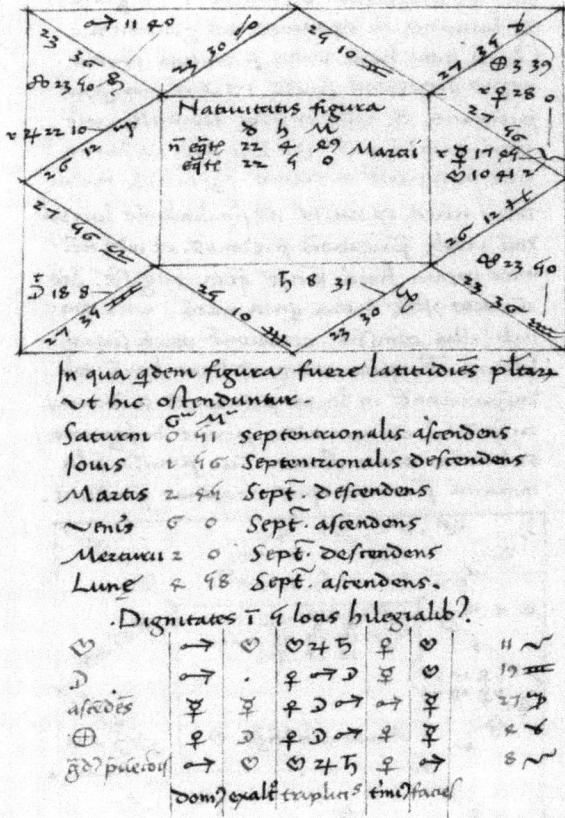

Figure 1.1. Emperor Maximilian I's geniture, calculated by Regiomontanus at the request of Eleanor, Maximilian's mother. "Epistola ad quandam Imperatricem judicium astrologicum de ejusdem filio contenens." Regiomontanus, "Epistola ad quandam Imperatricem judicium astrologicum de ejusdem filio continens," © Österreichische Nationalbibliothek, Vienna, cod. lat. 5179, 2v.

bach, modeling his geniture on the one his mentor had cast for Eleanor.[5] He brought his formidable intellectual talents and mathematical skills to bear on the task of interpreting the heavens and their influence on the infant Maximilian, combining precise astronomical calculations with rigorous analysis to read out of the stars a mixture of prediction and advice suitable for the Habsburg heir. In his letter to the empress, Regiomontanus emphasized the mathematical and logical method he employed, highlighted the precise data he had calculated, and made explicit his use of technical astrological rules.[6] The chart Regiomontanus cast and interpreted reflected the sophisticated and technical science the imperial family expected (see figure 1.1). This degree of skill and knowledge reflected an ideal that the young Maximilian would later expect from his own astrologers.

Although Maximilian's time of birth had been precisely reported, Regiomontanus stressed that he first had to confirm the accuracy of the time, a practice known as rectifying the geniture. Assuming that the time reported was only an estimate, Regiomontanus used two different methods to calculate the true time of birth and then compared the results to the record.[7] Based on his calculations, Regiomontanus concluded: "It is certain that this birth was at precisely 5 PM on the 22nd day of March in the year of our Lord 1459."[8] Regiomontanus expended this effort to determine Maximilian's precise birth time because the rest of his judgment depended on this initial step. He then presented tables of data, listing the precise positions of the planets, indicating the direction of their motions, their geometric relationships to one another, and their principal significations.[9] These tables provided the fundamental observational and calculated data for the geniture.[10] Having presented the raw data for his analysis, Regiomontanus turned to a detailed study of each horoscopic house. As was fitting for the Habsburg heir, Regiomontanus offered the most careful analyses of the first and tenth houses, those that revealed Maximilian's personality and constitution and the contours of his reign, his fame, and his honors.[11] According to Regiomontanus, the prominent place of Mars in Maximilian's geniture ensured that he would grow up to be strong, courageous, and virtuous in battle:

> On account of Mars it is foreshown that the child will be brave, powerful, strong, passionate, eager for weapons, an instigator of wars, courageous. He will confront dangers without fear of dying. He will hold no man before himself, nor humble himself before anyone. He will have confidence in his own power. He will often bring fury and vehemence against his enemies. He will destroy rebels. He will desire to take part in combat. In all these activities, however, he will obtain honor. He will bring aid to his friends and harm to his enemies. He will always conduct himself lawfully.[12]

Regiomontanus reassured the empress that Maximilian would not simply engage in battles and martial activities. Thanks to the influence of Venus, the imperial heir would also take particular pleasure in beautiful things.[13] He would be bold in his governance, courageous and prudent in battle, and gracious but strict in victory. He would be constantly active.[14]

Regiomontanus's letter to Eleanor and his horoscope for Maximilian locate astrology at the center of the Habsburg identity and imperial politics. As a systematic body of knowledge connecting the natural world to the social, astrology offered powerful political tools, both practical and symbolic, for the princes who controlled its production and use. In the realpolitik of Renaissance Europe, it presented a set of mechanisms for evaluating the risks and benefits inherent in any action. Princes consulted astrologers to learn when to engage in battles and to negotiate peace agreements, when to avoid expeditions or journeys, and when to make public proclamations. Nero and his mother Agrippina, who had poisoned Claudius, waited until the most "fortunate moment foretold by astrologers" to exit the palace and claim the imperial title.[15] Rulers also consulted the stars when evaluating possible marriages, as Frederick III had before he married Eleanor. The imperial astrologer Johannes Nihili of Bohemia accompanied Frederick to Rome for his coronation and marriage to Eleanor, where he recommended dates for both events. When the bride failed to arrive in time to celebrate the marriage on January 22, 1452, Nihili calculated the next propitious date to be March 16, 1452, which enjoyed a favorable Mercury-Venus conjunction.[16] Astrology also promised to provide insights into the inclinations, dispositions, and fates of a prince's enemies, allies, and subjects; Frederick III reportedly turned to it to learn such information.[17] Frederick's long-time enemy, the Hungarian Matthias Corvinus, had his own astrologer cast Frederick's geniture before invading Austria in 1485.[18] The Sforza dukes in Milan regularly used astrology to understand the fortunes of their allies and enemies.[19] Astrologers also served their patrons by identifying potential enemies and dangers. On a practical level, astrology could both guide and justify political action.

Beyond these pragmatic benefits, astrology also functioned symbolically to legitimize a dynasty or rule. Since the earliest years of the Roman Empire, emperors had deployed astrology to establish their legitimacy by portraying their rule as the realization of a greater destiny. Augustus had his zodiacal sign, Capricorn, stamped on coins and put on jewelry throughout the empire in an effort to associate his reign with a new era.[20] The political function of astrology was placed on a more robust intellectual foundation in the Islamic tradition. The most thoroughgoing effort to link politics and astrology was Albumasar's *On the Great Conjunctions*, which demonstrated the connection between the conjunctions of planets—especially between Jupiter and Saturn—and political dynasties and religions.[21] *On Great the Conjunctions* was translated into Latin in the twelfth century and quickly became the most important text to correlate large-scale historical changes to the motions of the heavens.[22] Princes located the basis for their authority in the stars, demonstrating the natural source of their sovereignty and legitimizing their rule as the realization of a divine plan. At the physical level, the celestial motions were the cause of change on earth. Although individual free will was protected from strict astral influence, astrologers had long argued that the motions of the planets inclined individuals to act in a particular way and influenced collective actions. Astrology invoked natural causes to explain societal changes. At the same time, the stars were considered signs from God. The skilled astrologer could read God's will in the motions of celestial bodies. This dual physical-theological understanding simultaneously undermined efforts by adversaries to reject a prince's authority and reinforced a prince's claims to power. Consequently, astrology's political significance grew during periods of instability or regime change.[23]

Horoscopes and astrological history helped to establish legitimacy in the person of the emperor. For Maximilian—who was obsessed with reinforcing his own claims to the imperial title as well as the authority of the House of Habsburg, particularly in the face of sustained opposition from the German princes and electors in addition to foreign kingdoms—reading the stars provided a powerful political and dynastic tool for locating legitimacy and authority in Maximilian himself.[24] The importance of astrology for Maximilian's self-presentation and his conviction that it played a fundamental role in his politics of representation is reflected in the emperor's sustained efforts to portray himself as a competent practitioner of the science as well as a generous patron of astrologers. From his earliest efforts to write his autobiography to his mature works, Maximilian repeatedly drew on astrology to structure and understand his own life and emphasized its importance for his own political program. Maximilian's autobiographical works functioned as "mirror of princes" literature. His emphasis on astrology, therefore, was an elaborate argument for its place in contemporary politics.

Autobiography as Memorial and Mirror

Maximilian's early encounter with astrology echoes throughout his autobiographical works. No early modern prince worked harder than Maximilian or expended more effort to shape and control his image, both for his contemporaries and for posterity. His *Gedechtnus* focuses on memorializing and institutionalizing his image and served as a form of self-aggrandizing propaganda for himself and the Habsburg family.[25] His massive autobiographical corpus—including his *Theuerdank*, *Weisskunig*, *Freydal*, and *Ehrenpforte*, his genealogical projects, and his various triumphal celebrations—was conceived and executed to project a carefully crafted image of the emperor.[26] To realize his goals, Maximilian employed teams of humanists and artists from Innsbruck, Augsburg, Regensburg, Nuremberg, and Vienna. He dictated the core of his autobiography to his personal secretaries Joseph Grünpeck and Marx Treitzsauerwein, who then along with humanists like Willibald Pirkheimer and Melchior Pfinzing edited, expanded, and polished the emperor's nascent ideas. Scholars such as Johannes Stabius, Jakob Mennel, and Conrad Celtis, as well as members of the faculty at the University of Vienna combed through monastery libraries and excavated ruins looking for source material to use. From these materials they recovered a mythical German past and reconstructed imperial genealogies, including the elaborate family tree decorating Maximilian's *Ehrenpforte*.[27] Maximilian also relied on networks of artists to illustrate his works, including Albrecht Dürer, Hans Burgkmair, and Albrecht Altdorfer. He rewarded these scholars handsomely, ennobling them, promising them large sums of money, awarding them incomes from mills, and appointing them to important and lucrative positions in his court. Maximilian's close oversight and management of these scholars and artists and his involvement at every stage of the process reveal his immense efforts to control and project his memorial.

In his *Weisskunig*, written as a mirror of princes for his grandsons Charles and Ferdinand, Maximilian justified his obsession with his image, claiming: "Whoever prepares no memorial for himself during his lifetime, has none after his death and that same person will be forgotten along with the sound of the bell that tolls his passing. And so the money I spend on my memorial is not lost; rather, to spare expense on my memorial is to suppress my future reputation. For what I do not produce towards my own memorial during my lifetime will not be celebrated after my death by you or anybody else."[28] Maximilian did not intend his autobiographical works merely to reflect his deeds; instead, he used them to construct an idealized portrait of the prince, his education, skills, interests, and his patronage practices. His *Weisskunig* was merely one part of a larger publicity program that projected his image to all levels of society, from the aristocratic and traditional spheres of authority to the *Burghers* and emerging bureaucratic class that he used to staff his offices and finally to the broad, scarcely literate citizens of the empire. A key aspect of Maximilian's self-portrayal was his own interest in, knowledge of, and support for astrology. On the one hand, it provided a powerful mechanism for structuring history and imposing regularity on seemingly contingent events by linking political and social events to the natural world, especially through the predictable motions of the heavens.[29] It could furthermore infuse Maximilian's various historical projects—his genealogies, his claims to Roman Imperium, his efforts to link the Habsburg family to the major royal families across Europe—with a natural necessity and divine authority. The thread of astrology linked Maximilian's autobiographical works, from his earliest efforts to dictate his life to his late, more mature efforts to portray himself as the idealized prince. The emperor used his texts to recount his familiarity with the subject, to gesture toward his patronage practices, and to indicate astrology's role in political practice. Artists illustrated these texts with images that reinforced the emperor's own understanding, either by portraying Maximilian at astrologically significant moments or by placing him in the context that revealed his knowledge and skills. Together, these texts and images represented an emperor deeply influenced by and knowledgeable about astrology.

Early Latin Autobiography

After experiencing the splendors of the Burgundian court, Maximilian returned to Austria convinced that he needed to dictate his autobiography in order to record his accomplishments and memorialize his character.[30] As early as 1492, a year before Maximilian was elected emperor, the humanist Heinrich Bebel had encouraged the young king to write one. Although pressing political and military engagements delayed the emperor's start, by the end of the decade he had begun to outline the contours of the work.[31] After his defeat by the Swiss troops at Dorneck in July 1499, Maximilian began dictating his *res gestae* to Joseph Grünpeck as they retreated across Lake Constance.[32] Grünpeck was a skilled Latinist from Augsburg who had been the emperor's secretary since 1496 and had been crowned poet laureate in 1498. These early dictations formed the source material that court historians and imperial secretaries were supposed to fashion into a respectable autobiography suitable for an emperor. Maximilian invested even this early, fragmentary work with astrological themes that would recur throughout his autobiographical corpus.

    In crafting his personal history, Maximilian sought to understand and explain the events of his life. Central to his efforts was his geniture, which connected his fortunes and misfortunes as well as his character to the motions of the stars.[33] Echoing Regiomontanus's interpretation of his geniture, Maximilian located key aspects of his personality in the configuration of the stars and planets at his birth. The malefic influences of his birth constellation had plagued him throughout his adventures, disastrously in his latest defeat at the hands of the Swiss troops. Fortunately, God's mercy was likewise part of his constellation and had frequently protected him from the worst physical dangers. The stars' positions at his birth was responsible for more than simply his corporeal health; Maximilian also traced his natural attraction to arms and his warlike character to the stars.[34] Later in this early autobiography, Maximilian again traces his unfortunate military losses to his inauspicious birth chart.[35]

Beyond recounting his own life in astrological terms, Maximilian gestured toward the importance of learning astrology, one of the seven liberal arts. As soon as he could understand Latin, the young king undertook the study of these seven arts so that he could comprehend and teach the deeds of great kings and princes.[36] In this context, astrology provided a means of structuring the actions of Maximilian's predecessors and organizing history. Just as Maximilian himself was never free from the influences of his birth constellation, the kings and princes in the past had acted under the beneficent and malefic influences of the stars. Understanding how and why they acted as they had, Maximilian implied, required knowledge of astrology.

Modeled on contemporary chronicles and especially the Burgundian examples by Olivier de La Marche and Jean Molinet, Maximilian's early autobiographical fragments concentrate on his military conflicts, peace negotiations, and other major political events.[37] Nevertheless, these early drafts contained themes that would assume ever increasing importance in the emperor's subsequent efforts to narrate his life, including the role of the stars for understanding Maximilian's own character and fortunes. Maximilian expected Grünpeck, Pirckheimer—whom he charged with correcting his "soldier's Latin"—and his other secretaries to use his Latin dictation as the basis for constructing a polished and complete work.[38] By 1499 Grünpeck had already capitalized on the emperor's interest in astrology to secure a position at the imperial court. It is not surprising then that when he turned to write the emperor's biography he emphasized and expanded its role. Whereas the Latin fragments only gesture toward astrology, subsequent works place it at the center of the emperor's birth, education, and political practice.

Historia Friderici et Maximiliani

In 1501 Grünpeck had to recuse himself from Maximilian's court when he contracted the French Disease.[39] The suffering was particularly acute for the young humanist who had recently railed against the moral corruption that had contributed the disease's spread through the Germans.[40] Separated now from the emperor's Latin dictation and the other material he had been using to compile a history of Maximilian's life and reign, Grünpeck had to suspend his efforts to write the emperor's autobiography. He did not, however, give up the idea of writing a history of Maximilian's reign. A few years later, after he seemed to have recovered from his illness, he once again used his literary talents to advertise his services to the imperial court. By 1507 Grünpeck had produced at least two new versions of the emperor's life story, his *Commentaria divi Maximiliani* and his *Gesta*. The *Commentaria* was largely a working draft while the *Gesta* was a more formal chronicle of Maximilian's reign up to the end of the War of Succession of Landshut in 1505.[41] Neither work appealed to Maximilian, who by this time had rejected Latin in favor of German for his autobiography. The emperor had realized that Latin was not the ideal language to propagate his image. His shift to German allowed him to reach a broader audience and to expand the basis for his political authority. Moreover, the dry, chronicle style of Grünpeck's *Gesta* failed to interest the emperor, who had already begun to outline his allegorical autobiographies, *Theuerdank* and *Weisskunig*.[42] Whereas Maximilian had previously focused on simply recording his deeds and achievements, he now wanted to glorify them as a means to memorialize and institutionalize the Habsburg dynasty.

Grünpeck apparently realized that Maximilian's interests had evolved and tried to adapt his Latin history to fit more closely the emperor's new interests. He presented his *Historia Friderici et Maximiliani*, an illustrated version of his history, to Maximilian in February 1516 in the hope of winning the emperor's approval and using it as a draft for a printed version.[43] Each chapter opened with a drawing by the young Albrecht Altdorfer, Grünpeck's fellow citizen of Regensburg.[44] Grünpeck's *Historia* along with Altdorfer's drawings highlight the emperor's interests and activities as well as those of his father, Frederick III. Although Maximilian never returned to Latin for his autobiography, he retained close oversight of Grünpeck's project, proofreading the text with Altdorfer's accompanying illustrations. He annotated various passages, indicating where the topic would be treated more fully in his *Weisskunig*, and crossed out certain illustrations that did not conform to his self-presentation. The emperor's close supervision was typical and ensured that his artistic and literary projects supported the image he wanted to convey.

In the *Historia* Grünpeck expanded the themes already found in Maximilian's early dictations, crafting a work that was less a chronicle and more a mirror of princes. Grünpeck projected his experiences at Maximilian's court back on to Frederick's and produced an ideal prince that was a composite of Frederick III and Maximilian. He built on Frederick's reputation for consulting astrologers and embellished Frederick's own skills at reading the stars. In one chapter, Grünpeck recounted how Frederick had spent his leisure time studying mathematics and the motions of the stars because he recognized that such knowledge provided not only a guide to an individual's health and future but also an understanding of the nature and habits of rulers and princes, including his own heir.[45] Astrology, therefore, promised both natural and politically useful knowledge. Further, it was a learned, princely activity that allowed Frederick not only to discuss the motions and influences of the heavens with his advisers and courtiers but also to cast and interpret horoscopes. Altdorfer's illustration for this chapter depicts the emperor discussing the influence of the stars with his astrologers and courtiers. Frederick points to a celestial globe with one hand and at the stars with the other while two astrologers point at the globe and another gestured toward the sky. Courtiers stand around listening. The text and the image portray an emperor who not only was sufficiently skilled in the science of the stars to evaluate competing advice from his astrologers but also could contribute to the discussion.

Following Maximilian's birth soothsayers and fortune-tellers offered various and often competing predictions about the young prince's future; in Altdorfer's drawing a palm reader inspects Maximilian's hand.[46] Grünpeck indicated the various groups of experts at Frederick's court vying for the emperor's attention and favor. The emperor was dissatisfied with their ambiguous and contradictory conjectures. Instead, he turned to the science of astrology and consulted experts in the subject. From them the emperor expected to learn about Maximilian's future, his successes, and challenges to his rule.[47] Once again Altdorfer's illustration reinforced the text, depicting Frederick and his astrologers standing in a vaulted hall. One holds the young Maximilian's hand and gestures toward the stars.

Although Grünpeck gave the impression that he was familiar with Frederick's reign, he had never been to Frederick's court. Instead, he drew on his experience as Maximilian's secretary to craft a portrait that emphasized the qualities and knowledge important to Maximilian—a thorough education and ultimately expertise in astrology, a preference for it over other forms of divination, and a reliance on the stars in dynastic and political matters. His portrait of Frederick, then, reflected the values and image Maximilian wanted to convey.

Grünpeck also portrayed Maximilian as an avid supporter of the sciences. A contemporary German translation of the *Historia*, probably completed by Grünpeck himself, underscores Maximilian's enjoyment of reading astrological and astronomical literature, especially the classic texts such as Ptolemy.[48] Of all his books, Grünpeck claimed, Maximilian preferred astrological texts. By reading such books, the emperor hoped to learn the science of the stars and to understand when and how to apply it to his political life. In Grün-peck's text, Maximilian was not content merely to rely on astrologers. Even more than his father, Maximilian wanted to possess the knowledge and skills necessary to assess their claims.

Early sixteenth-century learned astrologers sought to elevate their own mathematical sciences over the prophecies of inspired prophets,[49] and by the mid-1510s Maximilian had moved away from the more prophetic aspects of Grünpeck's text. The emperor crossed out the two chapters in which Grün-peck invoked prodigious phenomena. Much as Frederick had rejected the soothsayers and fortune-tellers, Maximilian also eschewed the subjective interpretations of prodigious events, interpretations that often relied on a charismatic individual claiming some special divine inspiration. Although Maximilian had capitalized on the Ensisheim meteor and Sebastian Brant's interpretation of it earlier in his reign, he now preferred the more certain forms of prognostication based on careful mathematical and historical anal-ysis, especially astrology.[50] Grünpeck's *Historia* failed to capture Maximil-ian's attention in part because it was written in Latin and in part because it resembled too closely Grünpeck's earlier efforts to chronicle the events of the emperor's reign. Nevertheless, Grünpeck dedicated the work to Archduke Charles, later Emperor Charles V, portraying Frederick III and Maximilian I as ideal emperors, whose education, action, and political practices should be emulated—including their knowledge and patronage of astrology.

Weisskunig

By the time Grünpeck had finished his *Historia*, Maximilian's conception of the project had evolved. At this time Maximilian began planning his great literary monuments, his multivolume German autobiography that would include *Theuerdank*, *Weisskunig*, and *Freydal*, his genealogical projects, and his *Ehrenpforte*. The emperor employed humanists and artists to realize his grand projects, which were directed by a new set of advisers and secretaries, including Jakob Mennel, Willibald Pirkheimer, Johannes Stabius, and Marx Treitzsauerwein. Maximilian was a master at deploying the power of print to construct and convey his image. He recognized that vernacular works would have a greater impact in disseminating and solidifying his image than would be achieved by works in Latin. Maximilian's mature autobiographical cor-pus—richly illustrated vernacular monuments—further cemented his reputa-tion as both a student and patron of astrology.

Maximilian's *Weisskunig* was a fictionalized autobiography modeled on courtly romances, especially those in the Burgundian tradition that he knew through his marriage to Mary of Burgundy. He divided it into three sections that recounted his parents' marriage, his own birth and upbringing, and his military campaigns during his reign. Maximilian portrayed his military campaigns as the necessary reaction to the aggressions of envious neighbors. His victories were made possible by his superior innate character—displayed through his pure, white armor—and his exemplary education and upbringing.[51] Maximilian's *Weisskunig* simultaneously provided an ideal model for his successors to emulate and was a powerful piece of propaganda intended for a broad audience. As a mirror of princes, the *Weisskunig* was intended primarily for other princes, his successor Archduke Charles, and the educated members of court.[52] This included the burgeoning bureaucratic class in the empire, which Maximilian had expanded considerably during his reign and drew on to staff offices at all levels in his government. Beyond this more elite audience, Maximilian clearly hoped to produce a *Volksbuch* version of his autobiography in an effort to disseminate his image through multiple registers of society.[53] His model *Volksbuch* was Johannes Hartlieb's *Alexander*, a vernacular translation of the romance that recounted the education and reign of Alexander the Great.[54]

As with his other projects, Maximilian relied on a team of scholars and artists to realize his ideas. The Augsburg humanist Konrad Peutinger, at the time one of Maximilian's most trusted advisers, coordinated the project.[55] For some time Peutinger had played an important role in the emperor's efforts to spread imperial news and propaganda, both in texts and images. Previously Maximilian had commissioned him to publicize news of his success in the War of the Succession of Landshut.[56] By 1516 Peutinger was involved in many of the emperor's publicity projects, including illustrations for Maximilian's three autobiographical works—*Weisskunig*, *Freydal*, and *Theuerdank*—as well as images for the *Triumphal Procession*. The emperor dictated the bulk of the *Weisskunig* to his secretary Marx Treitzsauerwein. The artists Hans Burgkmaier and Leonhard Beck completed most of the illustrations for the book. Although the emperor delegated the work to these artists and scholars, he retained close control over the final product, correcting versions of both the text and the images throughout the process. The fictionalized image in *Weisskunig* is a composite of the emperor's ideal and that of the artists and humanists who worked on the text.

In the *Weisskunig* Maximilian again set his autobiography within an astrological framework that traced his irreproachable character to the heavens and underscored his divinely ordained status. When he recounted his birth, Max-

Figure 1.2. The young Weisskunig in his nurse's arms while beneficent rays stream down from the heavens. © Österreichische Nationalbibliothek, Vienna, cod. lat. 3033, 92v.

imilian emphasized the fact that a comet appeared shortly before he was born, shone more brightly at the exact hour of his birth and then, shortly afterward faded and disappeared.[57] The comet was a portent that augured well for Maximilian's future reign and signaled his wonderful deeds. While it confirmed his warlike character, it also promised a kind and fair victor.[58] Just as in Grünpeck's *Historia*, in Maximilian's *Weisskunig* Frederick III turned to astrology to understand the young Habsburg heir's future. This time Frederick is portrayed as sufficiently skilled that he did not summon astrologers to his court, but rather he himself interpreted the heavenly signs to learn the general contours of Maximilian's future: "Now the old White King [Frederick] was extremely skillful in the knowledge of the stars and realized that through their influence and from the rule of the zodiac, under which the child had been born, that this same child would reach the highest rule, and through him many wonderful deeds and great battles would occur."[59] As he had in his Latin autobiography, Maximilian again attributes his bellicose character to the influence of the heavens. In the *Weisskunig* he added the comet as a signal of his divine selection. Although comets were often considered harbingers of doom, Maximilian appropriated this one to serve as the source of both his warlike behavior and his benevolence.[60] He had various contemporary models for this more positive association. Hartlieb's translation recounted how a comet had appeared at Alexander's birth.[61] Perhaps more useful for Maximilian were contemporary representations of Christ's nativity, which emphasized the corona at the time of his birth.[62] Any doubts about Maximilian's efforts to associate himself with the birth of Christ were dispelled by the woodcut that illustrated the chapter. Cradling the infant Weisskunig in her arms, a wet nurse stands within a columned room next to a crib marked with the Christogram for Jesus (see figure 1.2). Although the comet is not depicted in this woodcut, the beneficent influences of three stars and the moon stream down upon the newborn child.

The second section of the *Weisskunig* detailed the ideal education for a prince, devoting a chapter to each of the important subjects that the young Weisskunig studied. After learning how to read and write, the young prince should be tutored in the seven liberal arts as well as noble pastimes such as hunting, falconry, jousting, and painting. These early subjects provide the foundation for the most important princely activity, ruling well. To rule justly and fairly, Maximilian claimed, and to command the respect of one's subjects, the prince must receive a humane education and must know more than his subjects, both other princes and the people.[63] The political arts occupy a prominent place in the *Weisskunig*. The young prince studied books on human nature and disposition, on the estates and orders of society, and on law. He learned the science of politics and the art of negotiation. Success in politics, he claimed, required understanding the five different bases of political authority: God, the influence of the planets, reason, gentleness in rule, and restraint in war and violence.[64] For a prince, particularly the ruler of the fractious Holy Roman Empire, diplomacy and the art of negotiation was of greatest importance. However, diplomacy alone failed to provide the young prince with the tools necessary to succeed because it did not help him understand the hidden and natural inclinations that motivated people. Without a better understanding of why people behaved as they did, diplomacy and negotiations would ultimately fail. The science of astrology offered a way to decipher the hidden motivations and natural inclinations the influenced people's actions.

Following a chapter on rule, the various estates, and the five subjects a ruler should know—what Maximilian labels the *"gehaim wissen und erfarung der welt"*[65]—he made the case for learning astrology and stargazing, underscoring its importance in the opening lines of the chapter: "And now after the young Weisskunig had studied secret knowing [*das haimlich wissen der erfarung der welt*] and had learned it sufficiently, as previously mentioned, he then considered how in the future it would be necessary for him to recognize the stars and their influences as well as their effects. Otherwise, he might not correctly understand human nature, a topic that seemed to him almost completely absent from secret knowing [*dem haimlichen wissen der erfarung der welt*]."[66] Because the motions of the stars influenced human nature, which in turn guided people's actions in all aspects of their lives and especially in politics, Maximilian claimed that in order to succeed in politics the prince had to learn how to understand the stars. At first the young Weisskunig had tried to learn the science of the stars through reading and self-study, but no matter how diligently he applied himself he failed to make significant progress. He realized that the prince needed to secure expert astrologers to teach him the art. He sought the "most learned doctor in astrology," who along with tutoring him in the science of the stars also compiled a more extensive and accurate star catalog that would help explain human nature better. The young prince soon mastered the subject and relied on his own skills or those of his advisers throughout his reign whenever he confronted a serious event. The *Weisskunig* echoed the advice given in the *Secretum secretorum*, the most widely disseminated mirror of princes' text in the Middle Ages.[67] Written as a series of letters from Aristotle to Alexander the Great, the *Secretum* urged princes to consult with astrologers before undertaking any activity, however trivial.[68] In the *Weisskunig* the young prince confirmed his mastery of astrology by using it to determine the most propitious moment to engage in any activity and waiting until that moment to begin. This equanimity, which he claimed was contrary to human nature, distinguished him from his princely peers and his subjects and made him more suitable to rule.[69]

Maximilian's fictionalized history reflected his own life in important ways. The woodcut that accompanied the chapter on astrology placed the young Wiesskunig at the center of a scene in which the prince gestures toward a wheel of fortune while personifications of the planets Mercury and Mars occupy the upper corners of the sky, indicating that these two planets were important influences in the prince's life (see figure 1.3). The connection between Maximilian and Mercury is reinforced through the wheel of fortune in Mercury's hand. Although neither planet was on Maximilian's ascendant in his geniture, Mercury was the ruler of his ascendant and played an important and recurring role in Regiomontanus's interpretation of Maximilian's geniture. Similarly, warlike Mars figured prominently in Regiomontanus's analysis and had been implicated in Maximilian's earlier autobiographical fragments as the source of his bellicose nature.

FIGURE 1.3. The young Weisskunig learns the art of stargazing. Mercury and Mars, ruling planets from his geniture, stand in the heavens. © Österreichische National-bibliothek, Vienna, cod. lat. 3033, 22r.

Theuerdank

In the *Weisskunig* text and image functioned seamlessly as a mirror of princes, each reinforcing the other. They focused on the skills a prince needed to rule effectively and justified those skills by linking them to political practice. Astrology played a fundamental role in the prince's education and in his diplomatic and political efforts while astrologers at court served as tutors and counselors. Maximilian's *Theuerdank*, produced at the same time and by many of the same scholars and artists, shifted the focus from education to political praxis. Whereas *Weisskunig* focused on education, the *Theuerdank* offered a fictionalized depiction of the prince in the real world. The two texts were intended to complement each other and together offer a fuller picture of the role of astrology in Maximilian's world.

Maximilian's *Theuerdank* is an epic poem in which he narrates a fictionalized account of his journey from Vienna to the Netherlands in 1477 to claim his bride Mary of Burgundy. Like his other works, the *Theuerdank* was the complex product of Maximilian's own intentions and those of his humanists and artists who carried out the emperor's wishes. Melchior Pfintzing, Maximilian's chaplain, and Marx Treitzsauerwein, the emperor's personal secretary, were responsible for most of the supervision and composition of the text. The illustrations were completed by Hans Schäufelin, Hans Burgkmair, and Leonhard Beck. Although Maximilian had begun planning the *Theuerdank* as early as 1505, it did not near completion until 1512.[70] In 1517 the Augsburg printer Hans Schönsperger printed 340 copies of the *Theuerdank*—40 folio copies on parchment as gifts to German princes and other important members of the court, and another 300 on paper for wider distribution and sale.[71]

The *Theuerdank* recounts the many challenges and difficulties Maximilian encountered on his journey from Austria to the Netherlands. Throughout his adventures, Theuerdank, that is, Maximilian, was constantly accompanied by Ernhold, who aids the young hero in his struggles against the three adversaries Fürwittig, Unfalo, and Neidelhart—personifications of excessive passion, the dangers of chance, and envy, respectively. These three also symbolize the opposition Maximilian faced from the Italians, the Germans, and the Dutch.[72] In each case, Maximilian was able to avoid their plots and conspiracies through the timely intercession of his constant companion Ernhold, easily recognizable in nearly every woodcut by the wheel of fortune emblazoned on his tunic. Ernhold did not alter the course of events, either natural or social, so much as warn Theuerdank of impending danger and gave him the knowledge to avoid those events. He revealed for the hero when chance conspired against his success and when the winds of fortune would once again be blowing in his favor. As Melchior Pfintzing explained in his *clavis*, which was published with some of the early editions of the *Theuerdank*, Ernhold represented "the hint and evidence of truth that follows a person to the grave" whose role was to inform the young Theuerdank of impending favorable or dangerous circumstances.[73] Ernhold's struggle against Unfalo, Theuerdank's most common adversary, was a metaphor for the struggle of knowledge to understand the world and to provide the insight into how best to exploit it. Ernhold, in other words, was a fictionalized representation of the role of astrology in Maximilian's life. On the one hand, Ernhold represented the figure of the court astrologer, who was constantly at Maximilian's side offering advice on all matters from politics to health. On the other hand, Ernhold personified astrology as a body of knowledge that enabled the emperor to choose the most propitious time to act or helped him recognize when events and people conspired against his success.

That Ernhold represented astrology would not have been lost on contemporary viewers. His name, meaning Herald, is a clear reference to the planetary god Mercury, who was the herald or messenger of the higher gods, conveying their wishes to the lesser gods and to humans. Virgil, the source of much of Maximilian's political propaganda, had used Mercury to convey Jupiter's message to Aeneas.[74] Mercury's role as the messenger god provided a natural link to astrology, which enabled humans to read God's messages out of the stars. The connection between Mercury and astrology was reinforced through the rich tradition of *Planetenkinder*, pictorial representations of people engaged in the activities most closely associated with particular planets (see figure 1.4).[75] Among Mercury's children were stargazers, often using some instrument, as well as people constructing and using mechanical clocks and other time-keeping devices. Representations of *Planetenkinder* circulated in manuscripts throughout the fifteenth and early sixteenth centuries. Early printed texts ranging from single-sheet calendars to longer treatises often included woodcuts showing the various children of the planets. *Planetenkinder* also appeared in architectural reliefs, adorning palaces, churches, and other public buildings.[76] Early sixteenth-century viewers were familiar with these illustrations. More specifically, within Maximilian's autobiographical corpus the personification of the planet Mercury in the *Weisskunig* carried a wheel of fortune like the one decorating Ernhold's tunic in the *Theuerdank*, a visual connection that associates

FIGURE 1.4. A mid-fifteenth-century engraving of Mercury's *Planetenkinder*. Four astrologers stand in the middle gazing up toward Mercury and looking at an armillary sphere. Attributed to Baccio Baldini, London, British Museum, Prints & Drawings, Museum Number 1845,0825.475 © Trustees of the British Museum.

Ernhold with Mercury, the Lord of the Ascendant in Maximilian's birth horoscope, and emphasizes the important role that Mercury played in all aspects of Maximilian's life.

Theuerdank's travels to the Netherlands were fraught with dangers and perils, arising from the unfolding of events in the natural world as well as humans conspiring to impede his progress. Ernhold was constantly by the prince's side, offering advice and counsel, guiding the hero's actions and saving him from disasters of all sorts. When Theuerdank was hunting, Ernhold warned him against going into a particular part of the forest where imminent danger awaited. Armed with this knowledge, Theuerdank prepared for the attack and was able to defeat the bear that had been hiding among the trees.[77] Later Unfalo urged Theuerdank to venture out across an open field just as a storm approaches. Ernhold, providing one of the most common types of astrological advice, warned Theuerdank of the impending storm. This advice saved Theuerdank from harm as lightning struck the ground exactly where he would have been had he followed Unfalo's counsel (see figure 1.5).[78] When Theuerdank became ill, Ernhold was at his bedside offering advice to the physicians on how best to cure the ailing prince. Later, Theuerdank consulted with Ernhold before engaging in combat or jousting with another knight.[79] In each case, Ernhold offered the key for the prince to avoid almost certain disaster by steering him away from acting at inauspicious moments, by staying his hand, or by encouraging him to act quickly before the moment has passed. In this way, Ernhold, like astrology, helped Theuerdank "control" the wheel of fortune, not by stopping it but by giving him the knowledge to understand when it turned in his favor and when it might be moving against him.

The *Secretum secretorum* urged princes to consult with astrologers before undertaking any activity. The science of the stars, the author claimed, provided valid and useful knowledge about the future. In the *Theuerdank*, Maximilian realized the advice given in the *Secretum secretorum*. At every turn he relied on Ernhold to protect him, to guide his actions, and to reveal hidden dangers. This fictionalized portrait of the young prince traveling across Europe with his astrologer-adviser at his side reinforced the image of the ideal emperor. It was not sufficient merely to be trained in astrology, as Maximilian had indicated in his *Weisskunig*. The prince had to surround himself with and consult skilled astrologers who could provide advice and counsel on any activity and in any location.

Together, Maximilian's *Weisskunig* and the *Theuerdank* provided a complete picture of the place of astrology in Maximilian's idealized court. The prince should attract experts from whom he would learn the science of the

FIGURE 1.5. The illustration from chapter 52 in the *Theuerdank*, in which Ernhold's advice saves the young Theuerdank from a bolt of lightning. Ernhold is clearly identifiable by his tunic emblazoned with the wheel of fortune. Unfalo, on the left, had encouraged Theuerdank to proceed precisely when it was most dangerous. Maximilian, "Die geuerlicheiten vnd einsteils der geschichten des loblichen streytparen vnd hochberümbten helds vnd Ritters herr Tewrdannckhs," © Bayerische Staatsbibliothek, Munich, Rar. 325a, q5r.

stars. These learned astrologers would not simply teach the prince but also be actively engaged in advancing the science so that it could better serve politics. But education alone was not sufficient. These same astrologers should also serve as advisers to the prince, accompany him on journeys, offer advice at tournaments and jousts, and predict dangers in recreation as well as in combat.

* * *

Maximilian's unceasing efforts to shape his own image reveal a ruler acutely concerned about his authority and standing. The Habsburg claim to the title of Holy Roman Emperor was anything but certain. Although the family had been a powerful house within the empire, they did not succeed in gaining the title until Maximilian's father, Frederick III, was elected in 1452. But Frederick was never secure in his reign. He had to contend with opposition from the electors and was even besieged in Vienna by his brother's army. Maximilian's own claim to the throne owed more to his marriage to Mary of Burgundy and her family's immense wealth than it did to any Habsburg lineage. But even her wealth and heritage did not protect Maximilian. The early years of his reign were marked by failed efforts to reform government and to centralize authority in the empire. Moreover, he was forced to make concessions to a powerful association of princes and bishops who opposed his reforms and undermined his efforts to extend his control.

Maximilian's broad cultural and political program, manifest in his genealogical works, historical research, literary and publishing endeavors, and use of art, helped to reinforce his authority and bolster the Habsburg position in the empire and Europe more broadly. These efforts, no matter how impressive, remained historical contingencies. They could not offer necessity or a grounding in nature as a basis for their conclusions. Consequently, Maximilian sought sharper tools, which he located in astrology. For Maximilian the discipline offered a comprehensive and persuasive body of knowledge about the natural world, knowledge that was understood to be applicable to human experience. Maximilian used his autobiographical corpus to portray an ideal image of himself, emphasizing the characteristics, learning, and practices that distinguished, at least in his mind, the consummate prince from his peers. Central to this image was astrology. The ideal prince was himself a skilled astrologer, could use astronomical instruments, was arbiter of astrological knowledge, and patron and overseer of expert astrologers.

Maximilian's interest in astrology extended well beyond his self-presentation; it permeated his entire political program, from guiding political actions to shaping public opinion. Just as in his idealized courts of the *Weisskunig* and the *Theuerdank*, he attracted to his court and to the University of Vienna a number of talented astrologers, whom he appointed to important positions and whose expertise he enlisted in his various political projects. These included Joseph Grünpeck, the emperor's first biographer and personal secretary; Johannes Stabius, who created a number of astrological instruments and celestial maps for the emperor and edited and wrote various astrological texts; Stiborius, who constructed for Maximilian an astrolabe designed to work at the latitude of Vienna and wrote a set of canons that he presented to the emperor; and Georg Tannstetter, who produced yearly *practica* and *judicia* as well as the wall calendars that accompanied these texts.

Reading Tannstetter's *practica* in light of his position at the Habsburg court reveals the extent to which the two terms address the pressing political issues of the day. When Maximilian was embroiled in the War of the Succession of Landshut in 1503–5, Tannstetter's *practica* portrayed Maximilian's ally Duke Albrecht of Bayern-München in a favorable light and predicted wonderful successes for him. By contrast, Count Ruprecht was vilified in these same texts, as were the Bohemians who had supplied many of the mercenaries fighting in Ruprecht's army. Moreover, Tannstetter warned Kufstein, one of Count Ruprecht's strongholds, of impending disaster in the coming year—Imperial troops surrounded the fortress and forced it to surrender in October 1504. Tannstetter's student, Andreas Perlach, continued to support the Habsburg efforts through the 1520s and into the early 1530s. Like his mentor, Perlach produced pamphlets that disseminated a Habsburg message to diverse audiences.[80] Following the contours of astrology through the court, the university, and beyond into broader audiences reveals its central role in Maximilian's and later Ferdinand's political program and Habsburg efforts to profit from the science of the stars.

Chapter Two

Astrology as Imperial Propaganda

Emperor Maximilian I recognized the value of a well-established family. He also realized that the Habsburgs had only recently secured the claim of the Holy Roman Emperor and that their hold on the office was tenuous. To help secure his right to the title, Maximilian appointed a team of scholars to the task of reconstructing his family tree. In his *Weisskunig* Maximilian boasted:

> As a youth, the young Weisskunig often asked about his royal ancestry, for he would have liked to know the origins and history of such a royal and princely family, though as a youth he could not find out. Further, he was often annoyed that people paid so little attention to history. And so as an adult, he spared no expense but rather sent out scholars to do nothing other than to sift through the monasteries, cloisters, and books, and to ask all the learned scholars for information about the royal and princely house. And he had everything written down for the glory and praise of the royal and princely family. Through this research he traced his male lineage from one ancestor to the next back to Noah, which otherwise would have been entirely forgotten and the ancient records would have been lost because they were no longer studied.[1]

During the last two decades of Maximilian's life, scholars such as Jakob Mennel, Ladislaus Sunthaim, Johannes Trithemius, and Johannes Stabius produced no fewer than twenty different versions of Maximilian's lineage, tracing it back to Noah, Hector, and Aeneas.[2] At times, heated arguments broke out among these men over various branches of Maximilian's family tree. In the debate that revolved around whether or not it was appropriate to trace the Habsburgs back to the Old Testament prophets, the Theological Faculty at the University of Vienna considered it too controversial and so pruned it from Maximilian's family tree.[3] Other arguments focused not on major trunks of the Habsburg family tree, but on the minutia of relatively recent succession. The most colorful of these erupted between Stabius and Mennel over the succession of Frankish kings, though the real target of Stabius's attack was Trithemius.[4] Mennel had relied on evidence provided by Trithemius in proposing a genealogy from Clotharius to Otpertus. Stabius, in reply, criticized Mennel's suggestion and then charged Trithemius with inventing the evidence, calling him "no historian but the greatest author of fables."[5] No minor, antiquarian debates, the arguments about Maximilian's origins were important matters of state. Maximilian deployed his genealogy to establish the foundations of his authority in the Habsburg lands and the empire more broadly.[6] The emperor's genealogical projects found their most impressive and visible expression in his *Ehrenpforte* and, at the same time, figured prominently in learned works such as Mennel's *Die fürsterliche Chronik*.

Maximilian's efforts to find ancestors in both Greek and Roman antiquity played an important polemical and political role in his struggles to establish and confirm his claim to the title of Holy Roman Emperor. Maximilian's precarious hold on the imperial title was constantly threatened by both his subjects and his enemies. Consequently, he devoted much of his attention to establishing himself as the rightful heir to both the Western Roman Empire, centered on Rome, and the Eastern Roman Empire centered on Constantinople. By tracing his ancestry back in an unbroken line to Greek and Roman antiquity, Maximilian suggested that he was the embodiment of a continuous Roman Imperium. Visually, this association was reinforced on his *Ehrenpforte* by juxtaposing the succession of Roman emperors against the emperors in Maximilian's family tree.

Maximilian's ambitions extended beyond his own rule. He sought to bolster the political fortunes of the Habsburg dynasty by convincing his contemporaries that the Habsburgs were destined to hold the imperial title. His many historical projects conveyed prestige and authority for the House of Habsburg but failed to confer on Maximilian and the Habsburgs the inevitability that the emperor sought. Habsburg rule remained a historical contingency. Astrology could connect the emperor's historical analyses to the natural world, thereby bolstering Habsburg claims to the imperial title and generating support for his political programs. It offered a framework for understanding both how and why history had unfolded as it had—not simply a sequence of chance events but instead ordained by the celestial motions and ultimately God. Showing how past events had been caused by or at least signaled by the planets and stars not only revealed the pattern in history but also invested contemporary events with a greater significance because they were seen as part of some divine plan. Astrology, therefore, had a tangible role in politics. In the broadest sense, astrology could help legitimate Maximilian's claim to the imperial title, rooting it in the celestial motions and God's plan; more immediately, it could be used to support the emperor's many specific political struggles. Maximilian supported scholars at his court whose astrological analyses aligned with the Habsburg political agenda. Sebastian Brant and Joseph Grünpeck are two such scholars.

Early in Maximilian's reign, when the emperor faced strong resistance to his financial and military reforms, Brant lent his support to the imperial cause, interpreting a number of prodigious celestial phenomena as evidence of Maximilian's divine selection. Brant argued that the German princes and people had to support the Maximilian's reforms or risk destruction and ruin by the Turkish army. About the same time, Joseph Grünpeck interpreted celestial phenomena and the terrestrial prodigies that arose from them as confirmation that Maximilian was destined to be emperor. Capitalizing on Maximilian's interest in astrology, Grünpeck used his two short pamphlets on the cause and spread of the French Disease to help secure a position at the Habsburg court. Grünpeck's pro-Habsburg interpretations continued for the nearly three decades he served his Habsburg patrons.

Sebastian Brant's Astrological Propaganda

Modeling himself on Virgil, Sebastian Brant had assumed the role of imperial panegyrist and throughout the 1490s composed broadsheets that interpreted celestial or terrestrial prodigies as portents for Maximilian's political efforts to establish his authority within the empire and lead military offensives against the French and the Turks.[7] In 1492 Brant composed a broadsheet interpreting a meteorite that had fallen near Ensisheim on November 7. Maximilian was at the time in Alsace preparing for an offensive against the French. On November 26 he made an excursion to inspect the meteorite. The meteorite, Brant assured his readers, was a sign from God that Maximilian would be successful in his battles against the French. Brant's prediction seemed to come true when Maximilian defeated the French at Senlis two months later. Brant used the German victory as an opportunity to celebrate his own prediction and to produce another broadsheet on the meteorite. This time he claimed that Maximilian's victory over the French was just the beginning of his successful campaigns against his enemies.[8] In late 1493 Brant produced a third broadsheet on the Ensisheim meteorite. Because the French were no longer a threat—Maximilian had concluded the Peace of Senlis on May 23, 1493—Brant turned his attention to the Turkish armies assembling in Hungary. Once again, the meteorite was a sign from God indicating Maximilian's continued military successes, now against the Turks. Brant also claimed that the meteorite indicated that Maximilian would inherit the imperial crown and scepter.[9]

In 1495 Brant produced yet another broadsheet that supported Maximilian's political efforts. This time, he focused on the birth of conjoined twins near Worms. The two heads on the single body, the broadsheet said, presaged a unity between Maximilian and the princes, the two heads of the empire. The monstrous child symbolized the cooperation achieved between the German princes and Maximilian at the diet of Worms. Through this unity, the empire would be able to confront the Turkish problem. A year later Brant composed a pair of broadsheets interpreting first a pig with one head and two bodies and second a goose with two heads and one body. In both cases, Brant used these monsters to explicate the two most important political issues: unity within the empire, especially between the German princes and Maximilian, and the war against the Turks. Brant's method for interpreting these prodigies and monsters remained consistent. He searched for historical parallels and then compared them to his contemporary situation. Brant's historical review established his authority to interpret the new prodigies and to show that the contemporary situation presented unique threats.[10] Throughout these early works the key interpretive framework was provided by a prodigious event; astrology played little role in his analysis.

When the imperial diet convened at Worms in 1497, Sebastian Brant was hard at work composing a new poem for the second Latin edition of his *Ship of Fools*.[11] In this poem, titled "Concerning the annihilation of living things through the perversion of the natural order," Brant combined a prophetic interpretation of the past with an astrological analysis of future events. He then wedded these two traditions when he argued in support of Maximilian's contemporary political goals. Brant was not the first person to use astrology and prophecy to interpret moments of crisis. Throughout the medieval and early modern periods both played important roles in politics and religion.

This literature was often disseminated through pamphlets and other ephemeral texts that appealed to a broad audience across various strata of society.[12] Many prophetic and astrological pamphlets were written from the vantage point of some marginalized or neglected group and called for social or political reform. At the same time, they sought legitimacy for those reforms.[13] In contrast to this popular literature, astrological texts produced in princely courts and the rarified circles of the Church were written for a more narrow audience. Such texts were often longer and more complicated, making them both expensive to produce and inaccessible to most people. Such texts had only limited effectiveness as propaganda.[14] The popularity of Brant's *Ship of Fools* offered him the chance to adapt the prophetic and astrological aspects of the popular propaganda literature to the emperor's cause.

The prophecy from the Book of Daniel structured the first half of Brant's poem. He recounted human history by associating each of the four major historical empires with the four beasts from Daniel's dream. The Babylonian empire ended when the winged lion relinquished the imperium to the Persians. The bear witnessed the transfer of imperial authority from the Persians to Alexander's Greek empire, which ended with the advent of the four-headed leopard and division of the empire following Alexander's death. At that time the Romans had assumed the mantle of authority. Now the last beast, the ten-horned beast, threatened the disintegration of the Holy Roman Empire and the transfer of imperium from the emperor. According to Brant's analysis, the demise of each empire was brought about by insubordination and disregard for the natural order. At the time of Creation, God had established a natural and social order, which humans were obligated to respect. Failure to maintain this God-given hierarchy had caused the ruin and downfall of each of the previous empires.[15] Brant concluded his historical analysis by repeating his main point: disregard for laws and authority and failure to support the emperor had inevitably led to anarchy and ruin.[16]

Brant then turned his attention to the future and the planetary conjunction that would occur in 1503. He warned that this was a portentous conjunction:

> Believe me, a time full of dangers for Germany is soon coming,
> everything replete with great evils.
> Oh, how I fear lest impious fates reach us
> and carry away from us the scepter and empire.
> Consider, I beg you, the chart of the heavens which you now see:
> Extremely cruel stars join themselves to Cancer:
> Saturn, Mars and Jupiter conjoining in the variable sign of Cancer.
> It will become cold and wet,

and it indicates both old and young men and martial hearts
and that the clergy delights in instability.
They will begin to stir up everything, whatever is pleasant,
that will quickly crawl back like a crab.
Hence there will be grievous disasters: to our country
a common fury and evils will come forth:
gods, defend us from these threats!
In fact many celestial and savage fates threaten us.
No one knows how soon this day will come.
The time will come: the scepter will be taken from us
and pass far from us.
Ah, Teutonic land, at least lament your fate.
Who, who will shed tears for me so that I might be able
to deplore this common ruin or lament our destruction?[17]

Constantine had united the Christian and Roman empires when he trans-
ferred the seat of the empire from Rome to Constantinople. Charles the
Great, a German, had brought the Roman Imperium back to the west, where
through the cooperation of the electors and the emperor the empire would
stand forever. Now, however, the Turks, who had recently conquered Constantinople, were threatening to bring disaster and ruin to the Germans. Only
by aligning behind the emperor could the princes repel the Turkish threat
and avert certain ruin.[18]

Both astrology and prophecy are clearly represented in the woodcut that
illustrated Brant's poem (see figure 2.1). The four shields in the lower portion of the woodcut are decorated with beasts from the Book of Daniel: the
winged lion, the bear, the four-headed leopard, and the ten-horned beast. The
upper left corner is dominated by a horoscopic chart illustrating the positions of the heavens at 9:00 pm on October 2, 1503, when the three superior
planets—Saturn, Jupiter, and Mars—were ascending together in the zodiacal
sign of Cancer. The symbol for Cancer decorates the shield at the right of the
woodcut and links Brant's prophetic and astrological analyses. These were the
two causes for the anarchy portrayed in the center. The cart is placed before
the horses, which are being driven by a fool. Another fool, upside down and
with his torso turned around backwards, rides in the cart.[19] The woodcut visually summarized the poem and simultaneously located the source of current
affairs in the prophetic past and the astrological future.

Astrology stood at the center of Brant's poem and was the core of his analysis.[20] Brant and his contemporaries used the theory of great conjunctions
articulated by the ninth-century Arab astrologer Albumasar and summarized

CXLV

De corrupto ordine uiuēdi

pereūtibus. Inuentio noua. Sebaſtiani Brant.

Anno dni. 1503.
2. die octobris poſt
meridiem hora nona
aſcēdeñ. ad medium
vi. climatis.

FIGURE 2.1. The woodcut illustrating Sebastian Brant's "De corrupto." Sebastian Brant, *Stultifera navis* (Basel, 1498), Courtesy of the Special Collections Department, Bryn Mawr College Library, call # 11,377, 145r.

by Alcabitius to predict the effects of conjunctions between Jupiter and Saturn. The theory traced significant terrestrial events back to the conjunctions of the superior planets, especially Saturn and Jupiter and provided a cyclical framework for understanding history. Although the Arabic originals were inaccessible to Brant and his contemporaries, careful Latin translations had been circulating in manuscripts as early as the twelfth century.[21] By the early fourteenth century vernacular translations were available. Arnold of Friburg translated Alcabitius's *Introductorius ad magisterium judiciorum astrorum* into German in the early fourteenth century, a translation that was still being used more than a century later.[22] Albumasar's *Liber introductorii maioris ad scientiam judiciorum astrorum* and his *Conjunctio magnis* as well as Alcabitius's *Introductorius* were printed by the late 1480s and again in the early sixteenth century.[23] Albumasar and Alcabitius both asserted that conjunctions of Saturn and Jupiter, on the one hand, and Saturn and Mars, on the other, were the most significant celestial events and caused the greatest changes on earth. Albumasar had established a hierarchy of Saturn-Jupiter conjunctions in which the less frequent the conjunction the greater its effects. Great conjunctions, the most common, occurred every 60 years, greater conjunctions were separated by 240 years, and greatest conjunctions occurred after 960 years.[24] Albumasar had considered in detail the historical relationship between conjunctions of the superior planets and the transfer of rule as well as the rise and fall of religions. Alcabitius provided a schematic summary of this doctrine, which was imported wholesale into Latin translations of these works.[25]

Albumasar's and Alcabitius's texts were part of a broad corpus of texts and authoritative authors that refined the theory of great conjunctions. The Arab astrologers Haly Abenragel, Abubacher, Messahalah, and Almansor along with the Greeks Ptolemy and Dorotheus of Sidon all contributed to the larger body of astrological knowledge. By the end of the fifteenth century, Leopold of Austria and Guido Bonati were able to claim a place among the pantheon of astrologers.[26] Together, these authorities offered a broad spectrum of astrological doctrine that later authors wove together to suit their immediate needs.

Concrete applications of the theory further helped to spread the doctrine. Perhaps the most famous of such texts were Pierre d'Ailly's *Concordantia astronomie cum theologica* and its companion *Concordantia astronomie cum hystorica narratione*. D'Ailly had successfully used the Arabic theory of great conjunctions to explain the Papal Schism that divided the Roman Church from 1378 to 1417. Drawing on the theory of great conjunctions to interpret history, d'Ailly then applied it to the contemporary situation when he argued that the Church needed to heal its rupture or risk the devastating effects of the next conjunction, effects that would likely include the apocalyptic end of the world.[27] D'Ailly had shown how the theory of conjunctions applied to the Christian context. Moreover, he had carried out most of the difficult calculations and analyses, making the results available to a wider audience. His texts were widely circulated, often bound with other astrological tracts, including Alcabitius and Albumasar.[28] In the late fifteenth century Johannes Lichtenberger's *Prognosticatio* offered another application of the theory to history. Published in 1488, Lichtenberger's text distilled the theory of great conjunctions into its simplest form, which he had borrowed wholesale from Paul of Middelburg.[29] To this theory he added excerpts from the medieval Joachim and Methodian prophecies. Lichtenberger's work provided a convenient assortment of various doctrines. By the end of the fifteenth century Albumasar's theory of great conjunctions had become a standard astrological technique for interpreting the past as well as the present and future.[30]

When Brant composed his "De corrupto" he would have known these textual traditions. His own education had included at least basic astrology and most likely more advanced practice as well.[31] His interpretation of the 1503 conjunction repeated some of the standard tropes in the astrological literature. Most important for Brant's analysis, conjunctions of the outer planets caused significant changes in religions, the destruction of kingdoms, and the transfer of rule. These set pieces were found in the Latin translations of the Arabic texts and were the backbone on which d'Ailly and Lichtenberger had built their own texts.

According to Brant, however, the conjunction in 1503 had a few particularly sinister characteristics. First, warlike Mars, which influenced the fierce Germans, was suppressed by the malevolent Saturn, a point that was visually reflected in the woodcut by placing Saturn above both Jupiter and Mars. To interpret the significance of this fact, Brant drew on chorography, the astrological practice of associating particular regions of the earth with certain planets and zodiacal signs. Ptolemy had codified the practice in his *Tetrabiblos*, where he mapped the celestial bodies onto regions of his known world. Chorography became a flexible practice that could be modified to apply to different regions. Arab astrologers associated the planets and signs with cities and countries that were important to them; later astrologers in Europe further adapted the doctrine for places familiar to them. John of Seville, who translated Albumasar's works, offered two different methods: one that depended on the planet and the sign, and a second that related just the signs to different cities.[32] At the end of the fifteenth century, the chorographic tradition was in flux. Traditional relationships between the celestial bodies and cities or countries were being remapped.

When Brant interpreted the effects of the conjunction in 1503, the fact that malevolent Saturn would overpower Mars portended dire events for the

Germans. Second, the conjunction took place in Cancer, a sign symbolized by the crab. Ptolemy had divided the signs into fixed, mobile, and common, according to the motion of the sun in each one. Cancer and Capricorn were considered mobile or tropical signs because the sun reversed its path in them.[33] Just as the crab seemed to move contrary to nature, walking irregularly, sideways, and backward, conjunctions in the sign of Cancer were thought to disrupt the natural order and signaled radical changes on earth. More worrisome were Albumasar's specific predictions about such conjunctions. According to contemporary editions of Albumasar's *De magnis coniunctionibus*, Saturn-Jupiter conjunctions in Cancer were favorable to the Turks. Moreover, conjunctions of Mars and Saturn in the sign of Cancer indicated the transfer of rule from one place to another and the rise of the Turks.[34] In Brant's analysis, the Germans risked losing the imperial title, while the Turks, by contrast, would benefit from the planetary conjunctions and would assume the imperial title. He worried that the Holy Roman Empire of the German Nation was on the verge of disaster, and that the Turks were poised to capitalize on that situation.[35]

Brant was no simple fatalist, however. Indeed, to adopt such a stance would have precluded his real goal, which was to persuade the princes and electors to accept Maximilian's authority and to unite behind the emperor in his efforts to raise an army to defeat the Ottomans. For Brant, the conjunction of the superior planets revealed what would occur if the uncooperative Germans did not heed his warning. In the final section of his poem Brant abandoned his pessimism and suggested that there was still hope. He described the pleasant order that Maximilian had introduced into the empire. He urged the princes and electors to set aside their bitter internal conflicts and to unite behind the emperor.[36] He then addressed the German people. He assured them that by respecting Maximilian's authority, paying their taxes, and maintaining their proper station, they could avert the conjunction's sinister influences.[37] Through unconditional support for Maximilian's financial and governmental reforms, "Jupiter will look down on us with a kind face, and our forces will be stronger than Alcathoe."[38] Alcathoe was the legendary citadel at Megara, located just north of the Isthmus of Corinth. It had been impregnable as long as Nissus, its king, retained his lock of purple hair. The king's daughter, however, fell in love with the Cretan king and cut off her father's magical lock. The city immediately fell to an invading Cretan army. Alcathoe had been conquered through treachery and internal strife. Brant did not need to state the obvious message: if the German princes continued to foment strife, they would cause the downfall of the Holy Roman Empire. Conversely, if they united behind Maximilian by supporting his reforms, the empire would retain the Roman Imperium.

Brant's rhetoric echoed Maximilian's own efforts to secure both financial and military support. The emperor had seen his earlier reforms unravel at the hands of the German princes and electors, who had co-opted many of his efforts following the previous diet at Worms.[39] Maximilian's financial and military situation had worsened when the French invaded the Italian peninsula, forcing him to commit troops and spend his own funds to expel them from the peninsula. In 1497 he had once again convened the diet in Worms hoping to secure financial and military support for his campaign against the French and subsequently against the Turks. In such a climate, Brant's pro-Habsburg propaganda seems to have worked at least insofar as it helped secure him a stipend. At this time Maximilian apparently began paying Brant a salary and came to rely on him to direct the Alsatian humanists.[40]

Brant's production reached a fever pitch in 1497 and 1498. In addition to his "De corrupto" that he appended to the new edition of the *Ship of Fools*, he composed a dedicatory poem for his forthcoming *Varia carmina* that dealt with the Ottoman threat. He also edited the revelations of St. Methodius, which prophesied that a united Christian army would vanquish the Turkish, and he wrote "The Power and Terror of the Turks," an adaptation of the Methodian prophecies that identified Maximilian as the only person able to defeat the Turkish armies and ultimately convert the sultan.[41] What distinguished Brant's "De corrupto" from his other work was the central role that astrology played in his effort to convince his audience that Germany needed to unite behind Maximilian to have any chance of defeating the Ottoman armies.

Joseph Grünpeck's Early Astrological Texts

Brant's use of astrology drew on Maximilian's own growing interest in the science of the stars. In the mid-1490s the young humanist Joseph Grünpeck also recognized the emperor's interest and capitalized on it when he published two astrological prognostications and two astrological pamphlets on the French Disease. Unlike Brant, who was content to support the emperor's efforts without moving into direct imperial service, Grünpeck used his early astrological texts to attract Maximilian's attention and ultimately to obtain a position at his court.

In 1487 Joseph Grünpeck was admitted to the University of Ingolstadt as a pauper, where he studied rhetoric with Conrad Celtis and medicine with

Wolfgang Peyser.[42] In 1491 he went to Pavia to study medicine, before relocating to the University of Krakow to combine the study of medicine with astrology.[43] By this time, Krakow had a long tradition of astrology and was the only Central European university to have a chair in the subject. Many graduates obtained positions as physicians or astrologers at important courts throughout Central Europe.[44] Grünpeck remained in Krakow only a short time before returning to Ingolstadt in early 1495, where he tutored Latin rhetoric to students from the university. Always ambitious, Grünpeck had cultivated relations with important humanists such as Conrad Celtis, Willibald Pirkheimer, and Conrad Peutinger, whose connections could help him find more prestigious patronage.

When the plague broke out in Augsburg in the summer of 1495 Grünpeck traveled to Italy in the hopes of meeting the imperial troops and attracting Maximilian's attention. Instead of finding the emperor, Grünpeck ran into Maximilian's retreating army, which had been ravaged by the French Disease. Grünpeck returned to Augsburg and resumed his job as a Latin tutor, all the while searching for a better position. In the summer of 1496, he applied to Duke George the Rich, asking to be appointed court historian, but his efforts fell on deaf ears.[45] Undaunted by this most recent setback, Grünpeck redoubled his efforts to secure an appointment outside of Augsburg—he wrote to Celtis complaining that he found the city oppressive.[46] The last few months of 1496 were extremely busy for Grünpeck. In rapid succession he composed Latin and German tracts on the origins and causes of the French Disease and two astrological prognostications for the coming years—an ornate manuscript dedicated to Hans Langenmantel, the Burgermeister of Augsburg, and a printed one. Grünpeck's deft use of astrology to support Maximilian's political efforts attracted the emperor's attention and helped him secure an appointment as Maximilian's personal secretary.

The diet at Worms in 1495 had been filled with tension and disputes between Maximilian and the electors and German princes. Maximilian arrived in Worms with a broad reform agenda, much of which was eventually approved by the princes and electors. On August 7, 1495, he brought the diet to a close with high hopes of establishing unity and cooperation between himself and the German princes. Although he had been forced to compromise on a number of his reforms, he still succeeded in establishing the *Landesfrieden*, a perpetual peace in the empire, the *Kammergericht*, an imperial court for settling disputes, and the *gemeinen Pfennig*, a new system of taxation that would fund Maximilian's wars. He had also introduced the idea of a common imperial army, which he intended to lead against first the

French and then the Turks. The first months after the diet witnessed some of the greatest cooperation between the princes and Maximilian. However, this cooperation was short lived, and by the middle of the next year Maximilian was struggling with the German princes over control of the empire's finances and military, and imperial authority.[47] In 1496 as the emperor continued to fund his war with the French on the Italian peninsula these struggles were not abstract political contests. Into this context Grünpeck cast his early astrological pamphlets.

Grünpeck's two prognostications reveal common themes that recur throughout his writings. He asserted that there are three methods of predicting the future: history, astrology, and prophecy. A thorough understanding of history was the least specific method of predicting the future. Nonetheless, Grünpeck saw patterns in history and by comparing the contemporary situation to the past, could foresee future events. Astrology lay at the core of Grünpeck's attempt to understand present events and predict future ones. Drawing on Ptolemy and Aristotle, Grünpeck argued that all terrestrial matter was subject to celestial influences. The past and the future were, then, written in the motions of the planets. He was quick to point out, however, that this did not imply some sort of fatalism. Instead, the planets and stars were part of nature's machinery and, properly interpreted by the expert, could indicate likely futures. History and astrology were confirmed by and in turn confirmed the prophetic texts. Grünpeck brought these three approaches to bear on the contemporary situation, analyzing the prodigious celestial and terrestrial phenomena and predicting dire consequences.[48] Although Grünpeck was not the first to combine history, astrology, and prophecy in this way, or even the first to apply them to the Habsburgs—he certainly knew of Brant's political broadsheets and he silently borrowed Lichtenberger's approach—he took advantage of the urgent political situation and aligned his analysis with Maximilian's political goals.[49]

In these early works, Grünpeck identified the pope, the electors, and the German princes as the real threats to peace and prosperity in the Holy Roman Empire, in particular, and Christendom more generally. By 1496, Maximilian's position in the empire and with respect to the pope was once again on the decline. His successes at the diet in Worms had been short lived. He had watched the princes, led by Berthold of Henneberg, the Archbishop of Mainz, dismantle his various reforms.[50] In the meantime, Maximilian remained bogged down in his Italian wars, ostensibly against French incursions into the peninsula, but always wary of Pope Alexander VI, who had notoriously switched allegiance between the French and the emperor.[51] By mid-1496, when

Maximilian passed through Augsburg on his way back to Italy, his position was as precarious as it had been a year earlier. Grünpeck's decision to blame these factions for upsetting God's natural order seems to have resonated with Maximilian.

Grünpeck acknowledged that the pope occupied a special place among people on earth. Nevertheless, he reminded his reader that the pope was, like all people, subject to God's laws and judgments, which were handed down through the planets and stars. It was sheer arrogance to think that God would allow His only son to be subject to His natural laws and yet would absolve the pope from them. The current pope was no exception. Grünpeck implied that Pope Alexander VI had done little to suppress the internecine wars and malicious gossip that plagued the Church and empire. He insinuated that the pope had caused strife throughout Christendom by setting the French against Maximilian. Although the pope had since turned against the French, prompting Charles VIII's invasion of Italy in 1494, Grünpeck and many of his German contemporaries blamed the pope for the current situation. Grünpeck linked the pope's duplicity to the spread of the French Disease up the Italian peninsula and into the imperial troops, who had been sent into Tuscany to fight the French. According to Grünpeck, as long as the effects of the great conjunction in 1484 continued, Alexander and his successors were potential impediments to peace in Christendom.[52]

In addition to these conflicts with the pope and the French Maximilian had to confront domestic unrest and rebellion among the princes and German bishops and archbishops. Echoing Maximilian's hope for unity between the emperor and the German princes, Grünpeck accused the electors, princes, and all levels of the clergy of failing to show the proper degree of loyalty and deference to Maximilian. Grünpeck repeatedly blamed Maximilian's unruly subjects for the plagues, foreign conflicts, famine, and general decline in society.[53] In both his manuscript and printed prognostications, Grünpeck lamented the current state of affairs. Not only was the Church suffering from lax morals and an excessive focus on worldly affairs, but there was general disregard for traditional social hierarchies.

Grünpeck found comfort in his predictions for the future, predictions that rested squarely on his combination of astrology and prophecy and that reinforced the increasingly common belief that Maximilian was the prophetic last emperor. Millennial speculations in the middle of the thirteenth century in the Latin West had placed Emperor Frederick II in the messianic role of the Last World Emperor, who would battle the Antichrist and usher in the 1,000-year reign of Christ. Although Frederick died without fulfilling these expectations, speculations about a Last World Emperor did not die with him. Indeed, his death seemed to introduce a period of increasingly intense millennial fears, all revolving around a future messianic Frederick III. In the fifteenth century, eschatological fears took on renewed significance when, in 1452, a new Frederick was crowned Holy Roman Emperor and then, one year later, the Ottomans conquered Constantinople and threatened all of Christendom. These events seemed to confirm the popular Methodian prophecy of a last Roman emperor who would defeat the infidels and surrender the imperial insignia to God as a sign of the Second Coming. Frederick III had been the object of these speculations for much of his reign, but he continued to disappoint his supporters. Johannes Lichtenberger, the most famous of the fifteenth-century prophetic authors, had long considered Frederick the prophetic Last World Emperor. By 1488 when he published his widely read *Prognosticatio*, Lichtenberger had shifted his support to Maximilian.[54] Grünpeck knew of these efforts to identify Maximilian as the Last World Emperor when he composed his prognostication in the mid-1490s.

Initially, Grünpeck hoped that the worst had past. In his manuscript prognostication that he dedicated to Langenmantel, he claimed that the effects of the great conjunction in 1484 had caused Maximilian innumerable problems and setbacks. The stars indicated, however, that Maximilian would prevail and soon extend his power and authority over the entire world.[55] Ultimately, Maximilian was preordained finally to receive the imperial crown, which would, in turn, further enhance his authority and finally enable him to subdue the French and drive the infidel from the earth. In his printed prognostication Grünpeck was more precise. There, he claimed that Maximilian would continue to be beset with complications arising from the great conjunction until its effects abated, which would occur in 1499, when Maximilian turned forty.[56] Grünpeck had no doubt where to place his support. His two tracts on the advent and spread of the French Disease, show the same degree of pro-Habsburg rhetoric.

At the end of the diet on August 7, 1495 Maximilian issued his Blasphemy Edict, which was promptly printed in both German and Latin and circulated throughout the empire. In it, Maximilian explicitly connected the blasphemous practices of his citizens with the French Disease epidemic in the empire. He claimed that God was punishing humans with famines, earthquakes, pestilences, and other plagues "and now, in our time, as is obvious, many similar and worse plagues and punishments have followed. Especially these days, there is the disastrous disease and human plague, named the evil pox, which never before in human memory has existed or been heard of."[57] For

Maximilian, the disease was the culmination of a series of punishments God had sent, and was particularly frightening because it had never before been seen. He appropriated this new disease, making it part of his larger reform agenda and urging his citizens to amend their social mores in order to reduce the threat of the disease. Grünpeck seized on the text and spirit of Maximilian's Blasphemy Edict when he produced his two tracts, the Latin *Tractatus de pestilentiali Scorra sive Mala de Franzos* and the German *Ein hübscher Tractat von dem Ursprung des Bösen Franzos das man nennet die Wylden Wärzten.*[58] Grünpeck modeled his tracts on the plague *concilia* that were printed in great numbers in towns throughout Europe whenever plague returned.[59] Like his models, he relied on astrology to explain the origin, spread, and symptoms of the French Disease. At the same time, he used these pamphlets to promote himself to Maximilian's court.

Grünpeck had no doubt that the ultimate source of the new plague was the will of God. The Holy Scriptures attested to God having punished the sins of pride, greed, and lust with pestilences, bloodshed, and famine.[60] The French Disease was merely the latest of God's punishments, sent at this time when these and worse sins were common among people.[61] Although God's will was the ultimate source, Grünpeck sought to explain its physical causes.[62] Apparently sensing no tension in his account, he explained to his reader how the motions of the planets and planetary conjunctions caused all manner of physical change on earth. For Grünpeck, the French Disease had an ultimate, divine source and an astrological cause. A proper account had to incorporate both.[63] Of these, however, Grünpeck had more to say about astrology.

The central feature of Grünpeck's astrological explanation was the theory of great conjunctions. He devoted an entire chapter to elaborating on the theory of great conjunctions and other types of astrological causation, all of which he would draw on in accounting for the advent, spread, and symptoms of the French Disease. In order to employ the theory of great conjunctions, however, Grünpeck first had to establish two important conditions. He had to determine the horoscope for the earth that showed the configuration of the heavens at the moment of creation. From this he could then calculate when the first Saturn-Jupiter conjunction had occurred in Aries. Finally, he needed to establish the age of the earth, so that he could calculate the number of years that had elapsed from that first conjunction until the present. Only then could he use the theory to account for the appearance of the disease.

The tradition of trying to establish the horoscope for creation, the *thema mundi*, stretched back at least to the fourth century, when Julius Firmicus Maternus used the assumption that a planet's exaltation—the portion of the zodiac in which the planet was considered particularly efficacious—corre-sponded to its location in the zodiac at the moment of creation.[64] Maternus and those who followed his example placed Aries at the highest point in the sky, the *medium coeli*, and the sun in Leo. The Christian tradition, by con-trast, placed the sun near the beginning of Aries, the vernal equinox, so that the Creation, Annunciation, and Crucifixion could occur on the same day. Pierre d'Ailly offered a variation from these two traditions, which he pub-lished in his *Concordantia astronomie cum theologia*.[65] Grünpeck borrowed d'Ailly's *thema mundi* and then used it as the point from which to calculate the first great conjunction and all subsequent conjunctions.

The *thema mundi* was only the first half of the problem confronting Grün-peck if he wanted to use the theory of great conjunctions to explain the French Disease epidemic. He also had to determine the chronology of the earth. The mathematical details would have been out of place in Grünpeck's texts and unnecessary to establish the astrological framework and display his expertise. Consequently, he simply presented the chronology, once again borrowing from d'Ailly's *Concordantia astronomie cum theologia*: 2,242 years from cre-ation to the Flood; 942 years to Abraham; 940 years to King David; 485 years to the expulsion from Babylon; 590 years to Christ.[66] Grünpeck then listed the seven greatest conjunctions between Saturn and Jupiter from the first 320 years after creation to the latest 735 years after Christ. When he could not correlate these conjunctions to any significant historical events he invoked the more frequent conjunctions between the two outer planets. He identified the source of the disease in the most recent Saturn-Jupiter conjunction, a greater conjunction that had occurred on the evening of November 25, 1484. Yet even this most recent greater conjunction preceded the onset of the disease by a decade. He still needed an explanation for natural disasters and the advent of plagues, an explanation that would also help him bridge the ten-year gap between the last conjunction and the outbreak of the French Disease. For this he relied on two further forms of astrological causation. First, Saturn-Mars conjunctions, which occurred every thirty years, signaled both famine and war.[67] Second, citing Albumasar, Grünpeck pointed to the theory of ten rev-olutions of Saturn, which influence the advent of pestilences, dire sicknesses, and death.[68] Together, these different types of celestial causation played a role in Grünpeck's account of the advent and spread of the pox.

Having set out his astrological theory, Grünpeck summarized the his-torical details that confirmed it. Three different planetary conjunctions had caused the French Disease. The first was the greater conjunction at 6:04 pm on November 25, 1484, when Saturn and Jupiter had met at 23°43' of Scorpio, a sign ruled by Mars. The solar eclipse that followed on March 25, 1485, was the second conjunction, whose effects were made worse by its proximity to the preceding November's greater conjunction. Finally, Saturn and Mars came together at 9° of Scorpio on November 30, 1485. Like the eclipse, this conjunction was under the malevolent influence of the greater conjunction a year earlier.[69] These conjunctions had a cumulative effect and ushered in a period of increasingly dire catastrophes, including plagues, famines, war, and natural disasters. Grünpeck reminded his reader of these disasters: Germany had been suffering from droughts and famines since the middle of the previous decade; war between the empire and Charles VIII of France was raging in Italy; the plague had become endemic in many southern German cities, breaking out most summers; recently earthquakes had been reported in different parts of the empire; and worst of all, the disease that had spread from the French troops into Germany.[70] He then invoked the ten revolutions of Saturn, claiming that the most recent occurred in 1489 and had been followed by: "numerous great evils, at first the great famine, which lasted for a full seven years and has not ended yet. After that the cruel pestilence, which also still rages, and the great war with the King of France. Beyond these ills, there now comes the fearfully cruel sickness, and the aforementioned French Disease."[71] Grünpeck emphasized the celestial causes of the recent catastrophes, the most severe of which was the French Disease.

Using astrology, Grünpeck just as easily explained the spread of the French Disease and its symptoms. Jupiter was a hot and moist planet that ruled over France. Its normally beneficent effects had been suppressed by the conjunction of the Saturn and Mars in 1485. Consequently, the French were the first to succumb to the new disease.[72] He explained the disease's spread into Germany by pointing to Mars's influence during the conjunction—Mars had been the lord of the conjunction—and the fact that it influenced the warlike Germans.[73] Finally, Grünpeck described the disease's symptoms. The seeping black sores, rancid stench, fevers, and burning in the limbs resulted from an excess of two humors, melancholy and cholera, both of which he traced back to their celestial origins.[74] Saturn influenced the production and expulsion of melancholy while Mars exercised similar control over cholera. Their conjunction in the sign of Scorpio, which ruled over the genitalia, accounted for the concentration of the black pustules on and around the genitals.[75]

Grünpeck's *Tractatus de pestilentiali scorra* and *Ein hübscher Tractat* were more than just his attempt to account for the French Disease. Capitalizing on the fact that the disease was a serious social problem, one that Maximilian had raised at the previous diet in Worms, Grünpeck used his texts to demonstrate his expertise to Maximilian and the court and to show his support for Maximilian's political agenda. As it was in Latin, the *Tractatus de pestilentiali* was beyond the reach of most people but served to advertise Grünpeck's abilities to a select audience, such as the Habsburg court. His vernacular *Ein hübscher Tractat*, by contrast, was accessible to a much broader audience, disseminating the text and its pro-Habsburg message to many more people.

Grünpeck dedicated his *Tractatus de pestilentiali* to his long-time friend Bernard Waldkirch. By 1496 Waldkirch had become a familiar figure among Maximilian's courtiers and had recently been appointed canon of the Augsburg Cathedral. The next year he facilitated Grünpeck's admission into Conrad Celtis's *Sodalitas litteraria Danubiana*—a circle of scholars Celtis established in 1497 with close ties to the Habsburg court.[76] Grünpeck used his *Tractatus de pestilentiali* to demonstrate his astrological expertise and his eloquence, talents that Maximilian apparently valued. Laden with classical allusions and learned digressions, Grünpeck's text reflected his efforts to persuade Maximilian that his services could be useful at the Habsburg court. Grünpeck implied that his intellectual talents and erudition placed him among "the most distinguished men, endowed with the highest learning, virtue, and wisdom."[77] At the end of his work, Grünpeck compared his text to some of the greatest works of antiquity, including Phidias's Minverva, Apelles's Venus, Homer's poems, and Virgil's *Aeneid*, all of which had been cherished by ancient kings and emperors.[78] He considered his work worthy of the highest accolades.

In the late 1480s and early 1490s, the University of Ingolstadt had become a center for humanist education and learning, centered on Conrad Celtis and his efforts to introduce humanist scholarly techniques and to establish the first of his many sodalities.[79] Celtis's efforts were instrumental in gaining imperial favor and his being crowned poet laureate—a title Maximilian awarded for explicitly political purposes—in 1487. Like Celtis, Grünpeck larded his text with classical allusions and indications of his own humanist talents. Xenophilus, the long-lived musician, and the Praenestine oracles along with Egyptian sages, Greek gods, and the Roman author Valerius Maximus all played important supporting roles in Grünpeck's work, alongside more typical sources such as the Bible, Aristotle, Galen, and Ptolemy. Grünpeck mixed classical allegory with astrology, humoral medicine, and physiology without distinction.

The title-page woodcut situates Grünpeck's text within Maximilian's political program by reinforcing the emperor's Blasphemy Edict and highlighting his position as the divinely chosen ruler (see figure 2.2). In the woodcut the baby Jesus sits on the Virgin's lap and heals a pair of infected but

¶ Tractatus de pestilentiali Scozra siue mala de Franzos.
Oziginem. Remediaꝫ eiusdem continens. cōpilatus a vene
rabili viro Magistro Joseph Gzynpeck de Burckhausen.
sup Carmina quedam Sebastiani Bzant vtriusꝫ iuris pzo
fessozis.

Figure 2.2. Title-page woodcut of Grünpeck's *Tractatus de pestilentiali scorra*. The same woodcut adorned the title page of his *Ein hübscher Tractat von dem Ursprung des Bösen Franzos*. Courtesy of the Historical Medical Library of The College of Physicians of Philadelphia, call # HHc 99 1497.

supplicant women. The corpse of an infected man who had not been sufficiently pious lies in the foreground while Maximilian kneels to the Virgin's right and receives from her the imperial crown. The woodcut suggested that even those afflicted with the French Disease could be cured through proper religious devotion. Moreover, Maximilian's privileged position on the Virgin's right depicts him as divinely chosen, pure in his devotion and thus in his right to lead the empire.

Grünpeck pruned his vernacular *Ein hübscher Tractat* of the more learned allusions and discussions found in his Latin text. Consonant with his goals and the purpose of the text, this version was a simpler handbook on the source and cause of the pox, one that conformed to the patterns and expectations of the late fifteenth-century vernacular plague pamphlets that had become common. An important aspect of most plague pamphlets was their connection to the local city council or prince. During outbreaks local authorities tried to assuage the fears of the citizens who were forced to remain in the city and infected areas. Attempts to compel physicians to stay, through either higher payments or threats of punishment, generally failed. Town councils did succeed in persuading local physicians to write short, vernacular texts on the origins, spread, and prevention of the disease. These cheap, vernacular pamphlets were widely available and advised readers on ways to avoid the plague or treat themselves during outbreaks. Linked to the local political authorities, plague pamphlets were political tools aimed at consoling the fears of the populace and quelling any unrest that occurred during epidemics.

Grünpeck's *Ein hübscher tractat* functioned like the plague pamphlets. He offered his reader a description, in simple German, of the origins and causes of the French Disease. His text suggested to the city fathers in Augsburg that he could serve a particular function. Indeed, he seems to have anticipated a practice that became more common in the sixteenth century, when city councils retained physicians specifically to treat the French Disease.[80] Grünpeck even dedicated his German text to Hans Langenmantel, the Burgermeister of Augsburg. However, Grünpeck's text also supported his efforts to secure a position at the imperial court by serving as a piece of propaganda in Maximilian's political program. Maximilian was constantly struggling to build cohesion and support among his German subjects and to generate support for his reforms in all strata of society. Grünpeck's vernacular *Ein hübscher tractat* retained all the political, pro-imperial implications of his more elite text, including the disparaging references to the French and the title-page woodcut showing Maximilian as savior. It helped to portray Maximilian as concerned for his subjects' health and welfare. Like other pieces of Maximilian's propaganda that were intended for broad audiences and, consequently, often printed in both Latin and German and distributed widely through the empire, Grünpeck's *Ein hübscher tractat* was simultaneously printed in

Augsburg, Cologne, Leipzig, and Nürnberg, thereby spreading Grünpeck's pro-Habsburg message across much of southern Germany.

Whether by happy coincidence or shrewd planning, Grünpeck's pro-Habsburg stance and his combination of history, astrology, and prophecy appealed to Maximilian's tastes. In 1497 Grünpeck was invited to present a play before Maximilian and his court. On November 26 Grünpeck presented his *Virtus et Fallacicaptrix*, a comedy in which true strength and virtue struggled against deceit and sensual lust, allusions to Maximilian and to the French. Grünpeck's carefully orchestrated play cast Maximilian in the key role as the contest's judge who decided in favor of strength and virtue. The play was the latest attempt by German humanists to associate Maximilian with the Greek hero Hercules.[81] At about the same time an anonymous single-sheet woodcut depicted Maximilian as the "German Hercules" (see figure 2.3). Grünpeck reinforced such efforts when he cast Maximilian in the role of Hercules in his own play, an adaptation of the fable "Hercules at the Crossroads."[82] Maximilian rewarded Grünpeck by having him crowned poet laureate on August 20, 1498, at the diet in Freiburg and then immediately appointing him personal secretary. Among his other tasks, Grünpeck was responsible for collating Maximilian's autobiographical musings in order to produce a celebratory biography.[83]

Portents, Pamphlets, and the Pro-Imperialist Agenda

For the next few years Grünpeck enjoyed a close relationship with the emperor, who showed his appreciation by granting Grünpeck a substantial yearly salary of 161 guldens.[84] In 1501 when Grünpeck contracted the French Disease at a party thrown by his friend Celtis, he had to withdraw from the court and to return to his hometown and seek a cure. Despite his absence from the court, Grünpeck remained a strong proponent of Habsburg rule and evidently maintained his connections. The following year he dedicated his "Prodigiorum potentorum" to Blasius Hölzl, who was rapidly becoming one of the most important people at Maximilian's court. Hölzl's keen financial sense had immediately attracted the emperor's attention. By 1498 he had been appointed finance secretary of the newly created *Hofkammer*. By the turn of the century, he was negotiating the conditions of loans with the Fuggers and arranging most of Maximilian's financial affairs.[85] Hölzl also became an important patron of literature and poetry at Maximilian's court and helped humanists gain access to the imperial court.[86]

In the dedication to his "Prodigiorum potentorum" Grünpeck promised

FIGURE 2.3. Anonymous single-sheet woodcut depicting Maximilian I as the German Hercules, 28.3 x 17.8 cm. Vienna, Albertina, DG1948/224r.

to explain the monstrous events, both celestial and terrestrial, that seemed to be increasing every day. These various prodigious events had entered the popular consciousness in the rumors circulating at the time and had become set pieces in debates about political and social reforms.[87] For Grünpeck, they offered the perfect opportunity to assess the state of the empire, to assign blame for the current lamentable conditions, and to encourage the rebellious princes and cities to unite behind Maximilian and stave off the wrath of God.

Grünpeck opened his elaborately illustrated manuscript by listing the many prodigious and monstrous events that had occurred during Maximilian's reign (see figure 2.4). He did not lack material. Since Maximilian had succeeded his father, reports of celestial portents had become increasingly common. Grünpeck recounted stones raining down from the heavens, strange lights and stars in the sky, fiery rains, and apparitions of armies battling in the sky. These unusual celestial phenomena were accompanied by equally disturbing terrestrial events: a pair of twins conjoined at the head, a boy with two heads and four arms, and a goose with two heads and four feet were just part of the macabre army of monsters marshaled together. Grünpeck argued that these marvels arose from the contemporary social and political situation and threatened future disasters.[88]

The monsters and prodigies reported all over Germany at the end of the sixteenth century were just the most recent manifestation of a long tradition of portents. As Grünpeck explained and illustrated in the subsequent chapters, history revealed that monstrous events had augured the demise of nearly every significant empire or reign, including the Persians, Alexander the Great, the Medes, the Egyptians, and even Rome itself. Echoing Brant, Grünpeck worried that the latest series of portents threatened the transfer of power from the Germanies and the loss of the imperial title.[89] Grünpeck's argument by analogy was persuasive, frightening and, at the same time, a warning. If the fate of each empire, ruler, or city had been determined by the motions of the stars and planets, there would have been little anyone could do to change that fate. But this was precisely what Grünpeck did not want Hölzl or any other reader to conclude. Instead, he considered these prodigious celestial and terrestrial phenomena as portents of a possible future. Portents were, according to Grünpeck, intimately linked to human actions. In this case, they were signs from God expressing His displeasure with the German citizenry and its blatant disregard for traditional hierarchies and Maximilian's rule. The dire consequences that Grünpeck prophesied could be avoided if the German people amended their ways and improved the situation that had initially invoked God's wrath.[90]

FIGURE 2.4. A drawing from Joseph Grünpeck's "Prodigiorum potentorum" showing a youthful Maximilian confronting the many monsters and prodigies that had arisen since his crowning. Universitätsbibliothek, Innsbruck, codex 314, 4r.

A common theme linked all of Grünpeck's portents, both celestial and terrestrial: truculent German princes and rebellious cities were disrupting the natural order. The proper state of affairs would be for them to unite behind the emperor and support his imperial reforms and foreign campaigns. Grünpeck reminded his reader that immediately before the diet in Worms in 1495, twins conjoined at the forehead had been born in the town. They quickly became part of the pro-Habsburg propaganda and were taken to be a sign from God that Pope Alexander VI and Maximilian were finally about to establish a permanent allegiance.[91] The unity between the empire and the papacy did not materialize. Seven years later Grünpeck claimed that this missed opportunity was in part responsible for the increase in portents seen recently.[92] Grünpeck was unambiguous about the social and political importance of these signs: each portent revealed what sorts of monsters result when the natural order was overturned. Just as God had established an order for nature, He had likewise determined a hierarchy for the social and political world. If the laws of that social world were violated, the result would resemble the monsters and marvels currently terrifying people all over the empire. In the worst case, these monsters and prodigies signaled the downfall of the empire and the transfer of power from the Germanies.[93] This bleak prediction could be avoided if people heeded the warning signs and maintained the proper social hierarchies. In other words, only if the electors, princes, and cities ceased to oppose Maximilian and instead supported him could the empire successfully avoid Grünpeck's dire predictions and retain the imperial title.

When Grünpeck composed his "Prodigiorum portentorum," he was still suffering from the disease he had contracted and surely realized that he would not be able to return to the court, at least not in person, until after he had recovered. Nevertheless, he was unwilling to relinquish the intimate ties he had cultivated with Maximilian. His "Prodigiorum portentorum" helped him maintain these relations. In the end, he was successful both in recovering from the disease and in relocating back to the imperial court. By 1506 he was once again working on Maximilian's memorial projects.[94] A year later, he had rejoined the emperor's immediate retinue, had resumed work on Maximilian's autobiography, and was again producing pro-Habsburg pamphlets.

Although Maximilian was de facto the emperor, he was officially only King of the Romans. He felt acutely the lack of sanction for his position and hoped to journey to Rome to receive the imperial crown from the pope. An important ceremony in the transfer of authority to the German emperors, formal papal recognition of Maximilian's imperium had so far been denied to him. In 1506 and 1507 Maximilian renewed his plans to travel with his entourage to Rome and receive papal confirmation. Finalizing these plans was one of the main items on Maximilian's agenda when he convened the diet in Constance in 1507.[95] Maximilian had originally intended to begin the diet on February 2 but postponed it until March 25. The later date likewise passed without incident, and Maximilian did not arrive and convene the diet until April 27. Many of the delegates, however, had arrived in early April and had taken up residence in the city, where they remained during the delay. This presented Maximilian with a captive audience that he could try to influence before the diet opened a few weeks later. Maximilian had become an expert in wielding the printing press to sway popular opinion and capitalized on the occasion by flooding the town with pro-imperial pamphlets.[96] Once again Grünpeck produced pamphlets that supported Maximilian's political agenda.

In his *Ein newe außlegung der seltzamen wunderzaichen und wunderpürden* Grünpeck boldly interpreted recent prodigious celestial and terrestrial phenomena with a keen eye focused on Maximilian's imperial plans. Grünpeck addressed his work to the princes, bishops and archbishops, and delegates from the estates who were in Constance to attend the imperial diet. He warned that the recent comets, monsters, rains of milk and fire, and mis-births all portended grave disasters for the German nation. If delegates followed his advice, however, they could avoid the impending disasters or at least mitigate the consequences.[97] For Grünpeck, this was an easy case to present. He had, after all, made the same argument five years earlier.

Grünpeck enlisted the standard range of portents—comets, strange apparitions in the sky, unnatural rains of stones, crosses and blood, and monsters and mis-births—in his attempt to persuade the German princes to support Maximilian. He explained that God had sent each of these signs as a warning for a particular transgression. The headless child was a sign that the princes were wrongly opposing Maximilian in both word and deed. They should, instead, recognize that Maximilian had been chosen by God to lead the empire. The boy with two heads and four arms and feet indicated the lies and treachery common among the princes and estates. The conjoined twins born in Worms made yet another appearance in this work, though they now signaled a rapprochement between the estates and Maximilian. Grünpeck admonished the princes for their own sedition and that of their citizens. He called on them to root out the enemies of the empire, especially those within its borders who incited discord and resisted Maximilian's authority. Grünpeck made it clear that Maximilian was the divinely ordained emperor. By resisting his reforms and preventing him from traveling to Rome to receive the imperial crown, these princes had overturned the God-given natural and social hierarchy, thereby provoking His wrath. The German nation was, for the moment, the head of Christendom, but if the princes and cities did not unite behind Maximilian, they would cause the Germans to lose the Roman Imperium, which would pass into foreign hands. Furthermore, their insurrection would call down on them plagues, disasters, and deaths that had beset other kingdoms and cities that had failed to heed God's warnings.[98]

Whatever the effects of Grünpeck's text, the delegates in Constance withheld their unconditional support. Maximilian's requests for money and soldiers met with limited success, and his efforts to establish a route and secure the finances for his journey to Rome were all but thwarted.[99] In the end, Maximilian had to be content with the title of Emperor-elect, conferred with the consent of Pope Julius II on February 4, 1508, in Trent.[100]

Grünpeck continued to produce pamphlets that supported Maximilian's political efforts. In 1508 he produced his most successful works, his Latin *Speculum naturalis cœlestis & propheticæ visionis* and the German translation, *Ein spiegel der naturlichen himlischen und prophetischen Sehungen aller Trübsalen*. Once again, Grünpeck used the triad of history, astrology, and prophecy to interpret the contemporary state of affairs in the Holy Roman Empire and, more broadly, in Christendom.[101] Perhaps reflecting Maximilian's preference for astrology over portents and monstrous births, in these later works Grünpeck abandoned the terrestrial monsters and prodigies and focused instead on the various planetary influences and the effects of conjunctions and eclipses.[102] He once again assumed the role of expert, warning his readers of the disasters foreshadowed by the celestial configurations and counseling them on how best to avert those disasters. According to Grünpeck, the citizens of the empire and throughout Christendom either could renounce their divisive practices and enjoy prosperity or could continue down the path of ruin and disaster. Grünpeck pointed out that the current celestial phenomena resembled those that signaled the transfer of Christianity from the Jews to the Romans. He wove his astrological argument together with an analysis of the Old Testament prophets Ezekiel and Isaiah to show that the Roman Church was on the verge of losing its authority in Christendom. By heeding Grünpeck's warnings the Church could avoid this fate.[103]

As he had done in his earlier works, Grünpeck again blamed the uncooperative German princes for the problems besetting Christendom. He accused

them of infecting all levels of the Church with their worldly ideals. Grünpeck may also have had in mind Pope Julius II, who just the year before
had been instrumental in preventing Maximilian's trip to Rome to receive
the imperial crown. Julius was famous for having brought to the Holy See
the ideals and concerns of worldly princes. By 1508 he had already led one
military campaign to expand the authority of the Papal States and to regain
Bologna and Perugia, and he was constantly trying to extend his control over
Christendom. Moreover, he had become one of the most powerful patrons
in Italy. His efforts to remake Rome produced some of the sixteenth century's finest works of art and architecture, including Michelangelo's ceiling in
the Sistine Chapel, Raphael's Stanze in the Apostolic Palace, and St. Peter's
Basilica designed by Donato Bramante. Julius brought a single-minded goal
to the Papacy: reestablish the authority of the Roman Church in the person of
the pope himself.[104] For Grünpeck, Julius's actions put him in direct conflict
with Maximilian. The pope should attend to the religious state and leave the
worldly affairs to the emperor. Julius's unrelenting efforts to extend the secular control of the Church would, in the end, destroy it.[105]

For much of Maximilian's reign the popes had ranged from allies, always
promising more than they would deliver, to enemies, usually aligned with
the French to prevent expansion of imperial control in Italy. Grünpeck's pro-
imperial, anti-papal stance ensured that his *Speculum* appealed to Maximilian,
who shortly after it was published wrote to Willibald Pirkheimer asking him
to purchase a copy of the work and send it to him. Unfortunately, Pirckheimer
informed the emperor, there were no copies available. Grünpeck's *Speculum*
had sold out. Apparently, Maximilian was not alone in wanting a copy of
Grünpeck's most recent work.[106] The German edition was reprinted numerous
times through the early 1520s and became Grünpeck's most popular work.
In the 1520s and 1530s Grünpeck repurposed his *Speculum* for Maximilian's
Habsburg successors, Emperor Charles V and Archduke Ferdinand. His
subsequent works once again marshaled astrological arguments in support of
Habsburg political and dynastic goals.[107]

Grünpeck continued to produce astrological works for the remainder of
Maximilian's life for both the emperor and local city councils; in 1511 he produced a judgment for Regensburg.[108] Maximilian maintained close relations
with Grünpeck and rewarded his services with the income from the mill in
Steyer. When Grünpeck heard that the emperor was on his deathbed, he
rushed to Wels to be present during his last moments. After Maximilian's
death, Grünpeck shifted his support to Emperor Charles V. Although he
found Charles to be a more difficult and inconvenient patron, he continued

to sing the praises of the Habsburgs and the pro-imperial position. He also cultivated his relationship with the more accessible Archduke Ferdinand and in 1527 cast a laudatory horoscope for Ferdinand's son Maximilian.[109] Grünpeck died in 1532 as committed to the Habsburgs as he had been four decades earlier. In his last work, a prognostication for the years 1532–40, Grünpeck portrayed Emperor Charles V as the greatest hope for repelling the Ottoman threat. Although the Turks had recently been driven from the Austrian plains, the siege of Vienna was still fresh in everybody's mind, and the Turkish threat seemed poised to invade Europe again. Grünpeck found solace in the celestial portents and planetary positions, which led him to predict that by 1536, after significant struggles, the Christian armies would retake Constantinople and ultimately Babylon.[110] Grünpeck repeatedly enlisted the stars and planets and the terrestrial prodigies they produced in his pro-Habsburg propaganda.

* * *

Sebastian Brant was not as sanguine as Grünpeck about the efficacy of his propaganda. Although he had produced a number of pro-imperial broadsheets in the 1490s, by the turn of the century he was convinced that his efforts had failed to change the attitudes and practices of the German people. In 1504, Brant lamented that because his predictions about the ominous planetary conjunction had not been heeded the empire was on the verge of destruction. In a letter to his friend Konrad Peutinger, Brant said:

> My predictions cannot be unknown to you (I believe): Did I not foretell about a decade ago in general terms and explicitly in my *Ship of Fools*, that there can be no constancy in human affairs? Indeed, I constructed the horoscope of that dreadful conjunction of the superior planets [of Saturn, Mars, and Jupiter in Cancer on October 2, 1503]. Whether the stars really do influence earthly events, as many authorities claim, or their conjunctions and oppositions have no affect at all, as our Pico believes, I was, alas, only too correct in my prognostications.[111]

Internal conflicts, refusal to support Maximilian's centralizing reforms, and the lack of unity among the Germans had divided the empire and ensured that Germany would relinquish the imperium to a foreign kingdom. For Brant, printed propaganda had ultimately failed to save the empire.

Shortly after Brant's letter to Peutinger, Maximilian began looking beyond Brant's and Grünpeck's printed propaganda in his efforts to generate support for his reforms and, ultimately, to establish his authority in the empire. Increasingly he drew on other astrological tools to bolster his political and dynastic programs. Astrological instruments, a corporate body of astrological experts at the University of Vienna, and more concerted printing campaigns became important mechanisms in Maximilian's efforts to accomplish his reform agenda and to establish the House of Habsburg as one of the most important European dynasties.

Chapter Three

Teaching Astrology

B y 1495 humanists around Maximilian and masters at the University of Ingolstadt were encouraging Conrad Celtis to move to Vienna.[1] In 1497 Maximilian himself urged Celtis to come to Vienna to revitalize the university, reform the teaching practices, and help restore its former glory.[2] Although the emperor emphasized Celtis's role as a professor of rhetoric and poetry, he expected Celtis to attract the brightest scholars from around Germany, including the members of his sodalitas at the University of Ingolstadt.[3] Maximilian was not disappointed.

Shortly after arriving in Vienna Celtis established the *Collegium poetarum et mathematicorum* in order to provide an institutional home for the scholars coming to the university. The *Collegium* was a distinct faculty within the university that was directly under Maximilian's control, providing him with a corporate body of experts he could turn to for political advice and could enlist in his dynastic program. Maximilian realized that his own reputation was enhanced by the caliber and number of scholars he could attract to his university, and supported scholars whose intellectual activities had clear political uses. For Maximilian, the masters at the university were political advisers as well as academics who moved between the university and the court. Consequently, their activities at the university—their lectures,

FIGURE 3.1. Hans Burgkmair's *Allegory of the Imperial Eagle*, a woodcut from 1507. Dimensions: 12 ⅞ x 8 ⁵⁄₁₆ in. New York, the Metropolitan Museum of Art, Harris Brisbane Dick Fund, 1926, Accession Number 26.72.87, URL: www.metmuseum.org.

their publishing efforts, and their teaching practices—all occurred with an eye on the local Habsburg court. They succeeded at the university and in securing imperial favor to the extent that their work reflected Habsburg values.

When Maximilian revitalized the University of Vienna, he supported astrology and related disciplines because they offered a distinct and distinctly persuasive body of knowledge that had political instrumentality.[4] Under his patronage, the university became one of the most important in northern Europe, famous as a center for both humanist studies and mathematical disciplines, especially astronomy and astrology.[5] The masters in the arts faculty and at Celtis's *Collegium* who realized Maximilian's goals of revitalizing the university reflected the particular forms of astrological expertise that he found most useful. Maximilian's patronage of the university quickly became part of the iconography surrounding him and his achievements. In 1507, Hans Burgkmair worked closely with Celtis to design and produce a woodcut that was an allegory of Maximilian, depicting him not only in his own image of himself as a generous patron of the arts and sciences and the University of Vienna but also in Celtis and the university's ideal of Maximilian as generous patron (see figure 3.1).[6] The crowned, double-headed imperial eagle protects Maximilian, the nine muses, and the seven liberal arts. On the eagle's wings human invention of the seven mechanical arts mirror God's work during the seven days of creation. Maximilian sits enthroned above the muses, clearly identified as their source. Water flows out of their pool and nourishes the seven liberal arts.

Among the first scholars to come to Vienna and accept positions at the *Collegium* were two of Celtis's colleagues from Ingolstadt, Johannes Stabius and Andreas Stiborius, both renowned for their astrological expertise. Their former student Georg Tannstetter soon joined them. By the 1510s a core group of masters included scholars who had moved to Vienna along with a growing number who had been educated in the city and had stayed there to teach in the arts and medical faculties, including Andreas Perlach and later Johannes Vogelin. Together they created a vibrant astrological curriculum in private and extraordinary lectures that supplemented the official prescribed lectures. Stiborius, Tannstetter, Perlach, and other members of this group lectured on astrology at the university, where they enjoyed Habsburg patronage through stipendiary support or as counselors at the Habsburg court. Their various forms of astrological expertise reflected Maximilian's own predilections and his understanding of the political instrumentality of astrological knowledge.[7]

Andreas Stiborius and Astrological Instruments

As a young master at the University of Ingolstadt Andreas Stiborius was already lecturing on astronomical subjects including astrolabes and related instruments. When he accompanied Celtis to Vienna in 1497 Stiborius's position at the University of Vienna seemed uncertain.[8] Maximilian had just begun to lay the foundations for revitalizing the university but had not yet established the institutional structures that would support Celtis, Stiborius, and their colleagues. Nevertheless, Stiborius soon began offering his foundational lectures on various methods of stereographic projection and the uses of different types of astrolabes and sundials, his "Liber umbrarum," which he revised once he arrived in Vienna. He frequently referred to the south side of the university's tower when calculating the polar elevation, zenith, or latitude, or offering other celestial descriptions.[9] These lectures introduced various methods of determining astronomical information for any given location and then explained how to use this information to determine the time of day, the rising and setting of constellations, and the degree of the ascending sign. In his "Liber umbrarum" Stiborius also presented theories of stereographic project that undergirded both standard and universal astrolabes. Although Stiborius recognized his debt to Messahalla's standard text on the planispheric astrolabe, he emphasized his own innovations.[10] These were the first in Stiborius's series of lectures on astrology and astrological instruments at the university.

Stiborius began by explaining some basics, such as how to determine the elevation of the pole, how to locate the horizon and the zodiacal houses, how to tell time and to find the positions, and how to determine the rising and setting times of constellations.[11] Stiborius intended his lectures to be illustrated and expected students to have ready access to an astrolabe.[12] He instructed students to place the rule on the front of the astrolabe at a particular point and then to find where it crossed the edge of the instrument, and he told students to inscribe different hour lines on the face of the astrolabe.[13] A typical brass astrolabe was impractical as a demonstration device in lectures. Not only was it too small to be seen from a distance of more than a few feet, it was too expensive for each student to own. Further, it was difficult to modify a brass astrolabe by marking the reference points or inscribing the various hour lines on its face, as Stiborius instructed his students to do.[14] Instead, Stiborius probably used inexpensive, paper instruments to illustrate his lectures. He could have easily constructed a large, functional, paper astrolabe to show students how to mark or color the face of their instruments or how to move the various parts of the astrolabe.[15] Students could have modified their own paper astrolabes and followed Stiborius's instructions for how to manipulate the astrolabe or which portions to color. Such instruments could have been simply paper cutouts that were held together by a pin at the center, or the various parts of the astrolabe could have been printed, cut out, and pasted onto a wood support.[16]

Stiborius's "Liber umbrarum" established his domain of expertise—the construction of instruments and their use in astrological practice. At the same time it anticipated his subsequent lectures by providing the foundation for his more sophisticated and detailed treatment of astrolabes and astrological instruments. Stiborius compared the standard astrolabe to various universal instruments as well as different methods of stereographic project. He surveyed the relative usefulness of these different methods and the techniques for using them and promised a fuller discussion of these topics in later lectures.[17] In these early lectures Stiborius also established his characteristic teaching practices, using paper astrolabes as demonstration devices and explicitly linking his lectures to others at the university.

Stiborius's lectures on astrolabes renewed a long tradition at the University of Vienna of masters lecturing on astrological instruments, a tradition that stretched back nearly a century to Johannes de Gmunden's lectures on the astrolabe.[18] Stiborius considered himself the successor to Regiomontanus, Vienna's most famous master, and modeled his teaching and scholarly activities on his predecessor's. Shortly after arriving in Vienna, Stiborius began a systematic study of the heavens to improve the star catalog and established a tradition of designing innovative astrological instruments. He also began offering a series of lectures at the University of Vienna that taught students both the theoretical foundation for these instruments and how to use them. After Regiomontanus's departure from Vienna and a period of decline during the Hungarian occupation of the city, Stiborius's lectures on astrological instruments contributed to the university's return to intellectual prominence. His "Liber umbrarum," however, were not recorded in the official lecture lists compiled at the beginning of each year. They were instead private or extraordinary lectures. Masters at the University of Vienna supplemented their income through the fees students paid to attend such private or extraordinary lectures.[19] Consequently, masters chose topics that attracted students willing to pay those fees. Stiborius's "Liber umbrarum," with its focus on using astrolabes as astrological instruments must have been a topic that students were eager and willing to pay to learn.

Stiborius's lectures also appealed to Maximilian, who promoted Stiborius to more important and lucrative positions in the university. For Maximilian, astrological instruments were not merely ornate political gifts; rather they embodied astrological expertise. They represented a concrete and especially public application of astrological knowledge to political events, and thus Stiborius, with his expertise in both the theoretical underpinnings and design of instruments, played a key role in bolstering his public image as an emperor who sought expert advice. In 1501 Stiborius's status at the university improved rapidly when he was appointed to one of the chairs in mathematics and astronomy at Celtis's newly founded *Collegium*. Two years later he was promoted to a chair in the *Collegium ducale*. Appointments to both colleges were carefully controlled by Maximilian, who exercised influence on the university through the faculty he appointed to these positions. With chairs in both colleges and his pay coming directly from the imperial coffers, Stiborius had become one of the rising stars of the University of Vienna.[20]

By 1504 Stiborius had completed a new set of lectures on the astrolabe, the "Canones astrolabij."[21] He began by reviewing the parts of the astrolabe. Stiborius then explained how to use the astrolabe to carry out typical astrological operations: locating the location of the sun and the planets, determining the four cardinal points—the ascendant, *medium coeli*, descendant, and *imum coeli*—and the zodiacal houses. He surveyed the various time-telling conventions and how to convert between them. Stiborius focused on real-world applications rather than theoretical issues. In each example, he stated the canon, defined the terms, and explained how to solve the problem. Throughout he explicitly referred to an astrolabe, instructing the student on how to move various parts, where to place the rule, and how to read the results off the face or the edge of the instrument.[22]

After surveying the standard uses, Stiborius introduced more sophisticated operations that were particularly useful for the practicing astrologer, such as determining the true ascendant at the time of an election or conjunction, dividing the zodiac into mundane houses, or determining a horoscope's significator, the planet that exercised the strongest influence in the horoscope. Although astrologers typically relied on tables for these operations, such as Regiomontanus's *Tabulae directionum*, Stiborius claimed that an astrolabe was quicker and sufficiently accurate.[23] Stiborius also explained how to use an astrolabe to predict the motions of the humors. At least one of his students was excited by these applications and recognized their immediate use in astrology, noting in the margin that they "were not found in previous examples and will be useful for judgements."[24] Throughout his lectures Stiborius foregrounded instruments in both his own teaching and in the practical tasks of astrologers and physicians.[25] He also connected his lectures to other courses on astrology by introducing astrological topics students would encounter later at the university. When he pointed out that a physician could use an astrolabe to predict the flux and reflux of humors or to determine the critical days of a disease, Stiborius emphasized how an astrolabe facilitated these predictions and previewed a topic students would encounter later in Georg Tannstetter's lectures on medical astrology.[26]

Part of Stiborius's achievement and what made his work appeal to Maximilian was his codification and systematization of astrological knowledge, particularly as it related to astronomical instruments. For Maximilian, who had in his autobiographical works depicted himself using astronomical instruments often while consulting with expert astrologers (probably university masters), Stiborius's lectures—both their technical content and their focus on instruments—were important aspects of the emperor's self-representation. Another aspect was Stiborius's success in implementing a series of lectures that disseminated and reproduced his own expertise, which in turn reinforced Maximilian's image as generous patron of technical experts. This relationship between patron and astrologer reflected authority in both directions.

Stiborius collected and arranged canons that were otherwise scattered through the various texts and added to them various new canons of his own.[27] His "Canones astrolabij" provided his students with the foundation needed to use a variety of other astrological and horological instruments. Students working their way through Stiborius's lectures now advanced to more sophisticated universal astrolabes. Stiborius presented the first of these in his lectures "Canones Saphee." The saphea had first been developed by the eleventh-century Toledan astronomer Ali ibn Khalaf and had been described in King Alfonso X's *Libros del Saber de Astronomia*. Despite circulating in academic contexts for nearly three centuries, it had never enjoyed the popularity of the standard astrolabe. Stiborius recognized that his students would be unfamiliar with the saphea and so opened with a brief definition, history, and justification of the instrument.[28] He emphasized the advantages it offered, pointing out that when fitted with a rule and type of rete the saphea could be used to solve a wide variety of celestial and terrestrial problems.

Stiborius began with the operations that could be solved simply by using the face of the saphea (see figure 3.2). Astrological uses dominated these early canons: determining the true time, the *medium coeli*, the ascendant, and the location and latitude of the sun and various stars. He also explained how to tell time, determine the rising and setting time of the sun, and calculate the distance between towns. When fitted with a rete and moveable zodiac, the saphea became particularly useful for quickly calculating the rising time of

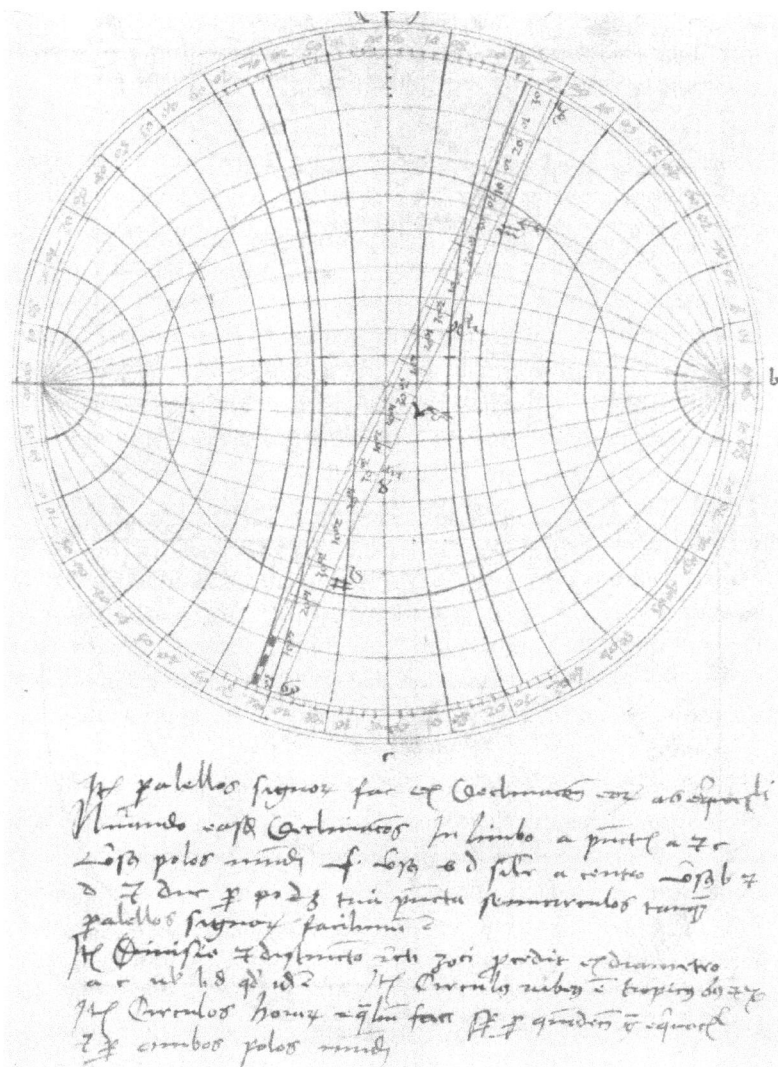

FIGURE 3.2. A diagram representing the face of Stiborius's saphea, from an illustration in one copy of his lectures "Canones Saphea." © Bayerische Staatsbibliothek, Munich, Clm 19689, 313r.

zodiacal signs, the mundane houses, and the ascendant, and for determining the *medium coeli* and the positions of the constellations. Stiborius concluded his lectures by extending the astrological uses of the saphea. He explained how to use the saphea as a universal *horologium* that was valid at all latitudes and presented a series of alternative methods for finding the four cardinal points of a horoscope, dividing the zodiacal houses, and adjusting the significator in order to determine critical days and to cast elections.[29]

Stiborius's "Canones super instrumentum universale quod organum ptholomei vocant" completed his cycle of lectures on different types of astrolabes. Here again he was concerned that his student would not be familiar with this type of astrolabe and so began with a brief description, relating it to the saphea.[30] Stiborius followed his well-established pattern, relating the instrument to more familiar types of astrolabes and then proceeding quickly through a set of canons. Finally, he indicated how the *organum ptholomei* enabled the astrologer to perform operations that the others did not. In this case, Stiborius pointed out that the *organum ptholomei* enabled the astrologer to calculate the sine of any arc. He concluded by offering two examples.[31]

As if fulfilling a series of prerequisites, students had to progress through Stiborius's lectures in order, from simpler to more complex. Stiborius's "Canones super instrumentum universale" were built on the foundation provided by his previous lectures on the saphea as well as his "Canones astrolabij," "Liber umbrarum," and "Liber horologium."[32] At times Stiborius explicitly referred his students to earlier lectures for a detailed treatment of some point or other.[33] Some students bound their lecture notes together in the appropriate order for easy reference.[34]

Although Stiborius's status at the University of Vienna continued to improve during his tenure there, his lectures did not appear on the official course lists.[35] Perhaps because he was lecturing on instruments rather than traditional texts and subjects, or perhaps owing to his close association with Maximilian, Stiborius's lectures attracted remarkably little attention in the normal university bureaucracy. Students, however, considered them an important part of the astrological curriculum. Jakob Ziegler, renowned for developing a historical geography, was interested enough in Stiborius's lectures to copy out the version he found, even though he feared that it was incomplete.[36] The famous Swiss humanist Joachim Vadian, likewise, guarded his copy of Stiborius's "Canones astrolabij," taking it with him from Vienna when he returned home to St. Gall.[37] More than forty years after Stiborius's death, Joachim Rheticus praised his accomplishments in developing various astronomical instruments.[38]

Stiborius had first established himself in Vienna by consciously modeling himself on Regiomontanus and by lecturing on astrological instruments. Stiborius praised his predecessor's work but emphasized how his own lectures and instruments surpassed Regiomontanus's. After he had gained Maximilian's support, Stiborius continued to pursue these topics throughout the remainder of his tenure at the university. Although Stiborius concentrated on different types of astrolabes, he recognized that instruments formed only one part of a broader astrology curriculum. Arguing that the science of the stars was unlike other forms of knowledge, in particular commentaries on Aristotle, Averroes, or Aquinas, Stiborius suggested that astrology proceeded through new observations, improved calculations and theoretical texts, and the development of better instruments.[39] He admitted that in order to understand his lectures, students also had to know and understand the theories and causes of celestial phenomena. His own lectures on instruments were based on the work of his younger colleague and one-time student Georg Tannstetter, whose interest in calculations and the theoretical facets of astrology complemented Stiborius's concentration on instruments and observations. In a prefatory letter to Tannstetter's edition of Regiomontanus's *Tabula primi mobilis*, Stiborius claimed that his work on instruments and any real knowledge of how instruments functioned depended on tables of data and calculations like those contained in Tannstetter's book: "Likewise, innumerable instruments depend on this knowledge of the *primum mobile*: the astrolabe, saphea, *organum ptolomei*, meteorscope, armilary sphere, torquetum, rectangle, equatoria, compass, quandrant and many other similar types. Oh how splendid, how noble, how necessary for all students of astronomy, just just like the alphabet and elementary education, without this nothing is finished, nothing accomplished in this preeminent discipline of astronomy. Therefore, you who are a true friend of the heavens, take these tables produced and printed with diligence and solicitously by my dear colleague Georg Tannstetter."[40]

Stiborius's closing statement was more than just hollow praise. By 1514 when he wrote this letter, he and Tannstetter had long been colleagues and close friends. Tannstetter had studied under Stiborius at the University of Ingolstadt and then had followed him to Vienna, where they worked together for nearly two decades.[41]

Georg Tannstetter, Editor and Lecturer

Georg Tannstetter arrived at the University of Vienna in 1501 as part of Maximilian's efforts to revive the university. Stiborius had helped secure Tannstetter's initial appointment in the arts faculty.[42] Tannstetter quickly established his expertise in astrology's theoretical and mathematical techniques. Tannstetter's role at the university combined lecturing with a vigorous publishing practice, thereby helping to realize Maximilian's goal of restoring the University of Vienna to its former place of preeminence, particularly with respect to astronomy and astrology. Maximilian, who exploited the power of print to construct and disseminate his own image, understood that printed texts were particularly powerful instruments for building and maintaining reputations, and by 1505 he had enlisted Tannstetter's particular form of astrological expertise for the task of producing yearly *practica* and wall calendars. That same year Tannstetter lectured on the *Theorica planetarum*.[43] In this way, Tannstetter's lectures connected his university courses to the almanacs, wall calendars, and *practica* that were expressions of the Habsburg political program and evidence of Habsburg authority.[44]

In addition to his wall calendars and *practica*, Tannstetter published numerous editions of astrological works, which he and his colleagues used as textbooks.[45] At the same time, Tannstetter used these editions to celebrate the revival of astrology at the university and to reinforce the intellectual filiation between his own contemporaries and the university's famous masters from the previous century. Tannstetter foregrounded this intellectual patrimony in his *Viri mathematici*—an academic genealogy of the most famous Viennese astrologers and astronomers, which Tannstetter appended to his edition of Georg Peuerbach's *Tabulae eclipsium*.[46] Alongside Gmunden, Peuerbach, and Regiomontanus stood Johannes Stabius, Andreas Stiborius, Stephen Rosinus, and a handful of Tannstetter's other colleagues at the University of Vienna. Like Maximilian's genealogical projects, which were meant to reinforce the emperor's claims to the imperial title, the *Viri mathematici* was intended to celebrate and reinforce the authority of the astronomers and astrologers at Maximilian's court.[47] According to Tannstetter, Maximilian so delighted in the astrological and horological instruments that Johannes Stabius and Andreas Stiborius had designed for him that he sponsored lectures on astronomy and mathematics: "Every day the unconquerable and illustrious Caesar Maximilian enjoyed his unique inventions. Admiring his and Stiborius' talent (about whom I'll speak a little later), he established at Vienna by means of a new stipend public lectures on astronomy and mathematics."[48] Tannstetter implied that the astrology curriculum was flourishing once again thanks to Maximilian's generous support. Thomas Resch, who had recently been crowned poet laureate by Maximilian, drove the point home in two dedicatory letters. He praised Tannstetter's efforts to produce new editions of astrological texts for the students to use at the university. He also lauded Maximilian for his efforts to restore the university to the prominence it enjoyed before Matthias Corvinus occupied the city. Maximilian's patronage had reestablished the university as an important center of astronomy, astrology, and mathematics:

> Hence the memory of lost authors owe you their eternal thanks. And the entire university of distinguished scholars, to which you are about to restore the fame and splendor through your immense renown, is in debt to you. Indeed, I often considered those scholars: both you and Stabius and Stiborius, Rosinus, Angelus, and Ericius. All famous mathematicians, shining with the great splendor of scholarship, were preserved by God within the borders of our Germany, so that they might be those through whom the glorious and noble mathematical subjects, once shamefully and barbarously cultivated, might again hold their ground and recover.[49]

Tannstetter and his colleagues understood this publishing program to be an important facet of Maximilian's efforts to revitalize the university and to establish his authority as a generous patron of the mathematical subjects.[50] As Resch indicated, the success of Tannstetter's project extended beyond the university. By glorifying Germany's contribution to astrology, astronomy, and mathematics, the revival of the University of Vienna complemented Maximilian's sweeping efforts to scour the past for evidence of a distinct German national identity that rivaled the Italians'.[51]

Tannstetter used many of his new editions in his own lectures. His edition of Proclus's *De sphaera* was an introductory text for his students. The various texts circulating under the generic title *De sphaera* introduced students to the basic arrangement of the celestial and terrestrial spheres and the relationship between them. The most popular of these texts was attributed to Sacrobosco and appeared regularly in the lecture lists at the University of Vienna. Proclus's text was more elementary and provided an introduction to Tannstetter and Vadian's own popular commentary on the sphere.[52] As Tannstetter remarked in his dedicatory letter to Vadian, the text introduced the basic terms and concepts required to progress further in the study of astrology: the names and positions of the five important celestial circles, the reasons for the rising and setting of stars, the meaning of terms such as horizon, color, and meridian, as well as the basic composition of the heavens. One student worked diligently through the opening portion of this text to improve his basic understanding of astronomy, underlining important passages, clarifying unfamiliar terms, and adding cross-references.[53]

At least twice Tannstetter agreed to lecture on the *Theorica planetarum*, in 1505 as a young faculty member and again in 1511 as an established master at the university. As with the *De sphaera*, various texts circulated under the generic title *Theorica planetarum*, including Gerard of Cremona's version, Campanus of Novara's, and Peuerbach's. Although sometime in the 1510s Tannstetter produced a brief summary of Campanus's version of the *Theorica*, he concentrated on Peuerbach's *Theoricae planetarum novae*.[54] Tannstetter carefully studied and annotated his copy of the standard commentary on the *Theorica*, Albert of Brudzewo's *Commentum in theoricas planetarum Georgii Purbachii*.[55] By the mid-1510s Tannstetter assumed that his students would be familiar with Peuerbach's *Theoricae*, casually citing or quoting the text a number of times in his own lectures.[56] In 1518 Tannstetter produced a small, affordable edition of Sacrobosco's *De sphaera* paired with Peuerbach's, intended for students who first heard lectures on Sacrobosco's text before advancing to Peuerbach's. Tannstetter's edition was still being used as a text in the arts faculty well into the middle of the century.[57]

Although Tannstetter was a master at the university for nearly three decades, only two of his lectures on Peuerbach's *Theoricae* ever appeared in the acts.[58] His lectures occurred outside the prescribed curriculum. Nevertheless Tannstetter's lecturing and publishing activities attracted Maximilian's attention and helped establish his reputation at the Habsburg court. He soon enjoyed growing Habsburg support through appointments at both the university and the court. In 1509 he was appointed *Professor ordinarius in astronomia*, one of four chairs Maximilian had established as part of his broader effort to restore and reform the university.[59] Maximilian oversaw the appointments to these chairs and awarded them to those masters whose interests reflected his own and contributed to his own dynastic and political programs. In addition to the titular honor, each *professor ordinarius* received a 50 gulden stipend directly from Habsburg coffers and was given considerable freedom in choosing lecture topics.[60] After Tannstetter moved into the medical faculty in 1513, Maximilian brought him into direct service of the court by appointing him imperial *Leibarzt*.[61]

The standard astronomical texts were well represented in the ordinary lectures. Nearly every year students could hear lectures on the *Theorica planetarum*, *De sphaera* or the *Sphaera materiala*, as well as philosophical texts such as Aristotle's *De caelo*, *Meteorologia*, and *De generatione et corruptione*.[62] Missing from these lectures was any attempt to connect the theoretical subjects to their more concrete, practical applications. Tannstetter's lectures on using ephemerides and his commentaries on Pliny's *Historia naturalis* bridged this gap between theory and practice. For more than a decade Tannstetter regularly offered lectures on how to use ephemerides.[63] In 1514 Vadian attended these lectures; when he returned to St. Gall he took his notes with him.[64]

Tannstetter's favorite student and successor, Andreas Perlach, modeled his own lectures on Tannstetter's and tried, unsuccessfully, to convince Tannstetter to publish his. Ultimately, Perlach assumed responsibility for these lectures and developed his own manual. Tannstetter's courses filled a gap in the official curriculum by offering immediate solutions to the many practical problems students encountered when making astrological calculations and drawing horoscopes.

Like Stiborius's, Tannstetter's identity and expertise were inseperable from his practices, the technical content of his lectures. And it was his identity and his expertise that made him valuable in the Maximilian's efforts to represent himself as a skilled and expert patron. In his *Weisskunig* the young emperor was convinced that inaccuracies in astrological predictions, particularly as they applied to understanding human behaviors, resulted from inaccurate star charts. He selected the best astrologers and demanded that they create new charts from new and improved observations. Maximilian was immersed in and preoccupied with technical, astrological details and their application to his world. He was also tireless in his efforts to be seen to worry about these seemingly esoteric details. In this context, Tannstetter's expertise was useful because it was technical and systematic. It made credible the emperor's efforts to depict himself as both expert and profoundly concerned with expert knowledge.

Tannstetter first guided his students through the information contained in an ephemerides and then offered detailed instructions on how to use this information when practicing astrology, such as calculating nativities or elections, or predicting the effects of a solar or lunar eclipse. He expected his students to arrive at these lectures with a basic knowledge of astronomy and astrology, perhaps acquired through his own course on Sacrobosco's *De sphaera*. He also expected students to be familiar with the general format and content of Johannes Stöffler and Jakob Pflaum's *Almanach nova*, the most common ephemerides in circulation. An ephemerides was a required reference book for practicing astrologers. Students who did not already possess a copy would have most likely purchased one soon after hearing Tannstetter's opening lecture. He began by listing the information found on the opening pages: important calendric numbers like the golden number, the solar cycle, the Roman *inditio*, as well as the symbols used to represent the planets and signs. He added to this discussion a history of why these calendric values had been developed, how they had changed, and why they were still important.[65] Tannstetter did not dwell on the basics or their history but progressed quickly to more pragmatic topics.

The first task for an astrologer was to determine the relevant celestial phenomena. He had to calculate the positions of the planets, their motions within the zodiac—whether progressing or retrogressing or ascending or descending in the ecliptic—the rising and setting times for the important constellations and fixed stars, and the divisions of the zodiac for a given time and place. In practice, astrologers typically relied on an ephemerides, which contained all this information. These ephemerides were always calculated for a particular city. Consequently, the information in them often had to be adjusted to reflect the astrologer's local position. Tannstetter devoted a great deal of time explaining the calculations required to translate the data in Stöffler and Pflaum's ephemerides to any new location. Tannstetter introduced each operation, explained how to calculate it, and concluded with an example.[66]

Tannstetter divided the remainder of his lectures into sections, each treating a different set of celestial phenomena. These latter sections extended the standard canons on ephemerides. Tannstetter treated eclipses in great detail because they were considered the most powerful of common celestial events. He reminded his students that solar and lunar eclipses for any year were illustrated on the opening page of an almanac. This illustration also indicated how much of the sun or moon would be occulted. However, it was not sufficient simply to know if there would be an eclipse in any given year. The astrologer also needed to determine the exact time it would occur and how long it would last. This was a relatively simple calculation, but again required the astrologer to adjust the information in the ephemerides to reflect his locale.[67] Tannstetter's lectures were entirely traditional. He followed closely the methods in Ptolemy's *Quadripartitum* for using eclipses to make predictions. The period of occultation revealed the duration of the eclipse's effects. Where the eclipse occurred—between the eastern horizon and meridian, on the meridian, or between the meridian and the western horizon—indicated when the effects would begin and when they would worsen or abate.[68] He explained how to find the time and location of the true conjunctions. Finally, he pointed out how to determine if an eclipse occurred near the *caput draconis* or the *cauda draconis*, two points at which the lunar path crossed the ecliptic.[69]

In the final section Tannstetter shifted his attention from natural astrology (predicting the general, wide-scale effects of eclipses) to judicial astrology (specific predictions based on nativities, elections, interrogations, and judgments). Judicial astrology required a more fine-grained analysis. In order to interpret the strength and duration of the planetary influences, the astrologer had to take into consideration not just the location of the planet but also its motion—whether direct, retrograde, or stationary—as well as its celestial latitude, speed, and distance from the earth. Tannstetter also introduced a new layer of sophistication, one that relied heavily on the theories of planetary Peuerbach presented in his *Theoricae novae planetarum*. Tannstetter considered Peuerbach's *Theoricae* a key astrological text that provided the theoretical foundation for his own lectures on astrology.

Tannstetter was not confident that his students would remember his lectures on Peuerbach's *Theoricae* and so began by carefully defining key astrological terms. He then provided a detailed discussion of how the observed motion of the planet was a combination of the motion of the center of the planet's epicycle and the rotation of the epicycle itself, and why the moon never seemed to retrogress. Tannstetter's exposition came directly out of Peuerbach's *Theoricae*. Although the practicing astrologer could find a planet's motion in the tables in the ephemerides, Tannstetter implied that the best astrologers understood why a planet moved in a particular manner. Such knowledge enabled them to interpret more accurately the effects of that planet.[70]

Tannstetter's subsequent propositions surveyed different aspects of celestial events that affected the planetary influence. Some of this information could be gleaned from the ephemerides, such as the planetary latitudes, but in each case Tannstetter provided a more thorough discussion of these points and explained why the astrologer should take them into account when making a prediction. According to Tannstetter, only by knowing the latitude of a planet could the astrologer determine precisely the true conjunctions between the planets themselves and between the planets and the fixed stars as well as their rising and settings with the fixed stars.[71] Other propositions required knowledge and information not found in the ephemerides, such as finding the center of a planet and determining the location of a planet on its epicycle.[72]

Tannstetter's students recognized the value of his lectures and continued to rely on these lessons long after Tannstetter had stopped lecturing on ephemerides. One student used Tannstetter's lectures as the foundation for his own calculations of a solar eclipse in 1523. He carried out each calculation in the order Tannstetter had set out in his lectures.[73] Another student added calculations for various solar eclipses through the 1530s, implicitly comparing the guidelines set out by Tannstetter to the approach of his two famous students, Andreas Perlach and Johannes Vögelin.

Tannstetter, like Stiborius, considered his different lectures a coherent body of knowledge, progressing from basic to more advanced topics. His lectures on ephemerides built on his previous lectures on the *De sphaera* and the *Theoricae* and, in turn, provided the foundation for more advanced lectures on astrology. In a passing remark in his lectures on ephemerides, Tannstetter referred his students to his recently published edition of Regiomontanus's *Tabulae primi mobili*.[74] Regiomontanus's text along with Peuerbach's *Tabulae eclypsium* were advanced astrological textbooks that were too expensive and sophisticated for the typical undergraduate student at the university. The dense presentation of material benefited from an expert teacher, as one reader realized as he tried to work through the problems.[75]

In the mid-1510s, Tannstetter also lectured on books two and seven of Pliny's *Historia naturalis*. The second book of Pliny's *Historia naturalis* had long been a foundational text for astrologers, thanks to its collection of received astronomical and astrological lore.[76] During the fifteenth and sixteenth centuries the number of complete printed editions of Pliny's text as well as printed and manuscript commentaries increased dramatically.[77] Along with humanist concerns for the language came a renewed interest in the content of Pliny's text, exemplified in the quarrel between Niccolò Leoniceno and Pandolfo Collenuccio.[78] Bringing his vast knowledge of Greek philosophical and medical texts to bear, Leoniceno argued that many of the errors in Pliny were not the result of careless copyists and editors, but stemmed from Pliny's own ignorance, particularly of his more ancient sources.[79] Nonetheless, for many scholars, Pliny's text still offered the most accessible introduction to classical astronomy and astrology and a useful source for their astronomical handbooks.[80] Like other commentators, Tannstetter considered the *Historia naturalis* a useful teaching text, once he had corrected the errors and clarified the confusing passages.[81]

Rather than produce a complete commentary on book two of Pliny's text, Tannstetter concentrated on those chapters most flawed by inaccuracies or confusing and incorrect planetary theories. Some of Pliny's errors were simple matters of precision: Saturn did not revolve about its orbit in 30 years, as Pliny claimed, but in 29 years, 163 days.[82] Tannstetter relied on various canonical sets of tables—the Alphonsine tables, or those by Gmunden or Ptolemy—to identify errors in Pliny's discussion of planetary motions and to supply missing information.[83] Tannstetter wasted little time on these factual corrections. He was more interested in the knotty problems arising from Pliny's account of planetary motions and his implicit planetary theory.[84] Tannstetter struggled to superimpose Ptolemy's epicycle-deferent system onto Pliny's model, which was based on a simple notion of apsides. Pliny's account failed to provide a useful explanation of the various observed motions—retrograde and direct motions as well as stationary points—and seemed to be in conflict with motions recorded in various ephemerides. Tannstetter occasionally returned to this point, finally renaming parts of Pliny's model to distinguish between the different apsides, and importing more recent theories of epicycles to account for the various planetary motions.[85] Tannstetter excused Pliny for not using epicycles because Pliny's own contemporaries either had been ignorant or had rejected them.[86] In addition to the problem of planetary motion, Tannstetter considered Pliny's discussion of astrology particularly lacking.

Tannstetter composed a long commentary on book seven, chapter 49 of Pliny's *Historia naturalis*. This chapter focused on the position of the stars at the moment of a person's birth, perennially the concern of both astrologers drawing natal horoscopes and physicians interpreting and predicting the course of a disease. Drawing on a number of examples, Pliny had claimed that the position of the moon at the moment of birth indicated the length of a person's life but stopped short of explaining how to calculate the latter. Tannstetter drew on Ptolemy and Arab sources to expand Pliny's account, focusing on how to determine the *hilech*, the planet that most influenced the length of a person's life, and the various signs that threatened a person's health and longevity. In particular, astrologers had to watch for the malefic planets, Saturn, Mars, or Venus when tinged by the harmful rays of Mars or Saturn, and had to note the harmful planetary aspects, opposition, or square, which intensified the effects.[87]

More than simply an idiosyncratic commentary on Pliny, Tannstetter's lectures helped bolster the university's reputation as an important center of astrological and astronomical learning. The influence of Tannstetter's lectures, both locally at the University of Vienna and farther afield in the empire, reinforced Tannstetter's own reputation and expertise. In turn, Tannstetter's growing reputation promoted the emperor's own image, thus serving Maximilian's political program and his efforts to portray himself as a generous patron of the university and its masters. Maximilian seemed to embody Machiavelli's advice: the prince's reputation is judged first by the counselors he has around him. He understood that his own authority was enhanced with the success of the university and its faculty. The emperor was, therefore, keenly interested in Tannstetter's authority and reputation, which were linked to how other masters incorporated and expanded his lectures.

Tannstetter's commentary found an eager audience among the masters and students at the University of Vienna, where it quickly became the standard astrological commentary on Pliny's text. Other masters at the university deferred to Tannstetter's expertise and relied on his commentary in their own lectures. In 1519 Phillip Gundelius, formerly Tannstetter's student and now a master in the arts faculty at the university, lectured on book seven of Pliny's

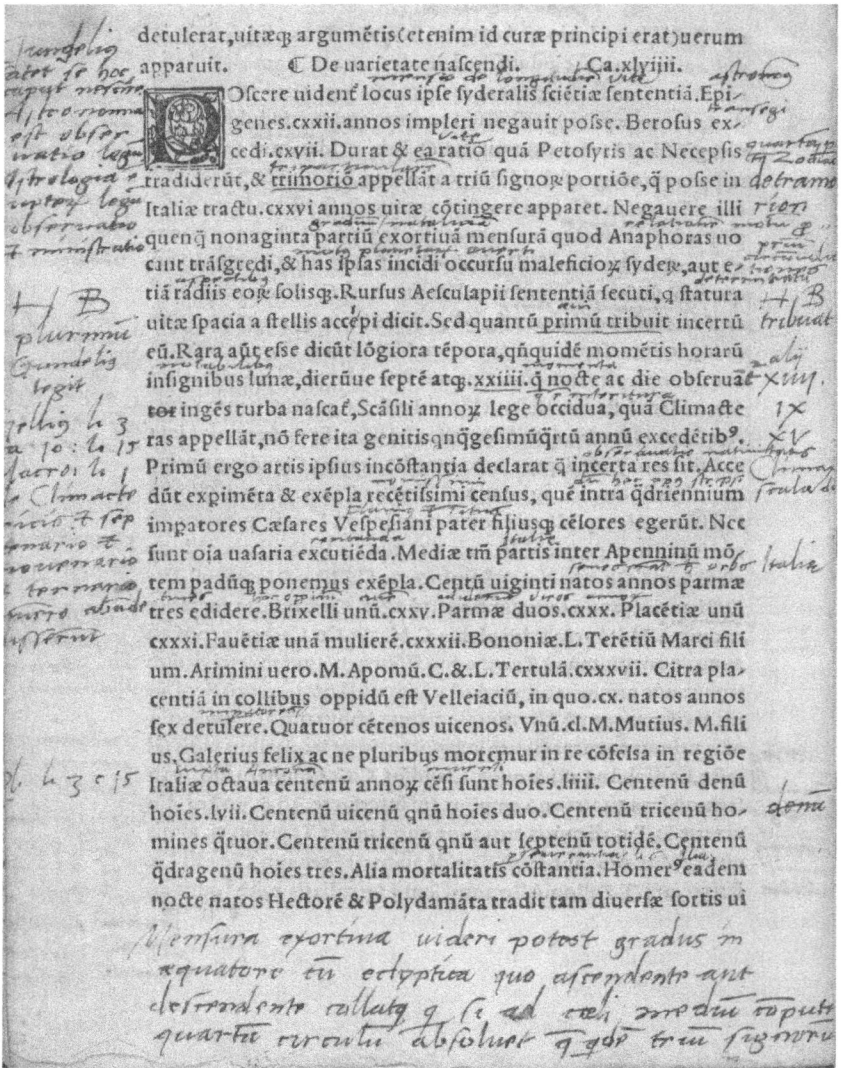

FIGURE 3.3. Philip Gundelius's comments on book 7, chapter 49 of Pliny's *Historia naturalis*. Houghton Library, Harvard University, Call # GC5.G9554.519p, e2v.

text. One student took careful notes in his own copy of Gundelius's edition of this text, including chapter 49 (see figure 3.3). Gundelius ultimately referred his students to Tannstetter's commentary for a more complete analysis of this chapter. This same student dutifully copied out Tannstetter's commentary onto the

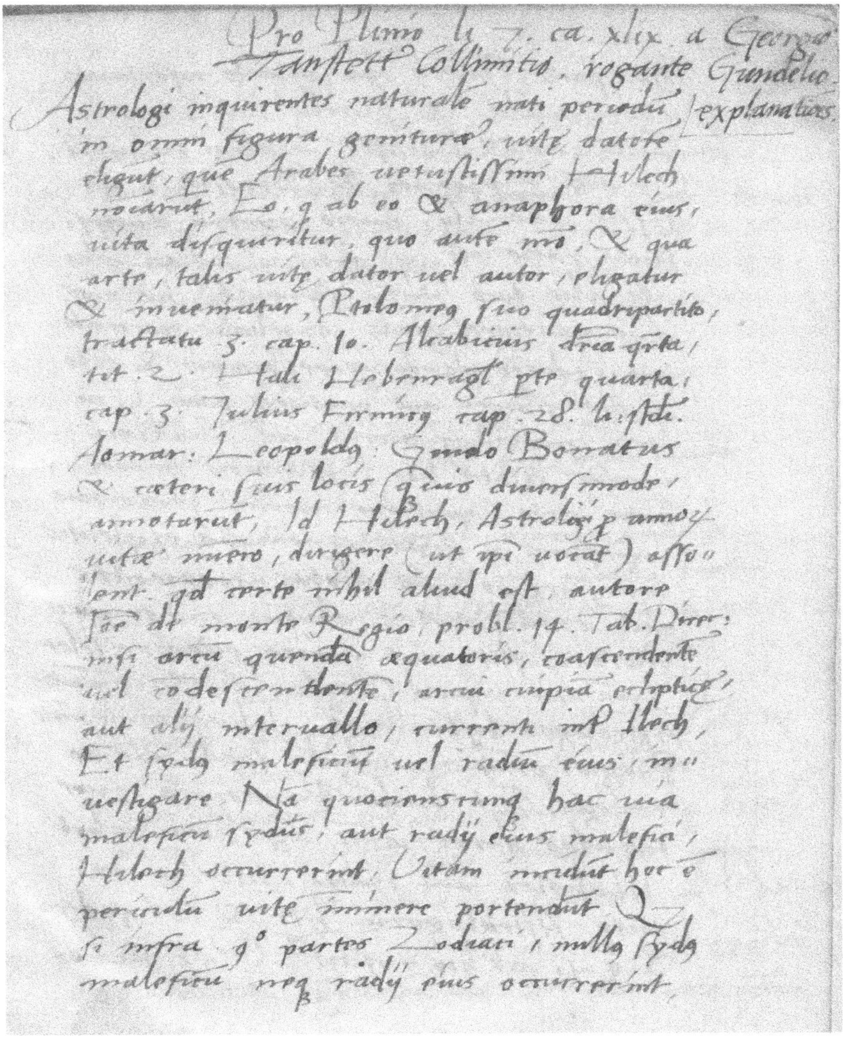

Pro Plinio li. 7. ca. xlix. a Georgio Tanstett Collimitio, rogante Gundelio explanatio

Figure 3.4. Tannstetter's commentary on book 7, chapter 49 of Pliny's *Historia naturalis* that an anonymous student copied into his copy of Gundelius's edition of this text. Houghton Library, Harvard University, Call # GC5.G9554.519p, np.

blank pages at the back of his copy of Gundelius's text (see figure 3.4). A note across the top of the first page of Tannstetter's commentary made the hierarchy between Tannstetter and Gundelius clear: "Georg Tannstetter Collimitius's explanations on Pliny, book 7, chapter xlix, in response to Gundelius's request."[88]

Tannstetter's efforts to correct Pliny's text were intended to make the *Historia naturalis* useful to his students and an effective teaching tool. These goals shaped his identification of errors in the text and his methods of correcting them. He mined his sources for specific corrections, taking from them what he needed and ignoring the rest.[89] Tannstetter drew on classical poets, classical, Arab, and medieval astrologers, and his own immediate predecessors. Ptolemy provided the epicycle-deferent system while medieval Arab astrologers, such as Alfraganus, Alcabitius, and Haly Abenragel, along with the Latin astrologers Guido Bonatus and Leopold of Austria, provided the mechanisms that Pliny had omitted. Tannstetter's own predecessors at the University of Vienna loomed large in his work. He drew on Regiomontanus's *Tabulae directionum*, Peuerbach's *Theoricae novae planetarum*, and Gmunden's tables of planetary data. Tannstetter found a strange ally in Pico della Mirandola. On the one hand, Tannstetter relied on Pico's attack on the doctrine of the 10 revolutions of Saturn, to dismiss Pliny's analogous belief that 100 revolutions of the moon would cause fires and storms. In both cases, certain planets were thought to return to the same place in the zodiac after a specified number of orbits. Pico had shown that this did not obtain for Saturn. Tannstetter implied that the same argument could be used in the case of the moon. On the other hand, Tannstetter claimed that Pico erred in rejecting the influence of the five planets as well as the sun and moon. According to Tannstetter, the motions of the planets, their relations to the fixed stars, and their distance from the earth all caused changes on earth.[90]

For some of Tannstetter's students his commentary was more than simply an introduction to astrology. Book two of Pliny's *Historia naturalis* could be used to replace basic Aristotelian physics and natural philosophy. Jacob Milich studied mathematics and medicine at the University of Vienna when Tannstetter lectured on Pliny. After leaving Vienna, Milich moved to the newly founded University of Wittenberg, where he produced his own commentary on book two of Pliny's text.[91] Although Milich was a proponent of astrology, his goal was to produce a commentary that could be used to replace Aristotle in the undergraduate curriculum. His efforts reinforced Philip Melanchthon's, who by the mid-1540s had established book two of Pliny's *Historia naturalis* as the introductory text in lectures on natural philosophy at Wittenberg.[92] Masters at the University of Wittenberg continued to lecture on Milich's commentary on book two throughout the middle part of the century.[93]

In some ways, Tannstetter's trajectory resembled that of other astrologers who found employment at local courts. Commonly, these court astrologers enjoyed academic positions in medical faculties or had been trained as and identified themselves as physicians.[94] By 1508 Tannstetter had begun studying medicine at the University of Vienna. Five years later, when he received special dispensation from the bachelor's exams, he assumed increasing responsibilities in the medical faculty, holding lectures and enjoying greater prestige associated with the higher faculty and his new position as imperial *Leibartz*.[95] Although he continued to lecture to arts students, perhaps in the remnants of Celtis's *Collegium poetarum et mathematicorum*, after 1513 Tannstetter increasingly turned his attention to the problems and issues in medical astrology. He brought the same pragmatic approach to his lectures on medical astrology that characterized his other lectures.

Tannstetter and Medical Astrology

Tannstetter was lecturing on medical astrology by the mid-1520s. His lectures were collected and printed in 1531 under the title *Artificium de applicatione astrologiae ad medicinam*.[96] His lectures probably supplemented the more advanced medical curriculum during the last years of instruction. In 1520, Martin Stainpeis published his *Liber de modo studendi* as a guide for medical students at the University of Vienna, listing the books that should be read and outlining a course of study for medical education at the university.[97] Along with Galen's *De criticis diebus*, most likely as part of the *Articella* that was studied in the first year, Stainpeis listed astrological works in the final three years of study. In the third year of study students read Hippocrates's *Astrologia medicorum*—prognostications based on the motions of the moon and planetary aspects—as well as his *De aere, aquis, locis* and an anonymous *De astronomia*.[98] In the final year they studied texts on using ephemerides and calendars.[99] Although Stainpeis did not mention Tannstetter or his works— the two certainly knew each other—Stainpeis must have been thinking of texts like Tannstetter's lectures on ephemerides when he composed his *Liber de modo studendi*.[100] Tannstetter, in turn, must have been aware of the curriculum Stainpeis outlined when he composed his own lectures on the uses of astrology in medicine. In his *Artificium de applicatione astrologiae ad medicinam* Tannstetter codified the astrological techniques used by the learned physician.

By far the most important task for the physician was constructing a correct geniture for the patient. The physician's predictive authority was grounded in the patient's natal horoscope. For this reason, Tannstetter reviewed at some length the various techniques for rectifying a geniture and the most accurate methods for determining the horoscopic houses. Although he admitted that using tables might result in marginally more precise house divisions, he recommended using an astrolabe or similar instrument from which the physician could simply read off the horoscopic houses. Tannstetter also recommended that the physician look up the locations of planets in an ephemerides in order to place the planets in the horoscope and then determine the interplanetary aspects.[101] The physician then used this natal chart to identify the five important signs: the ascendant, the part of fortune, the moon, the sun, and the *medium coeli*.[102]

By using the most sophisticated tools and techniques available, the physician constructed better natal charts and made more accurate prognoses. These tools also contributed to the physician's authority.[103] As Stainpeis had implied in book six of his *Liber de modo studendi* where he discussed the day-to-day practices of a physician, a physician who was surrounded by technical instruments and reference manuals, such as astrolabes and tables of planetary motions, conveyed confidence and authority.[104]

In addition to the geniture, physicians constructed astrological diagrams to chart the course of the disease. Prognosis was built around consulting these two horoscopes to determine the critical days, the days when a patient's suffering would worsen or begin to abate. Tannstetter expended considerable time and energy correcting Galen's method and explaining his own technique for calculating the critical days. Central to any method was the moon's motion. Tannstetter warned his students of the difficulties that arose from the moon's aberrant motions. Although he claimed originality, he largely echoed Pietro d'Abano's *Conciliator*.[105] D'Abano noted the problems in determining the medical month, which was measured from the point when the moon first became visible after a new moon. Unfortunately, sometimes the moon was visible on the second day after the actual new moon, sometimes the third day. This variability affected the physician's attempt to determine the critical days, which depended on the moon's first visibility. Tannstetter explained why the moon's motion was irregular and why in some months the first visibility occurred on the second day while in other months on the third day, and even occasionally on the fourth day.[106]

Tannstetter then introduced a different method of calculating the critical days. Galen's techniques used the mean motions of the sun and the moon. Relying on mean motions caused the difficulty in determining the beginning of the medical month. The solution lay in using the true motions of the planets rather than the mean motions. Once again Tannstetter celebrated the work of his predecessor Regiomontanus, who had published tables of true motions and had provided canons for using those tables to calculate the actual motions for any given day. Tannstetter claimed that Regiomontanus himself had indicated how critical days were related to the true motions of the luminaries. By using Regiomontanus's tables and his own technique, Tannstetter claimed he could determine the true first day of the medical month, avoiding the ambiguity introduced by trying to determine the moon's first visibility.[107]

Having identified the source of the problem and the tools needed to solve the problem, Tannstetter then explained how to construct a special chart to determine the critical days and how to use the chart to predict the course of an illness. He adapted Ptolemy's claim in the *Centiloquium* that the critical days could be determined by using a sixteen-sided figure and constructed a chart that resembled a horoscope with sixteen houses. The physician first determined the true location of the moon at the moment the disease appeared and recorded it on the chart, at a point analogous to the ascendant. Each critical day occurred when the moon had progressed one quarter of the way around the zodiac from its initial location. On Tannstetter's chart these points were analogous to the cardinal points in a horoscope. The physician first located the position of the moon in the zodiac at each cardinal point and then used a calendar to determine the date. Using this chart and an accurate calendar, such as the ones Tannstetter produced each year, the physician could determine the true critical days for any disease.[108] The physician then interpreted the effects of the individual planets and interplanetary aspects. Tannstetter provided a concise list of favorable planets and interplanetary aspects and concluded the section with lists ranking the favorable horoscopic houses, planets, and aspects.[109]

Physicians also interpreted the effects of recent eclipses when making a prognosis. When considering eclipses, the most important information was the magnitude of the eclipse and the Lord of the eclipse, the planet that most influenced the effects caused by the eclipse. Tannstetter's list of favorable planets served as a handy reference here. The magnitude of the eclipse allowed the physician to predict the severity and duration of its effects.[110] Finally, Tannstetter tried to give the physician a set of prognostic tools to determine when the disease would cause the patient's death. He pointed out that it was important for the physician to know when a disease was beyond his control, for he harmed his own reputation and that of the profession when he tried and failed to cure patients.[111] Professional realities of medical practice played an important part in Tannstetter's lectures on medical astrology.

For Tannstetter, his lectures on astrological medicine were the logical extension of his lectures in the arts faculty. He underscored the relationships between his lectures through frequent recommendations that physicians use an astrolabe and astronomical tables to determine the relevant astrological information. He expected his students to have heard either his own lectures on ephemerides or those of Andreas Perlach, his former student and now colleague. Perlach assumed responsibility for lecturing on ephemerides in the late 1510s and continued to do so for the next three decades. When Tannstetter encouraged his students to use astrolabes or other instruments, he drew on the strong local tradition in the arts faculty of developing and lecturing on astronomical instruments, exemplified in Stiborius's lectures. Stiborius and Tannstetter seemed to have coordinated their lectures. Stiborius frequently emphasized the medical applications of his various instruments, telling students how to color their astrolabes to make it easier to determine the motions of the humors.[112] Tannstetter also encouraged his students to modify astrolabes and similar instruments specifically for medical use.[113]

Tannstetter's lectures on medical astrology focused on pragmatic applications rather than theoretical topics. As a successful physician, he recognized that the point of a medical education was ultimately practical. He understood that astrology offered the physician a powerful predictive tool that could be used to explain the history of a disease and to predict its course and final outcome. Astrology also employed a sophisticated set of instruments and reference books that conveyed both specialized knowledge and authority. Deploying astrology—its instruments, books, charts, and techniques—demonstrated an expertise that gave the physician a competitive advantage in the medical market place.

Tannstetter's efforts to reinvigorate the astrological curriculum at the University of Vienna were not intended solely for his students—his own reputation and status benefited from his focus on astrology. His lectures and publishing program established his astrological expertise both within the Viennese academic community and at the Habsburg court. His Habsburg patrons, first Emperor Maximilian I and later Archduke Ferdinand, rewarded him handsomely. For two decades he enjoyed the rewards of being both a professor at the University of Vienna and *Leibartz* first to the Emperor Maximilian I and, following his death, to Archduke Ferdinand and his family.[114] In 1531 Ferdinand ennobled Tannstetter.[115]

Andreas Perlach and the Continuation of a Pedagogical Program

Tannstetter's success at both the university and the Habsburg courts was a model for other masters at the university. His favorite student, Andreas

Perlach, adopted the same approach and built his reputation around similar astrological content when he became a master at the university in the late 1510s. Perlach arrived in Vienna in time for the beginning of the fall term 1511, attracted to the university, at least in part, by its growing prestige following Maximilian's reforms. Within two years he began working with Tannstetter and continued to do so the next two decades.[116] In addition to attending Tannstetter's lectures, Perlach collaborated with him on various projects, including efforts to reform the calendar in 1514.[117] Following in Tannstetter's footsteps, Perlach as a young master lectured on the *Theorica planetarum* in 1517. And like Tannstetter, he appeared in the official lists of ordinary lectures only one more time, in 1524, when he again lectured on the *Theorica planetarum*.[118] Perlach used Tannstetter's recent edition of Peuerbach's *Theoricae novae planetarum* for his lectures.

In another way Perlach's trajectory at the university mirrored his mentor's. He worked to establish his expertise in mathematical astrology, especially as applied to everyday practice. He developed a series of lectures that codified operations and rules for using ephemerides in astrological practice, ranging from drawing horoscopes to interpreting specific questions and details. As with Stiborius's and Tannstetter's, Perlach's identity as an expert was bound up with his practices and his efforts to disseminate astrological knowledge. The emperor's image was, at least in part, a composite of the identities of the counselors and experts he supported at his court. What made Perlach legible as an expert to Maximilian was the technical content of his lectures.

By 1519 Perlach was lecturing on how to use ephemerides in daily astrological practice. On Thursday, March 3, 1519, Perlach began simply enough by defining the term *almanac*: "The Arabic *almanac*, Latin *diale* or *dirunale* or Greek *ephemerides*, is a book in which the planetary positions are written for each day."[119] He explained the symbols used to denote the zodiacal signs, presented the traditional qualities of the planets, and then set forth a schematic description of the terms and symbols used in an ephemerides. He then introduced the various time-keeping practices and how to convert from one to another. Perlach's initial lecture merely introduced the basics.[120]

Thursday afternoons at the University of Vienna, like most medieval universities, were reserved for extraordinary lectures, in which the master was largely free to choose any topic, especially one that would attract paying students.[121] As a young master at the university, Perlach depended on the income from these lectures, just as Tannstetter had previously. Although Perlach continued the university's tradition of lectures on ephemerides, his lectures escaped any note in the university's official records. Nevertheless, Perlach's lectures on the ephemerides filled a gap in the curriculum. As early as 1518 Perlach had recognized that Tannstetter's lectures on ephemerides were popular among the students. He had produced a short manual, his *Usus almanach*, in the hopes of prompting Tannstetter to publish his longer work.[122] Tannstetter remained unmoved and never published his lectures. A decade later, Perlach lamented the absence of a comprehensive work on using ephemerides and promised that he was working on just such a text.[123] Perlach, like Tannstetter before him, also failed to produce the promised text, which finally appeared shortly after his death in 1551. His students and colleagues collected his lectures from the preceding three decades and published them as a comprehensive text on using ephemerides, the *Commentaria Ephemeridium*, so that the university would continue to benefit from Perlach's efforts. Over the three decades Perlach expanded his lectures considerably, adding both propositions and examples throughout. He devoted the entire first section of his *Commentaria Ephemeridium* to explaining in detail the meaning and use of the symbols and the calendrical and celestial data found on the first page of an almanac.[124]

The next two sections of Perlach's *Commentaria Ephemeridium* detailed the tables of planetary data that made up the bulk of an ephemerides. Typically, an ephemerides included two tables of planetary data for each month. The left-hand page contained a table listing the positions of the seven planets for each day of the month. On the right-hand page was a table showing the aspects between the moon and the other planets as well as between the planets themselves. Perlach explained how to read and understand these tables. He also presented various calculations for manipulating this information, such as finding the speed or true motion of a planet, determining whether or not a planetary conjunction occurred above the horizon and whether or not planets eclipsed important fixed stars. These calculations often required his reader to use some other instrument, such as a sundial, or other tables of information, such as tables of stellar positions.[125] Perlach expanded both the range and depth of Tannstetter's lectures, often adding supplemental material and tables to assist the student in making a particular calculation. Moreover, he frequently divided propositions into more specific propositions and then added various examples in his effort to clarify a particularly difficult point.

In the fourth and final section of his *Commentaria* Perlach turned his attention from the general information and calculations needed to use an ephemerides for such tasks as predicting the weather and diseases to the specific tasks of calculating horoscopes. Here he addressed explicitly the daily concerns of the practicing astrologer and offered a complete guide to constructing horoscopes. He began with the basic framework of the chart, how to determine the cusps of the mundane houses, and how to locate the planets within this chart. He offered solutions for determining the four cardinal points and explained how to calculate the rising and setting times of planets and constellations.[126] These were general problems that an astrologer would need to attend to regardless of why he was constructing the chart. Perlach then shifted his attention to specific types of horoscopic charts, beginning with genitures. He explained carefully the complicated rules for determining and interpreting the *hilech*, which indicated both length of life and potential hazards.[127]

Perlach reviewed various methods of rectifying a geniture and stressed for his students the importance of determining an accurate and precise time of birth.[128] Far from a technical detail, the particular technique an astrologer used and was seen to use was an important marker of expertise and authority. An astrologer's preferred technique could have profound consequences, as the Polish astrologers Martin Bylica and Jan Stercze learned during their debate before the Hungarian diet.[129] By the early sixteenth century, astrologers commonly surveyed the various methods for rectifying a geniture before adopting the method they thought most accurate. The debate between Girolamo Cardano and Luca Guarico revealed how by the mid-sixteenth century rectifying a geniture had become an important part of the astrologer's professional identity. An astrologer's authority and credibility was bound up in displaying some technique, making some effort to rectify a geniture.[130] As Perlach's focus on these techniques made clear, rectifying a geniture both allowed the practicing astrologer to demonstrate his expertise and enabled him to calculate a more accurate natal chart. Perlach also discussed general elections and the *revolutio anni*, a horoscope for the coming year, and then concluded with a series of tables that facilitated drawing horoscopes, including tables determining the cusps of the mundane houses, the exact times that the sun would enter various zodiacal signs, and a range of other useful data.[131]

Perlach's lectures were pragmatic and utilitarian, addressing the real-world concerns of practicing astrologers. As extraordinary lectures, Perlach's courses on ephemerides had to be practical and to reflect students' preferences and perceived needs in order to attract paying students. For more than three decades Perlach's lectures attracted a body of students, reflecting a tradition of astrological and astronomical instruction at Vienna that spanned the first half of the century. But Perlach looked to his lectures as more than simply a source of additional income. Modeling himself on Tannstetter, Perlach used his astrological expertise to foster ties with and attract the favor of the local

Habsburg court. Initially, Perlach tried to establish a relationship with Maximilian's court. When the emperor died in 1519, Perlach shifted his attention to Archduke Ferdinand. Ferdinand, like Maximilian before him, looked to the university to find expert astrologers who could advise him on matters of state and could help influence public opinion. With his technical expertise, Perlach was the model astrological counselor. By the mid-1520s he was dividing his time between the university and the local Habsburg court, advising the archduke and producing pro-Habsburg astrological propaganda.[132]

* * *

Shortly after Maximilian recaptured Vienna from the Hungarians in 1490 he began trying to revitalize the University of Vienna by extending its privileges and giving grants to support new faculty. Maximilian sought to restore the University of Vienna to its former position of preeminence in the fields of astrology, astronomy, and mathematics. He attracted and supported a number of faculty members whose expertise lay in the various mathematical subjects. Stiborius provided an expertise in observations and astrological instruments. He was largely responsible for introducing a curriculum focused on instruments and their use in astrology. His students included some of the most important instrument makers of the sixteenth century. Peter Apian, Georg Hartmann, and Johannes Werner all received their initial exposure to instruments from Stiborius while they were students at the University of Vienna. When they left the university, they took with them an interest that would shape the ways instruments were designed and used throughout the remainder of the century.[133] Tannstetter's interests in ephemerides complemented Stiborius's lectures on instruments. His attention to practical details prompted him to correct Pliny's *Historia naturalis* and to produce basic textbooks for teaching astrology. Similar concerns motivated his lectures on astrological medicine in the medical faculty. As Tannstetter moved out of the arts faculty Perlach assumed responsibility for lecturing on almanacs. Together, Tannstetter's and Perlach's lectures on ephemerides provided the foundation for a number of sixteenth-century astrologers and physicians who studied at Vienna, including Joachim Vadian and Johannes Vogelin. Despite the enduring popularity of these lectures on instruments, almanacs, and astrological medicine, they were part of an astrology curriculum that flourished in the extraordinary lectures and outside the bounds of officialdom.

For Maximilian, the revitalization of the university, especially the astrological faculty, promised to provide him with local experts who could be consulted on political matters. He enlisted the faculty as counselors and turned to them for advice on various matters. At one point Maximilian turned to Stiborius's *Clipeus Austrie* to solve an immediate and pressing diplomatic problem. Tannstetter, by contrast, had a more sustained role, serving as *Leibartz* over the last decade of the emperor's life. He offered advice on issues ranging from political matters to personal health concerns; Tannstetter was summoned to treat the emperor during his fatal illness and was at his side when he died in Wels. The local Habsburg court regularly drew on Tannstetter's and Perlach's expertise to determine the effects of the planetary positions and the celestial motions on Habsburg political, dynastic, and military actions.

Beyond these ad hoc tasks, Maximilian enlisted the university in his broader political agenda. His own glory and reputation was reflected in the strength of his university and the caliber of scholars he could attract to Vienna. The emperor asserted and maintained his expertise and authority through the scholars he supported at his university and the counselors he brought to his court. At the same time, Maximilian enjoyed the support of those counselors who viewed the emperor and his university as their patron. In this way, the University of Vienna and its masters became important instruments of political authority.[134]

Chapter Four

Instruments and Authority

In the summer of 1562 Andreas Schöner dedicated his *Gnomonice*, a book describing various sundials and their uses, to Archduke Maximilian. Although the archduke was interested in medicine and natural history, he had not yet demonstrated much interest in the mathematical sciences.[1] In the preface to his *Gnomonice*, Schöner celebrated contemporary German princes who had supported the construction of astronomical instruments, suggesting that the most highly praised and honored princes were both patrons and practitioners of the astronomical sciences.[2] "Now there are many distinguished princes who also excel in the knowledge of the mathematical arts. Through their own efforts and public activities they amuse themselves in these studies and recover a certain relaxation of the mind. They design instruments, and I have seen many of this sort designed by princes. They observe the motion of the stars and perform duties similar to mathematicians."[3] Schöner drew attention to recent efforts by Protestant princes to develop their own mathematical skills in order to emulate the practices established at the Habsburg courts. What had distinguished the Habsburgs from other German princes was their active participation in both the design and use of such instruments and the long Habsburg tradition of supporting instrument makers. In a survey of instrument makers at various Habsburg courts, Schöner singled out two from Emperor Maximilian I's court: Andreas Stiborius and Johannes Stabius.[4]

Stiborius and Stabius had spent many years at Maximilian I's court and at the University of Vienna. Their technical expertise in making astrological instruments attracted attention and praise even during Maximilian's reign. In 1514 Georg Tannstetter, then master at the university and Maximilian's *Leibartz*, wrote an early biographical history of astrologers—*Viri mathematici*. By the time it was written, both Stabius and Stiborius had dedicated various instruments to the emperor along with canons on their use, and the two astrologers both received detailed entries. Tannstetter listed the numerous astrological and horological instruments that Stabius and Stiborius had designed for the emperor's use, and attributed Maximilian's support of the astrology curriculum at the University of Vienna in part to the success of these instruments with the emperor.[5]

Like Grünpeck's and Brant's pro-Habsburg propaganda, astrological instruments reinforced Maximilian's claim to the imperial title and confirmed him as heir to the Roman Imperium. Instruments were also functional devices that could be used to solve concrete problems. They enabled the emperor or his advisers to cast horoscopes or to determine propitious moments to engage in important actions, and also underscored the emperor's prestige and authority by celebrating his own knowledge and ability as well as the talents and skills of the astrologers and astronomers he was able to attract to his court.

Andreas Stiborius's *Clipeus Austrie*

Andreas Stiborius understood that instruments functioned on multiple levels when he produced his *Clipeus Austrie*, an astrological instrument designed for Maximilian to use in Vienna that grew out of Stiborius's lectures at the University of Vienna. Stiborius's appointment as master of astronomy and astrology in the *Collegium ducale* at the University of Vienna owed much to the coincidence between his activities and the interests and goals of Emperor Maximilian. Stiborius lectured on the uses of various instruments, especially universal astrolabes, and often spent nights observing and cataloging the fixed stars in an effort to improve the existing star catalogs.[6] As a friend of Conrad Celtis, Stiborius would have known of Maximilian's efforts to legitimate his claims to the title of Holy Roman Emperor and understood his work at the university as contributing to the emperor's political and dynastic program.

In the summer of 1506, Stiborius was presented with a specific opportunity to contribute to Maximilian's political efforts. Maximilian was in Vienna during the final months of a conflict with the Hungarians and to conclude peace negotiations.[7] Stiborius took advantage of Maximilian's presence in the city to design an instrument that had an immediate political function: guiding Maximilian's military campaign and peace negotiations. Stiborius designed a particular type of astrolabe instrument for use in Vienna, *Clipeus Austrie*, to determine important astrological information. It was simultaneously a practical instrument that could guide the emperor's actions and a symbolic instrument that reinforced Maximilian's authority.[8]

In the opening lines of his preface, Stiborius returns to Greek and Roman antiquity to find prototypes for his *Clipeus Austrie*. He locates these in the mythic past of Homer's *Iliad*, Hesiod's *The Shield of Heracles*, and Virgil's *Aeneid*.[9] Symbolically, Stiborius could not have found more important models. Each of these shields had been forged by the gods and given to a hero, a deity of mortal origin. Further, Stiborius's reference to Hercules would have immediately resonated with Maximilian, whose courtiers frequently associated him with the Greek hero—as noted in the earlier discussion of Grünpeck's play *Virtus et Fallacicaptrix*. Stiborius reinforced recent efforts to portray Maximilian as a modern, German Hercules (see figure 2.3) when, alluding to the labors of Hercules to rid the world of monsters, he praised Maximilian for his unceasing efforts to cleanse the world of the monsters of tyranny. Although few readers in 1506 would have missed the allusions to the Papacy in the south, the French in the west, the Ottomans in the east, and the Bohemians in the north, Stiborius made the association explicit: "However, divine King Maximilian, since the revolutions of the heavens tracked from long ago and your most illustrious, wonderful and courageous deeds, the most outstanding triumphs [ever] proclaimed to whole world, from which in every direction, east, west, north and south, you do not cease to root out the monsters of tyrants from whenever they lurk, [these things] confirm you to be another Hercules for the human race, brighter than the moon, as if you had been sent down from the heavenly host by Jove himself as a helper to mortals."[10] Throughout his preface Stiborius repeatedly linked the emperor to Hercules, playing on the importance of this equation in Maximilian's political rhetoric. But Stiborius advanced a more compelling argument for his *Clipeus Austrie*. Unlike the mythical shields of antiquity, Stiborius's *Clipeus Austrie* did not merely display fabulous pictures or the history of a single transfer of imperium, but instead unveiled the true sequence of the imperium throughout history.[11] In other words, Stiborius's *Clipeus Austrie* both revealed the true succession of emperors and conferred the authority of that succession on Maximilian.

The transfer of imperium from Rome to the German emperors was a common theme in histories written during the fifteenth century. For Maximilian, this theme acquired increased significance as he struggled to solidify his position as Holy Roman Emperor and to marshal support for his crusade to regain both Constantinople and, eventually, the Holy Land. The challenge facing historians in the fifteenth and early sixteenth century was to show that the Roman and the German empires constituted an unbroken sequence and thus that the German emperors were the rightful heirs to the imperial title, the same title worn by the Roman Caesars.[12] Maximilian's frenetic search for ancestors in both Roman and Greek antiquity as well as his efforts to discover an ancient German ancestor were one part of his project. The different columns of Maximilian's *Ehrenpforte* visually depicted this mythical genealogy of Greek, Roman, and German ancestors as well as imperial predecessors.[13] However powerful these historical projects, they were grounded in the vagaries of the historical past and were, therefore, always contingent. Stiborius's *Clipeus Austrie* complemented these historical arguments by adding the force implicit in astrological analysis. No longer was it simply the case that Maximilian was the emperor; rather, he had been destined to become the emperor.

Stiborius situated his *Clipeus Austrie* in the political sphere and endowed it with political and dynastic ambitions. As he pointed out, it remained wedded to his astrological curriculum in the Arts Faculty at the University of Vienna.[14] Maximilian had appointed Stiborius to the faculty to hold lectures on astronomy and astrology and continue to develop astrological instruments. Stiborius did not disappoint. The *Clipeus Austrie* gave Stiborius the chance to express his appreciation to Maximilian and allowed him to celebrate his activities in the *Collegium ducale*. Stiborius linked *Clipeus Austrie* to his university lectures when he described it as a new instrument "by a certain new invention, by which the circles of the celestial globe are variously extended in straight lines onto a plane, just as they are depicted, projected and stretched out in my fifth book of shadows (which should soon be dictated to Your Majesty)."[15] In addition to showing Maximilian how to use the *Clipeus Austrie*, Stiborius indicated that he would instruct him in methods of stereographic projection and different types of astrolabes, all described in his lectures *Libri umbrarum*, which were themselves part of Stiborius's series of lectures on astrolabes and related instruments that he held regularly at the university.[16] In this way, Stiborius implied that he was familiar with Maximilian and that he expected the relationship to continue.

The canons for the *Clipeus Austrie* open with a brief description of the instrument, far too brief to make much sense without some sort of visual aid.[17] Paper instruments were not uncommon in university settings, as it would be too expensive and time-consuming to make an instrument out of metal for a one-off use; thus, Stiborius and Tannstetter seem to have made a practice of teaching with paper instruments. In addition, Stiborius was not an instrument maker, though he designed paper instruments. This combined with the description in the canons suggests that the *Clipeus Austrie* was likely a type of paper instrument with no rete or, apparently, any moving parts.[18] Stiborius had designed the *Clipeus Austrie* to be easy to use. Moreover, in keeping with his preference for instruments, Stiborius claims that he designed it specifically to avoid the laborious and manifold calculations astrologers typically had to carry out when using tables of directions, ascensions, and declinations.[19] To facilitate its use, Stiborius designed the *Clipeus Austrie* specifically for use in Vienna and drew only the lines and information needed to determine right and oblique ascensions at that latitude.[20]

Stiborius adopted the same methodical approach in his canons that characterized his lectures at the university. Just as he did in his university lectures, Stiborius divided his subject into specific topics, presented each as a proposition, and then explained in clear, simple Latin how to perform that operation. He concluded most of them with a worked example. His pragmatic approach to problems dictated both how he structured and presented his canons and what topics those canons covered. He stayed close to the basics. Every horoscope had four cardinal points, the ascendant, the *medium coeli*, the descendant, and the *imum coeli*. Each of these points revealed important information, but the ascendant and the *medium coeli* received particular attention. Stiborius presented a number of different ways to determine these points. The precision of his methods varied rather widely, from determining simply the ascending sign to identifying the degree of the sign.[21] Adapting a practice that was common when using a standard astrolabe, Stiborius explained that the cusps of the mundane houses, those divisions drawn on a horoscope itself, were marked on the face of the instrument. Finally, after having established the mundane houses, the necessary first step to drawing a horoscopic chart, Stiborius concluded with canons explaining how to use an ephemerides to determine the planets' positions in the horoscope and determine whether or not the planet was ascending or descending in the zodiac. To this point, Stiborius had provided all the basic calculations needed to construct basic horoscopic charts.[22] He designed an instrument that offered Maximilian a quick and reasonably accurate method of drawing a horoscope that he could use when the situation required it.

Maximilian had an immediate use for the *Clipeus Austrie*: bringing his conflict with Hungary to a close. Although he agreed to a ceasefire on July 9, 1506, Maximilian then entered a period of negotiation with the Hungarian king. The emperor, who prided himself on always waiting for the best moment to undertake any action, spent the next ten days in Vienna receiving Hungarian emissaries. The humanist and Habsburg diplomat Johannes Cuspianian recorded the various meetings between the emperor and the Hungarian ambassadors in July.[23] Finally, at 8:00 pm on July 19, Maximilian signed a treaty with the Hungarian king Ladislaw that not only ended hostilities between the Habsburgs and the Hungarians but also laid the foundation for the Habsburg Double Marriage in 1515 and the formation of the Austro-Hungarian Empire.

Stiborius had brought Maximilian's interests in astrology down to earth and located them in the practical tools of the art: its instruments. He had used astrology to complement Maximilian's broader historical and genealogical projects. Whereas history and genealogy provided an account of the unfolding of history in all its vagaries, astrology conferred a necessity on that sequence of events.[24] In the end Stiborius returned to the University of Vienna, his lectures, his students, and his observations. Maximilian rewarded Stiborius for his services when, in 1507, he appointed him canon of St. Stephan's Cathedral in Vienna. Although Stiborius maintained his ties with the imperial court, he did not leave his position as master at the university. Nevertheless, Maximilian continued to rely on Stiborius for expert opinions, most famously in 1514, when the emperor solicited Stiborius's and Tannstetter's assessment of Pope Leo X's proposed calendar reform.[25]

Johannes Stabius's Paper *Horoscopion*

Johannes Stabius also produced astrological instruments for the emperor, as well as his immediate counselors. Stabius's instruments were printed, broadsheet instruments that could be reproduced and distributed easily, extending their effect well beyond the emperor himself. Placed in the hands of the emperor and his advisers, they combined practical benefits with symbolic value, enabling the emperor to determine auspicious times to engage in different activities while also portraying Maximilian as both a generous patron and skilled practitioner of astrology.[26] Further, by distributing Stabius's horological instruments as gifts to nobles and bureaucrats, Maximilian asserted his authority to standardize or at least exercise some control over time-telling practices, which complemented his other efforts to centralize various mechanisms of government, including establishing central courts and the post.[27]

Maximilian had invited Johannes Stabius to the University of Vienna to take up one of the newly founded chairs in mathematics and astronomy in

Celtis's *Collegium poetarum et mathematicorum*. For Stabius, this appointment was transitional, between his lectures at the University of Ingolstadt and his appointment at the court. In all likelihood Maximilian too considered the appointment temporary. In 1503, shortly after arriving in Vienna, the emperor signaled his intention to bring Stabius into direct service of the court when he crowned Stabius poet laureate, a title Maximilian awarded for explicitly political purposes.[28] Stabius soon left the University of Vienna and moved into imperial circles. Initially, this required him to adopt a rather peripatetic existence as he traveled with the emperor's entourage. During these early years he produced few works.[29] Sometime around 1510 Stabius took up a more permanent residence in Nuremberg, where he directed a team of artists—including Albrecht Dürer, Hans Burgkmair, and Hans Springinklee—to produce some of Maximilian's most impressive monuments, such as the *Triumphzug* and the *Ehrenpforte*. These massive printing projects functioned in important and complex ways in Maximilian's efforts to reconstitute and legitimate imperial authority in the character of the emperor himself.[30] Unlike Grünpeck's propaganda, which was aimed at a broad, literate audience, these works were elaborate gifts intended for a select group of recipients. As political tools, they were simultaneously rewards for loyalty and attempts to encourage it. They also spread Maximilian's presence and fame geographically and temporally. In this way, the *Ehrenpforte* and the *Triumphzug* were both political instruments that served his immediate need and monuments that ensured Maximilian's legacy.[31]

By the time he relocated to Nuremberg Stabius recognized the efficacy of prints and woodcuts as political tools and was aware of Maximilian's preference for the new medium of print. He was also aware of the emperor's growing interest in scientific prints. One of his early projects was a set of celestial maps of the northern and southern hemispheres that he completed with Conrad Heinfogel and Dürer as well as a terrestrial map he and Dürer produced.[32] Although in 1515 these maps were ultimately dedicated to Maximilian's brilliant and ambitious adviser Matthäus Lang, Bishop of Gurk, an early version of them seems to have been sent to Maximilian's secretary Jakob Bannisius in 1512—in April of that year Cuspinian wrote to Bannisius asking if he had received a set of celestial charts.[33] Stabius also understood the authority that astrological instruments could confer on the imperial title. Traditional metal instruments were, however, limited by the fact that they were expensive and difficult to produce, and they could only be circulated within very limited, if elite, circles.[34] Printed instruments were a more efficient mechanism for disseminating the imperial message. They could be reproduced rather cheaply in

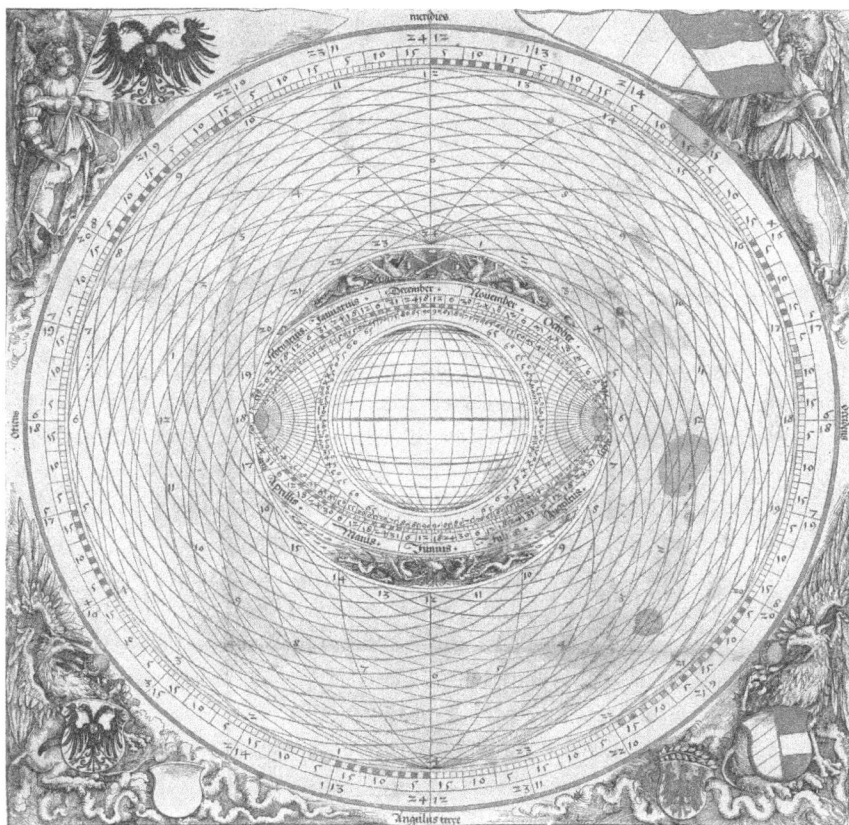

FIGURE 4.1. Johannes Stabius's *Horoscopion universale.* The instrument was hand-colored and dedicated to Emperor Maximilian I in May 1512. Note the double-headed eagles and the Austrian coats of arms in the corners of the print. Hand-colored woodcut, 46.8 x 46.3 cm. Albertina, Vienna, DG 1950/219.

large numbers and were much easier to circulate among important audiences. These printed instruments were meant to be hand-colored and fitted with moving parts. They could be used to tell time, determine important astrological information, and select propitious times to engage in various actions.

Stabius's first instrument was the *Horoscopion universale*, completed by May 31, 1512, and dedicated to Maximilian (see figure 4.1). It was printed on a large single sheet without any explanatory text; thus, the dedication and canons for this instrument must have been printed separately or copied out by hand and included with each copy of the *Horoscopion*, perhaps to increase the value of the instrument.[35] The lack of obvious explanatory canons enhanced the *Horoscopion*'s impact on a viewer and gave Maximilian the opportunity to display his erudition as he explained its uses to suitably impressed viewers.[36] The connection to the emperor was made clear, despite the lack of an obvious dedicatory epistle, by the imperial double-headed eagle and the Austrian coat-of-arms that adorned the two flags in the upper corners of the print, and griffins in the lower corners that carried shields with the same emblems. The placement of these figures and the general layout of the *Horoscopion* echoed Stiborius's preface to the *Clipeus Austrie*, where Stiborius had suggested that his *Clipeus Austrie* allowed Maximilian to comprehend in a single glance the past, present, and future, subjecting time to the emperor's gaze.[37]

Stabius's print, likewise, suggested Maximilian's control over the world by subjecting the terrestrial globe, at the center of the print, and time itself to imperial power and Habsburg rule. Stabius intended viewers to have that impression, which he reinforced in the dedicatory letters to his other instruments. According to Stabius, another *horoscopion*, "imitating the orb of the world, it exists of a circular form. This I produced at the command of earth's most authoritative, divine Caesar Maximilian. Since it is a sphere, it is perfectly suited for every mathematical figure. Therefore the greatness of His Majesty Caesar either embraces or surpasses every height of the earth's orb and whatever is outstanding in it or excellent in the opinion of mortals."[38] Stabius elevated Maximilian above all other nobles in the empire and all other humans. This passage also reinforced Maximilian's divinity, by metaphorically raising the emperor above all mortals and by suggesting that Maximilian's authority was an earthly numen, or divine will.[39] Such sentiments, as well as those accompanying Stabius's other instruments, placed the *Horoscopion* squarely in the middle of Maximilian's political projects and echoed imperial attempts to legitimate Maximilian's authority.

Stabius's *Horoscopion universale* represents a substantial increase in sophistication from Stiborius's *Clipeus Austrie* and reflects Maximilian's increasing knowledge and familiarity with astrological instruments. Whereas Stiborius's *Clipeus Austrie* was valid for only one elevation, Stabius's *Horoscopion* was a universal instrument, equally applicable at any latitude. Having traveled extensively with the emperor, Stabius recognized the need to convert between local time-telling conventions and to calculate horoscopes in a variety of locales. In his dedicatory epistle, Stabius pointed out that his *Horoscopion* made it easier for Maximilian to travel through his widely dispersed lands and immediately determine the time according to local convention and the polar elevation.[40] Overcoming the vagaries of local convention, Stabius implied in the same passage, was closely linked to Maximilian's efforts to build cohesion throughout the empire and to assert his authority. The *Horoscopion* was, like the *Clipeus Austrie*, a political tool as well as an astrological one.

Recognizing that the instrument could appear daunting at first glance, Stabius began by describing its parts. Using the instrument was aided by the fact that its various sets of lines were supposed to have been colored to help distinguish them from other lines. On Maximilian's own copy the lines representing the planetary hours were colored red. Such attention to detail and presentation certainly contributed to the value of this *Horoscopion* as an instrument of display and a cherished gift.[41] Stabius then stepped through the conversions between the four common time-keeping conventions—measuring the hours from sunset, sunrise, noon, or midnight—though he did not mention which regions or cities employed which practices. Accurately determining the time of day had obvious astrological implications. In fact, knowing the time of day was the starting point for any further astrological investigation, and many horoscopes reported the time according to local conventions and astronomical time.

Maximilian was constantly on the move and would have had ample opportunity to construct elections—horoscopes for particular events and actions. For Maximilian, the chance to correlate his entrances into various imperial cities to astrologically propitious moments, or to determine the best time to sign a treaty would have been extremely useful, or at least to have an instrument that indicated that he possessed the skills to cast elections and select the most propitious times for such actions.[42] This task, however, depended on being able to construct a horoscope, which, in turn, required knowing the time of day. For this, Stabius's *Horoscopion* was particularly useful. Perhaps more useful for the emperor were the planetary hours clearly marked on the front of his instrument. According to a fixed pattern, each hour of the day was ruled by a specific planet. These planetary rulers influenced any activity undertaken at that time. Consequently, astrologers tried to correlate activities with a ruling planet that was most likely to produce a favorable result. Planetary hours were especially common in selecting dates and times for journeys and voyages. The planetary hours clearly marked on the face of the *Horoscopion* would have been particularly useful. On the one hand, these lines enabled the user to quickly and easily elect a day and an hour ruled by the planet that was most favorable to his activity. On the other hand, they suggested that Maximilian recognized the value of such knowledge and its use.

The planetary hour lines on the face of Stabius's *Horoscopion* also facilitated drawing actual horoscopes. By the sixteenth century, planetary hours had

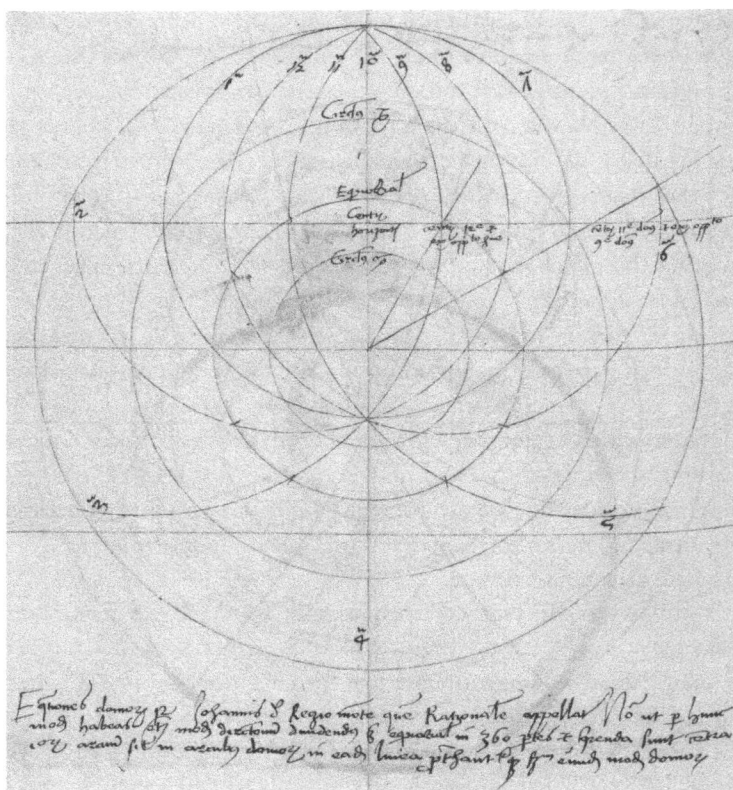

FIGURE 4.2. A diagram showing Regiomontanus's division of the zodiac into the twelve mundane houses. This diagram was probably drawn by the same person who copied Stabius's "Horoscopion," which precedes the diagram in the codex. © Österreichische Nationalbibliothek, Vienna, cod. lat. 5280, 55v.

largely been abandoned as means of telling time but continued to function as important astrological tools, both for determining ruling planets and for finding the divisions between the horoscopic houses.[43] Astrologers commonly used the planetary hours lines that instrument makers continued to inscribe on astrolabes to divide the ecliptic into the twelve mundane houses and thus to construct horoscopes.[44] And while Stabius dwelt on how to determine the unequal hours, it was a rather simple operation to convert planetary hours into the cusps of the mundane houses. Although Stabius passed silently over the fact that he is using Regiomontanus's method for dividing the zodiac, it was certainly no accident that he too adopted this approach (see figure 4.2).

Stabius's two other instruments from 1512 were variations on this same theme. In theory, both were applicable across a broad range of latitudes and could be used to convert between the various time-keeping conventions. These two instruments also alluded to Regiomontanus in that they both adapted his innovative design that he first published in his *Kalendarium*. Once again Stabius worked with Springinklee, who cut the woodblocks for his next two instruments, the first and second *Horoscopion omni generaliter*.

The first of these instruments was dedicated to Matthäus Lang. From the mid-1490s, Lang had been one of Maximilian's most important political advisers. He was also a proponent of an independent German church, in the same manner that the French church had achieved autonomy from the Papacy in the Pragmatic Sanction. Therefore, in 1511 he had been a supporter of the Council of Pisa and had been one of the architects behind Maximilian's fanciful plans to become pope, plans that briefly seemed realizable when Pope Julius II suddenly became seriously ill in August 1511. Lang was, however, a shrewd politician and when Julius II recovered, Lang traveled to Rome to negotiate directly with the pope. His efforts produced a renewed union between the pope and the emperor, succeeded in gaining admission for the imperial legates to the Lateran Council, and helped heal the schism in the Church. In 1512, Maximilian appointed him governor of imperial holdings in Italy, where he took up residence in Verona.[45]

Stabius dedicated his *Horoscopion omni generaliter* to Lang on July 30, 1512 (see figure 4.3). This instrument paired two Regiomontanus-style dials that allowed the user to calculate the local time at any latitude from 0° to 66° and to convert between common time-telling conventions. Stabius printed the canons as well as tables of useful information along the bottom of the print. The canons were similar to those he had composed for his *Horoscopion universale*. Although this new instrument was designed very differently, it could be used to perform many of the same operations. It is unclear how Stabius intended Lang to use the *Horoscopion omni generaliter*. The instrument's elongated form made it unwieldy, and the canons printed horizontally across the bottom made it difficult to consult them while using the instrument. One explicit use of this instrument was calculating the planetary hours, an operation that required converting from local time to the planetary hours. Stabius's canons on using the instrument to determine local time were at times too brief and omitted information that would have clarified the use of the instrument, but his explanations on how to determine the planetary hours were more explicit.[46] Not only were his instructions clear and precise, he included a chart of the planetary hours. Although the *Horoscopion omni generaliter* could be used to find the time, it required suspending the instrument by one end. By contrast, converting between different time-telling conventions or finding the planetary hours could be accomplished while the instrument hung horizontally on a wall. However it was displayed, its size and complexity made it an impressive display of intellectual and political authority.

As governor of the imperial holdings in Italy, Lang was an important recipient of Maximilian's patronage and a key figure in the emperor's efforts to establish his authority in Italy. On the one hand, Lang represented part of the growing class of newly empowered bureaucrats on whom Maximilian increasingly relied for his authority and on whose loyalty the emperor could depend. Stabius's *Horoscopion omni generaliter* linked Lang to the emperor, referring to him as "chief counselor to his majesty the Caesar." Stabius himself was clearly identified in the dedicatory letter and by his coat of arms at the right end of the instrument. In this way, Stabius's *Horoscopion omni generaliter* served as an instrument of imperial propaganda, the sophistication of the instrument extending both the emperor's reputation as a powerful patron by highlighting the scholars who served him and as a skilled practitioner. Stabius's instrument also reflected the emperor's efforts to standardize time by at least symbolically subordinating local variations to his convention. Lang's role as governor in Italy ensured that Maximilian's efforts to extend his control would reach an influential audience.

Stabius dedicated his last instrument, another *Horoscopion omni generaliter*, to Maximilian's personal secretary Jakob Bannisius.[47] Sometime in 1511 Maximilian had appointed Bannisius his personal secretary and had later promoted him to the position of director of the emperor's Latin correspondence. From his initial position as deacon of the cathedral at Trent, Bannisius quickly assumed greater responsibility, negotiating between the emperor and the pope, gaining considerable influence with the emperor, and eventually becoming deacon of the cathedral in Antwerp.[48] In the summer of 1515 Bannisius helped assemble the princes and nobles in Vienna to participate in the Congress of Vienna and to celebrate the Habsburg double marriage.[49] Although the *Horoscopion* was dedicated to Bannisius, once again its connection to the emperor was immediately apparent. The female figures adorning the instrument carried the imperial eagle and the Austrian flag and resembled not only those on Stabius's first instrument but also the figures on the *Ehrenpforte* (see figure 4.4). Further, across the bottom of the instrument Stabius printed the imperial privilege protecting his *Horoscopion* from pirating: "By imperial edict it is forbidden, under the penalty of 50 gulden, that anyone else reprints or tries to sell this instrument and its canons."[50] Even the casual observer would have immediately recognized Stabius's work and associated it with the emperor.

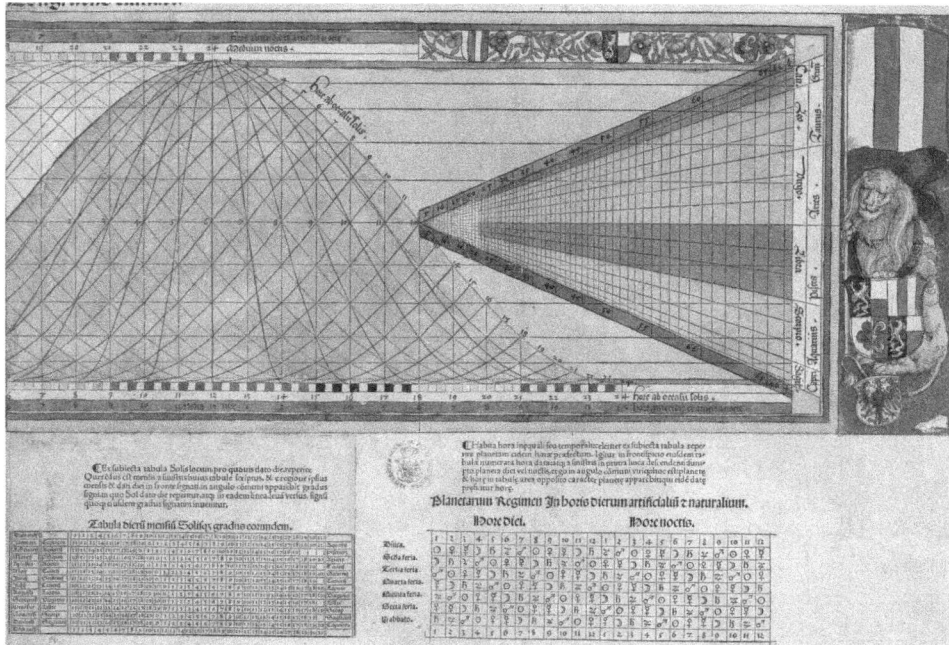

FIGURE 4.3. The right half of Stabius's *Horoscopion omni generaliter congruens climati* (1512) dedicated to Matthäus Lang, showing the chart of planetary hours. Hand-colored woodcut from two blocks and printed on two sheets, total size: 36.8 x 105 cm. Kupferstichkabinett, Staatliche Museen, Berlin, Germany (Volker-H. Schneider, Art Resource, NY), Inv. 16-1999.

Stabius's *Horoscopion omni generaliter* was closely modeled on Regiomontanus's universal dial, a point Stabius makes in the text that accompanied his instrument. This last *Horoscopion* was more convenient to use than the one he had dedicated to Lang. It was both smaller and oriented such that it could easily be used to determine the local time. Like his other instruments, this one functioned not only to determine the time but also to convert between the common time-telling conventions, subordinating the local practices to a centralized system. Equally important were the planetary hours, which were supposed to be hand-colored to distinguish them from the other lines on the face of the instrument. As with his other instruments, these planetary hours both highlighted this form of knowledge and served as a handy guide, allowing the user to determine auspicious times for undertaking different activities.

FIGURE 4.4. Stabius's *Horoscopion omni generaliter congruens climati* (1512) dedicated to Jakob Bannisius. Note tabs at the top suggesting that the Horoscopion was to be cut out and hung on a vertical surface. Hand-colored woodcut, 24.6 x 39 cm. Germanisches Nationnalmuseum, Nuremberg, HB 25805.

On July 25, 1515, Stabius dedicated a second astrological instrument to Bannisius, Stabius's *Astrolabium Imperatorium*. Stabius's timing coincided with Bannisius's involvement at the Congress of Vienna and the Habsburg double marriage that established the Habsburg monarchy. Stabius included with his *Astrolabium* a fuller explanation than he had with his previous instruments. He began by describing the various parts and areas of the instrument itself (see figure 4.5). Along with scales for converting between the various time-telling conventions, this instrument included lines that enabled the user to map the zodiac onto the mundane houses as the sky rotated throughout the day, and the important fixed stars. It was, as Stabius boasted in the dedicatory letter, a simplified astrolabe represented on a single sheet that did not require a rule, or any other moving part. By inspection alone the user could determine the important information. He then explained in detail how to use his *Astrolabium* to draw a horoscope for any given day.[51] While drawing a horoscope did not exhaust the possible uses for his instrument, Stabius considered it one of the most important astrological uses.

As well as making the astrological uses of his *Astrolabium* explicit, Stabius also suggested how his printed instruments functioned and why they were as effective as political devices. Paper instruments could easily be transported along with the emperor on his many journeys, a point Stabius made in the text at the bottom of his instrument,[52] and could be distributed as gifts in various cities.[52] For Maximilian, who was constantly on the move, such portability would have been welcome. Stabius's instruments were also meant to be used to determine the local time and identify the ruling planets for any given day and time, and they overcame various local conventions, converting local practice to a common standard.

Stabius's instruments were intended to reach recipients well beyond their select group of dedicatees. In the text accompanying his *Astrolabium* Stabius wrote: "Then through the art of printing I distributed numerous copies, so that other people too will enjoy contemplating the rotation of the heavens."[53] He had perhaps as many as several hundred copies of his instruments produced for distribution.[54] Stabius's three *Horoscopion* and his *Astrolabium* were meant to be gifts distributed to important bureaucrats, nobles, and princes. Each *Horoscopion* and the *Astrolabium* was to be hand-colored, effectively making each one a unique gift. In addition, each *Horoscopion* had a blank shield where the recipient's coat of arms would be added, further personalizing these gifts. To produce astrological instruments that were as attractive to look at as they were functional, Stabius had worked closely with Dürer and Springinklee, both of whom also worked on the emperor's more elite projects,

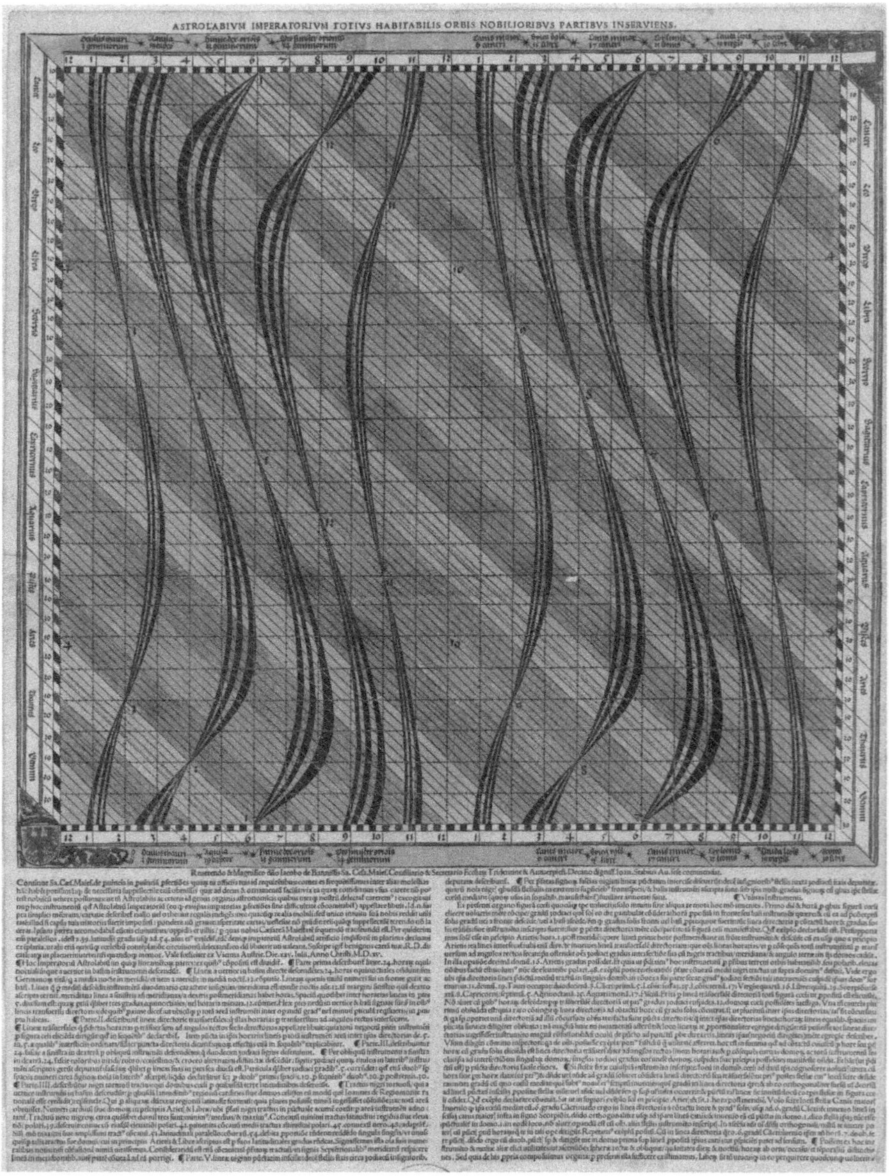

FIGURE 4.5. Stabius's *Astrolabium Imperatorium* was the second instrument dedicated to Jakob Bannisius. Stabius dedicated this instrument on July 25, 1515, during the Congress of Vienna that established the Habsburg monarchy. Hand-colored woodcut, 45 x 61 cm. © Bayerische Staatsbibliothek, Munich, Einbl. VIII, 12.

including his *Ehrenpforte* and his *Triumphzug*. Like those projects, Stabius's instruments were tools in the emperor's political program. They were not, however, memorials, looking back to Maximilian's past. Instead, Stabius's instruments were testimonies to both Maximilian's generous patronage of astrology and his own command of that body of technical knowledge.[55]

Stabius's efforts were clearly appreciated by the emperor. Maximilian had visited Nuremberg in February 1512 and, at that time, had officially charged Stabius with overseeing the production of his *Ehrenpforte* and *Triumphzug* projects. In 1514, long before these projects had been completed, or even completely planned, Maximilian ordered that Stabius be paid 1,000 Rheinisch gulden. This was an enormous sum, and in the absence of the available funds, Stabius was supposed to receive an advance.[56] Stabius apparently received only 200 gulden as an advance, prompting Maximilian to order his *Rat- und Schatzmeister* Jakob Villinger to obtain the balance from the Fuggers.[57] Maximilian's esteem for Stabius was reflected in the size of his stipend and his insistence that Stabius be paid promptly. In 1514, Dürer was still complaining that the emperor had not paid him for his services. When Dürer finally received payment, in late 1515, he was paid 100 gulden a year from the taxes collected in Nuremberg, half of Stabius's stipend.[58] Stabius's position as *Hofhistoricus* and *Poeta laureatus* had become more than an honorary title.[59] The emperor's willingness to pay Stabius such a large sum reveals Maximilian's appreciation for Stabius's services, which up to this point had largely focused on astrological instruments. In the summer of 1515, Maximilian granted Stabius a title of nobility. It was probably no coincidence that Matthäus Lang's companion and secretary, Ricardo Bertolini, gave the celebratory speech before the Congress of Vienna.[60]

* * *

In 1514 Georg Tannstetter cataloged Stabius's and Stiborius's various astrological instruments in his *Viri mathematici*. In addition to Stabius's various *Horoscopion* from 1512, Tannstetter seems to have known about Stabius's *Astrolabium Imperatorium*, which he lists as "*Instrumentum ascendentis cum domibus & stellis fixis ad diversas elevationes.*"[61] As an important member of Maximilian's court, Tannstetter had likely seen copies of Stabius's instruments. He was particularly impressed by the *Horoscopion universale*, which he considered an amazing instrument. Tannstetter probably knew about Stiborius's *Clipeus Austrie* through working with him at the University of Vienna. The two had a close relationship and explicitly referred to each other in their lectures at the university.[62] Regardless of how Tannstetter knew of these instruments, he clearly identified them as important pieces of Maximilian's imperial project. Tannstetter's emphasis on Stiborius's and Stabius's astrological instruments probably echoed the emperor's own recognition that such instruments were politically useful technologies. On the one hand, these astrological instruments were functional devices that could guide Maximilian's political decisions, enabling him to determine quickly and easily the most appropriate time to engage in political action, whether signing a peace treaty or entering a city. On the other hand, these instruments combined practical, astrological utility with political, symbolic value. Whereas Stiborius's *Clipeus Austrie* was tied to a particular place, Stabius designed instruments that could be used across the empire. Copies were personalized and sent to specific recipients.

Printed astrological instruments offered Maximilian a mechanism for asserting and extending his authority. As gifts distributed among the newly empowered bureaucratic classes, these instruments reinforced the emperor's authority by glorifying his knowledge and celebrating his ability to attract expert astrologers. Maximilian's interest in these instruments was also guided by his desire to establish uniformity throughout the empire. Stabius had suggested in his canons for the *Horoscopion universale* that by observing a common time-telling system Maximilian would be able to build cohesion among his constituencies. Through his different horological instruments Stabius offered Maximilian a mechanism for establishing that common convention, or at least a set of symbols that demonstrated his efforts to establish that convention. Stabius's instruments reflect Maximilian's broader efforts to centralize governance and strengthen the emperor's position in the empire through the *Landsknechte*, the *gemein Pfennig*, and *Hofkammer*, as well as other councils and courts.[63]

Stiborius's and Stabius's instruments also reflected a growing awareness by the emperor that skilled experts could play an important role at his court. Shortly after Maximilian had come to power he began trying to revive the University of Vienna in order to establish a body of experts he could rely on for practical, political advice.[64] Stiborius's *Clipeus Austrie* exemplified the astrological consultation that Maximilian expected masters at the university to provide. Stiborius stated more than once that his *Clipeus Austrie* relied on Regiomontanus's rational method for determining the horoscopic houses.[65] Regiomontanus's rational method was considered the most accurate method of dividing the zodiac into the mundane houses. In Vienna, invoking Regiomontanus did more than simply refer to the best method of drawing a horoscope, however. Such a reference also linked Stiborius's and his colleagues' works to the intellectual flowering that had occurred at the university in the middle of the previous century. Allusions to Regiomontanus reinforced in their own minds and those of their readers this intellectual lineage. For Maximilian, references to Regiomontanus recalled his father's confidence in and reliance on the astrologers at the University of Vienna and confirmed that his efforts to reinvigorate the university were succeeding.[66]

The authority of the University of Vienna depended on its intellectual genealogy—a genealogy that, a decade later, Tannstetter tried to codify in his *Viri mathematici*. In turn, Maximilian's own reputation and status as a patron benefited from the university's improved status and the caliber of scholar he could attract to it. But the emperor looked to the faculty at the university as offering more than isolated advice. As a body of experts they could contribute regularly to his political program.[67] Increasingly Maximilian turned to the masters at the University of Vienna for support in his struggles with German princes, Italian Popes, and French monarchs.

Chapter Five

Wall Calendars and *Practica*

The importance of wall calendars in the sixteenth century is made evident in a particularly vituperative passage from Philipp Melanchthon:

> One day in the hall at dinner time, I had an argument with a certain doctor, who began to disparage the study of mathematics. Since he was sitting next to me, I asked whether it was necessary to know the divisions of the year. He replied that it was not really necessary, for his peasants even knew when it was day and night, when it was winter and summer, and when it was noon without the knowledge of such things. I in turn said: "That response is certainly not fitting for a doctor." Oh that is a fine doctor, a rude fool. Someone should shit a turd in his cap and put it on him. What insanity is this! God's great gift is that anyone can have the letters of the Calendar on his wall.[1]

Melanchthon was not alone in this opinion. In the sixteenth century, calendars enjoyed unprecedented popularity. By drawing on shared bodies of knowledge and practices, calendars were an ideal vehicle for reaching a broad audience that spanned all levels of society and for fostering both a uniformity of social practices as well as a sense of community. The saints days, prominently displayed on every calendar, tied them to local communities and religious observation, and gave them a local identity. The astrological guidelines that governed when to bathe, take medicines, plant, or cut hair and nails shaped behavior and encouraged common practices. The conventional symbols used to represent the astrological guidelines guaranteed that even illiterate audiences would understand the information displayed on a calendar. More than just being accessible to a broad audience, calendars were well within the reach of most people. The spread of printing considerably reduced the costs and difficulty of producing calendars. Printers took advantage of the large market for calendars.[2] By the time Melanchthon ridiculed his colleague, calendars had become an important aspect of everyday life.

German princes realized that calendars and related texts offered them a way to assert their authority by being seen to have identified expert astrologers and enlisted their predictions at court. These ephemeral texts allowed princes to shape public opinion through the promulgation of official interpretations of periodic celestial phenomena such as eclipses and planetary conjunctions. In this way, the selection of astrologers for favor at court and the materials that these court astrologers then produced functioned as a means for a prince to centralize and control the production of astrological knowledge. Princes encouraged masters at their local universities to produce annual wall calendars that were grounded in technical astronomical and astrological knowledge.[3] This practice allowed princes to celebrate local scholars, thereby enhancing their own reputations. At the same time, wall calendars gave princes a mechanism to regulate everyday life. Along with these schematic wall calendars, masters produced *practica*, annual pamphlets that offered more thorough guidelines as well as interpretations of celestial phenomena. *Practica* provided the prince with a powerful political instrument.[4] As with wall calendars, *practica* were grounded in the authority of astrology and promised to interpret significant celestial events for the benefit of the reader. *Practica* offered to guide people's actions in matters of health, family, travel, and business, and provided detailed predictions about diseases, war, and the weather. By choosing masters at the university to produce both wall calendars and *practica*, the prince could exercise some influence over the meanings ascribed to natural phenomena and could appropriate those phenomena and enlist them in his political program.[5]

Maximilian was one of the first princes to recognize the political instrumentality of this ephemeral literature and to exploit it systematically. In his effort to enlist the broadest possible audience in his dynastic program, Maximilian appropriated the mechanisms that ensured him access to that audience. Wall calendars and *practica* were an important part of his program. Having revitalized the University of Vienna, in part as a means of ready access to a group of experts to consult, he enlisted the masters at the university to realize his project. For nearly two decades Georg Tannstetter had the responsibility of producing wall calendars and annual *practica*.⁶ His *practica* and wall calendars served to publicly express princely control over contemporary and future events, as well as assert and reinforce princely authority, shore up social and political hierarchies, and stabilize social order.⁷ For Tannstetter, the popularity of the wall calendars and the *practica* made them ideal vehicles for establishing his expertise and developing his reputation as a skilled astrologer. These ephemeral texts linked Tannstetter's position as a master at the University of Vienna to his role as *Leibartz* to Emperor Maximilian I and later Archduke Ferdinand. His calendars and *practica* reflected the political and social concerns most important to Tannstetter and his Habsburg patrons.

From the end of the first decade of the century, Tannstetter was inextricably linked both to the University of Vienna and to the Habsburg court. Tannstetter's fame—evidenced by his opportunity to produce *practica* and calendars for other cities as well as the popularity of his texts—reflected favorably on the emperor, who as patron benefited from the authority of his astrologers. Through Tannstetter, the emperor also exercised control over the print production of certain forms of astrological knowledge. That this astrological knowledge was embodied in cheap, accessible pamphlets made it particularly useful for the emperor precisely because it was not confined to the intellectual and political elite. Maximilian used these popular texts to reach and to enlist a broader segment of society in his political program than had previously been possible through more elite forms of political propaganda. Tannstetter identified the astrological causes for pressing social and political issues, whether war, disease, natural disaster, or diplomacy. By invoking the shared authority of astrology, Tannstetter grounded his pro-Habsburg analysis in a framework of natural causes. He used his *practica* to disseminate his message to multiple registers of society, ranging from university-trained astrologers and physicians to scarcely literate citizens.

Georg Tannstetter's Early Wall Calendars

Single-sheet calendars were some of the earliest printed works.⁸ Until the latter half of the fifteenth century, calendars were most commonly found in books of hours or used as reference tools for physicians to determine propitious times for bloodletting, bathing, and taking medicines.⁹ Printing greatly expanded the production, sale, and spread of calendars, which found an eager audience each year. Astrologers capitalized on this development and soon began producing annual large, single-sheet calendars for different cities.[10] By the end of the fifteenth century calendars had largely acquired a standard content and format. They typically included basic calendrical information—the year, golden number, dominical letter, and a wealth of astrological content—full and new moons, propitious times for bloodletting, cupping, weaning children, and planting or sowing.[11] Woodcut illustrations depicted the rulers of the year and any conjunctions or eclipses for the coming year—eclipses and conjunctions were important celestial events, and the rulers of the year determined the weather and general characteristics of each season. The position of the moon in the zodiac was given for every day of the year, enabling the reader to draw on the tradition of *Monatsregeln* to determine propitious and inauspicious times for different activities.[12] Increasingly both famous and obscure astrologers signed their calendars.[13] They used their calendars to display their expertise and establish their reputations in a community. Attribution also connected the calendars to the astrologer's patron or university.

Georg Tannstetter began producing calendars shortly after he arrived at the University of Vienna.[14] Previously, other masters at the university had produced the local calendar. Johannes Muntz had composed wall calendars from 1495 until his death in 1503. The following year one of Tannstetter's colleagues, Stephan Rosinus, produced the wall calendar for Vienna.[15] After Tannstetter took over the task in 1505, he held a near monopoly for the next two decades.[16]

Tannstetter's wall calendars conformed to the conventions that had developed over the previous fifty years. A large woodcut decorated the top of the calendar often illustrating significant astrological information such as the planetary ruler of the year or the seasons. Immediately below that Tannstetter was identified as the author and as a master at the University of Vienna. He listed important calendrical variables: the golden number, the dominical letter, the solar cycle, the *inditio romanorum*, and the number of days between Christmas and the *Estomihi* Sunday.[17] Next, Tannstetter provided a legend of the symbols used in the body of the calendar. These symbols indicated when it was astrologically propitious or hazardous to let blood, take a bath or medicines, wean children, sow seeds, and, occasionally, make clothing (see figure 5.1). The calendar itself was divided into columns, each of which included a letter representing the day of the week, the saint for that day, the position of the moon in the zodiac, and the various symbols denoting propitious times for engaging in the activities noted.

Across the bottom of the wall calendar Tannstetter added further astrological information. He often explained how the moon's position in the zodiac

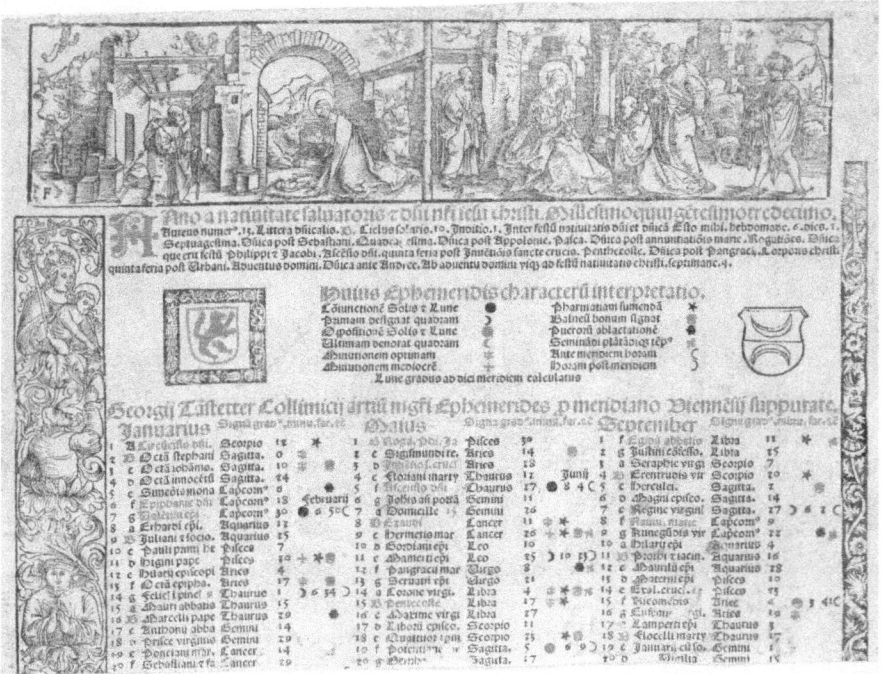

FIGURE 5.1. The upper fragment of Georg Tannstetter's 1513 wall calendar. A woodcut typically decorated the top of a wall calendar. The legend of explained the symbols used in the calendar. © Bayerische Staatsbibliothek, Munich, Einbl. Kal. 1513a.

affected parts of the body and when to treat ailments. A central woodcut illustrated either the zodiacal man or important celestial phenomena, such as conjunctions, eclipses, or the planetary rulers for the various seasons of the coming year. Finally, Tannstetter concluded with a detailed description of the conjunctions or eclipses that would occur during the year (see figure 5.2). Tannstetter culled his data from Stöffler and Pflaum's popular *Almanach novum*, correcting the times listed there to reflect local time in Vienna.

Pope Leo X and Calendar Reform

Initially Tannstetter did not seem to invest too much effort into his calendars. He was perhaps more interested in the additional income and notoriety that accrued from producing them.[18] In 1514, however, Tannstetter had a reason to consider more carefully the calendar and its problems. Late that year Max-

FIGURE 5.2. The lower fragment of Tannstetter's 1513 wall calendar. The woodcut shows the Lord of the Year and the corulers, Mars and Venus as well as a solar eclipse in Pisces. The text on the right gives the exact time and location of impending conjunctions as well as the solar eclipse. © Bayerische Staatsbibliothek, Munich, Einbl. Kal. 1513a.

imilian asked Tannstetter and Stiborius to evaluate Pope Leo X's proposed calendar reform. The two colleagues wrote a short pamphlet outlining their response to the pope's proposal and presenting their own solutions to the problem.

Scholars faced various problems when they confronted calendar reform, such as determining the length of the tropical year and reconciling the conventional dating systems with what they observed in the heavens. Astronomers and astrologers in antiquity had recognized the precession of the equinoxes that slowly caused a disjunct between the spring equinox and the beginning of the tropical zodiac, shifting the first degree of Aries further from March 21, the conventional date of the vernal equinox. By 1500 the vernal equinox had drifted about twenty degrees and corresponded roughly to the tenth degree of Pisces. As the equinox drifted further from March 21, the date established by the Council of Nicea in 325 and used to calculate the date of Easter, scholars became increasingly concerned with reconciling the astronomical equinox with the calendrical. As early as the thirteenth century the problem of calculating the date of Easter had occupied Church councils and scholars, attracting a number of proposed reforms.[19] Regiomontanus had raised similar problems in his *Kalendarium*.[20] In the fifteenth century the issues expanded to include related problems: What era was appropriate for determining the date of the initial equinox: the origin of the earth, the birth of Christ, or the Council of Nicea? What was the correct meridian for calculating the date of Easter, Rome and Jerusalem were the two obvious choices? These were not trivial questions, and they resurfaced repeatedly.

Early in the sixteenth century Paul of Middelburg was arguing most forcefully for calendar reform. In 1513 he wrote a long work on the subject titled *Paulina de recta Paschae celebratione*, which he addressed to Pope Leo X, Emperor Maximilian I, and the College of Cardinals. The following year, Leo invited him to the Fifth Lateran Council and placed him in charge of the commission on calendar reform. Middelburg's proposals for reform were far from revolutionary. He proposed to set the equinox to March 10 and to shift it one day every 134 years. He suggested adjusting the lunar cycle and periodically resetting the lunar calendar to bring it back into agreement with the calendrical equinox.[21] In preparation for considering these proposals in the tenth session of the council in December 1514, the papal curia solicited the opinions of the Christian monarchs throughout Europe, requesting the advice of their learned astronomers.

Emperor Maximilian forwarded a copy of Middelburg's proposal to Tannstetter and Stiborius, asking for their expert opinions. For Maximilian, there was more at stake than just correcting the errors in the calendar. Tannstetter and Stiborius's evaluation of Middelburg's proposal was part of the emperor's broader political project to glorify himself and Germany. The problems with the calendar were real issues that the Church and the various states had to contend with and had been struggling with for centuries, and a reform constituted an important and far-reaching change. A solution originating from Maximilian's court astrologers would thus confer prestige and renown to the empire. Maximilian expected Tannstetter and Stiborius to devote their full attention to the matter of devising a reform: "We most eagerly entrust this to you so that when you receive this you, with exacting care for this enquiry, might apply yourselves very carefully to it, and so that through your judgment and estimation you might increase and amplify the glory and fame of our

Germany. You will convey to us by the hand of our counselor and secretary Jacob de Bannisius a written account of your judgment and plan regarding the Pope's request, if you will be unable to come to the . . . tenth session, on 1 December."[22] The emperor's letter was dated October 4, 1514. Stiborius and Tannstetter did not have much time to evaluate the proposed reforms and to formulate their own.[23] They set to work immediately, and composed and published their *De Romani calendarii correctione*. Despite their best efforts, they missed the deadline; they handed their text over to Jakob Bannisius on December 16, 1514.[24]

Tannstetter and Stiborius opened their work with a brief account of the two most significant inaccuracies in the calendar—the fact that the equinox no longer occurred on March 21 and errors in the lunar cycle. These errors, they asserted, had increased over the centuries because the Church continued to use the mean motion of the sun and the moon rather than the true motions to calculate the date of the vernal equinox and subsequent full moon. They rejected mean motions as being imprecise and susceptible to error.[25] They cataloged a number of likely errors that resulted from the Church's reliance on the mean motions. In one particularly telling example, Tannstetter and Stiborius indicated how using the mean motions could yield a date for Easter that was up to a month later than the actual date. Their advice was simple: the Church should rely on the true motions of the sun and moon and, therefore, the true conjunctions and oppositions. Moreover, they urged the Church to exercise more care when calculating the dates of the equinoxes and the subsequent full moon.

Stiborius and Tannstetter devoted the remainder of their work to the mechanics of correcting the calendar. In the first place, they needed to establish the true date of the equinox. They rejected Middelburg's suggestion because it would cause too many difficulties, and opted instead for a fixed date of March 8. They proposed adjusting the calendar so that the equinox would no longer drift from this fixed date. Here they accepted Middelburg's suggestion of eliminating one day from the calendar every 134 years. Finally, they addressed the geographic question: Which location should be given primacy in determining the vernal equinox and the subsequent full moon? In other words, should Easter be celebrated according to local phenomena, which could cause Easter to fall on different Sundays in different cities, or should a single city be chosen for calculating the date of Easter? Tannstetter and Stiborius illustrated the difficulty by considering the example of Lisbon and Canton. These two cities were diametrically opposed, and so noon in Canton was midnight in Lisbon.[26] Although this was an extreme example, in principle the same held true for cities separated by only a couple hours. They conceded that this presented an insuperable problem. They were not particularly troubled by this difficulty because it would rarely occur and if it did the Church should simply determine the appropriate date for Easter.[27] In the end, Tannstetter and Stiborius adopted a position similar to Middelburg's and settled on Rome as the location that should be used for determining the date of Easter.[28]

According to Tannstetter and Stiborius, calendar reform should have been easier than ever before. They were optimistic that in 1514 it would be easier to implement the changes they were proposing thanks in large part to the proliferation of almanacs, ephemerides, and other instruments, which had helped spread astronomical and astrological knowledge and awareness throughout the population.[29] By the 1510s, they claimed, even "to such an extent that even the poorly learned and ignorant and almost worthless are able to see and read what centuries before the most learned were hardly able."[30] Thus, not only were these tables available to a greater number of people, but they were accessible to them as well. Moreover, contemporary astronomical tables were both more accurate and more common than ever before.[31] Their reference to ephemerides and other astronomical tables called attention to the calendars and astronomical tables that Tannstetter and Stiborius themselves had produced and anticipated the ephemerides that their student Perlach began composing by 1519.

Georg Tannstetter's Later Wall Calendars

Tannstetter immediately saw the connections between his work on calendar reform and his wall calendars, connections that were made clear in the *De calendarii Romani* through various references to almanacs and ephemerides. From 1515 until 1527 when he stopped producing calendars, Tannstetter approached his own wall calendars with an increased attention to detail.[32] Beginning with his wall calendar for 1515, Tannstetter seems to have used a different value for correcting the times listed in Stöffler and Pflaum's ephemerides. His task became easier when he delegated the work of carrying out the calculations to his gifted young student Andreas Perlach. In 1519 Tannstetter drew his information from Perlach's almanac when he composed his calendar for that year.[33]

Tannstetter's reputation quickly spread beyond the walls of the city. In addition to producing calendars for Vienna, he composed wall calendars for a number of other cities, including Passau, Buda, Salzburg, Olmutz, and Krakow.[34] Although most of the cities for which Tannstetter produced calendars had no university, Krakow was home to a famous university with a long tradition of astrology. Tannstetter must have enjoyed a considerable reputation for the city to have looked to him for the 1517 calendar, rather than one of their own faculty. Along with spreading geographically, Tannstetter's fame was also spreading through various strata of society as the purchasing audience for calendars continued to expand. Tannstetter and Stiborius had argued in their *De calendarii Romani* that ephemerides and almanacs were available to nearly everybody. Similarly, in his argument with his colleague, Melanchthon suggested that everyone had a calendar decorating their wall. Tannstetter's calendars were regularly printed in both Latin and German. He certainly profited from the fact that calendars were accessible, affordable, and popular.

Like most calendars in the early sixteenth century, Tannstetter's drew on a visual vocabulary that made them accessible to a broad audience. The common astrological symbols used in wall calendars had first been used in the *Holzkalender* of the fifteenth century.[35] In the latter part of the fifteenth century the *Holzkalender* was quickly replaced by the printed *Bauernkalender*, which grew into its own genre of calendar in the sixteenth century.[36] Both types of calendars relied on little or no text, using instead a rich set of symbols whose meanings resided in oral traditions and were shared by literate and illiterate people alike.[37] Like other printed wall calendars, Tannstetter's drew from and adapted these common symbols. In addition to being accessible to nearly everybody, perhaps as many as 400 to 500 copies were printed each year, suggesting a wide and eager audience.[38] At the turn of the sixteenth century, Vienna had a population of approximately 22,000.[39] Enough wall calendars were produced to put them in the hands of most literate people and on the walls of many households in the city.[40] Finally, wall calendars sold for a price that most people could afford. In the late fifteenth century, a single-sheet calendar cost approximately one day's wages for a reaper in southern Germany.[41] The relative prices remained consistent over the next few decades: by 1530 a farm worker in Austria still paid about one day's wages for a calendar, while a more skilled laborer had to part with only about one-half of a day's wages.[42] Wall calendars, with their practical layout, accessible content, and affordable pricing, would have been common sights in houses and market squares throughout the city.

The legend that decorated each wall calendar explained the symbols used in the calendar and cataloged its astrological content. The primary focus was on the regulation of health through bloodletting, bathing, taking medicinal plants and herbs, and weaning children.[43] As a reference tool, wall calendars provided easy guidance for common activities. In the case of bloodletting, Tannstetter calculated both the best and mediocre times for bleeding. The importance of these medical uses was reinforced by the text across the bottom of the wall calendar, where Tannstetter explained in greater detail when to bleed from different parts of the body and when to take medicines. Tannstetter's text repeated the standard relationship between the location of the moon in the zodiac and the parts of the body, beginning at the head and progressing to the feet and also included additional information about taking medicines and treating illnesses.

Tannstetter's wall calendars provided the literate and skilled audience more than simply the guidance found in the symbols and recommendations printed on the calendar. When paired with one of the many Latin or vernacular handbooks in circulation, his calendar became convenient and useful reference tool. Much of the daily astrology outlined in these handbooks depended on the position of the moon in the zodiac and the relationship of the other planets to the moon, precisely the information that was detailed in most wall calendars. Using a handbook such as Perlach's *Usus almanach seu ephemeridium*, Tannstetter's literate audience could use his wall calendar to make more sophisticated predictions or guide a wide range of personal activities. Perlach provided detailed canons for the standard uses of a calendar—the activities related to health, planting and sowing seeds—as well as numerous other uses. In each case, Perlach's canon used the exact location of the moon along with the person's complexion or temperament to determine the most propitious time for a particular activity or treatment.[44] Vernacular handbooks made this information available to people who could not read Latin. The German *Temporal*, attributed to Regiomontanus, included advice on purging, taking medicines, bathing, and various other daily activities.[45] Tannstetter's calendars also included the basic information needed to make general and specific weather predictions. The phases of the moon were indicated throughout the calendar itself, while the ruler of the year and the rulers of each season were frequently indicated in the woodcuts across the top or bottom of the calendar. Finally, Tannstetter often reported the exact time and location of any eclipse or conjunction and included a brief discussion of their effects. For the literate audience, the trained astrologer, and learned physician Tannstetter's wall calendars contained the basic information needed to make many common astrological predictions.

Tannstetter's wall calendars were a versatile medium, projecting his own astrological knowledge and communicating astrological information to diverse audiences. Drawing on a rich set of standard symbols, his calendars provided a wealth of astrological information that could be read off the face of the calendar. This superficial level reflected those aspects of life that were most important to the broad audience and structured them according to a set of astrological practices, including guidance on personal health and well-being, planting, and family and household management. While these commonplace predictions relating to daily life were not in themselves critical to advancing the Habsburg agenda, they served key rhetorical functions. On the one hand, they allowed Tannstetter to establish himself in popular opinion as a knowledgeable and trustworthy expert astrologer. On the other hand, these everyday forecasts, especially when paired with the guidance in the *practica*, were what made wall calendars and *practica* so popular—and it was this ubiquity and widespread use that enabled Tannstetter and the emperor to capitalize on them as a means to influence a wide audience when predictions took a more overtly political turn.

By the 1490s the plague had become endemic in southern German and Austrian cities. A few years later French Disease spread through the empire, killing many and terrifying many more. These particular diseases along with the typical challenges for remaining healthy turned people's attention to medical matters. At the same time, southern Germany and Austria were suffering from a prolonged drought and food shortages. Any advice on planting was surely welcome—one reader was so concerned with the upcoming harvest that he covered the back of his calendar with calculations and notes about it.[46] His specific concern was the fate of the staple crops—wheat, barley, and rye—suggesting that he had a yearly *practica* at hand. This reader's use of a wall calendar alongside a *practica* was typical, even expected.

Practical and Political Uses of *Practica*

Tannstetter's wall calendars offered a schematic set of predictions, guidelines, and useful astrological information, but they did not provide a wide range of predictions. The person wanting more comprehensive predictions for the coming year turned to his yearly *practica*. Tannstetter signaled the connection between his calendars and *practica* through explicit references and implicit visual clues. At the end of his discussion of eclipses on his wall calendar for 1523 he referred the reader to his *practica*: "The portents and their significance are set out in the yearly prognostication."[47] His wall calendars and *practica* were linked implicitly through the use of the same woodcut images, as in 1513 (compare figures 5.2 and 5.3). Booksellers reinforced this connection by displaying them next to each other.[48] *Practica* found an eager audience and

FIGURE 5.3. The title page from Georg Tannstetter's *Judicium astronomicum* for 1513. The title-page woodcut matches the woodcut found on his calendar for that year. Compare with figure 5.2. © Bayerische Staatsbibliothek, Munich, Sig.: Res/4° Astr.p. 510, 25.

were so popular that they attracted numerous parodies. In the 1530s François Rabelais lampooned the *practica* literature.⁴⁹ His satirical *Pantagruéline prognostication* was translated into English perhaps as early as the late sixteenth century, where it joined other comic prognostications.⁵⁰ Similarly, in Italy Pietro Aretino produced a number of satirical prognostica, which he often used to attack a specific astrologers—Tommaso Giannotti Rangoni and Lucas Guarico were his two most common targets.⁵¹ Three decades before Rabelais penned his satire, however, the yearly *practica* were no laughing matter— Tannstetter and his contemporaries took them very seriously. Tannstetter's *practica* formed part of a rich but relatively recent tradition. Unlike calendars, which had existed in manuscript form for more than a century, *practica* were largely a product of the late fifteenth century.⁵² Inexpensive and immediately useful, the genre quickly became popular. While Tannstetter certainly knew of this broader tradition, he needed only to turn to his own teacher Johannes Stabius to find a model. As a master at the University of Ingolstadt, Stabius had produced a number of *practica* as early as 1499.⁵³

In 1498 Stabius had succeeded Johannes Angelus as lecturer in mathematics at the University of Ingolstadt but already hoped to move to Vienna.⁵⁴ He aligned his early *practica* with pro-Habsburg politics. Stabius began his *Practica Teutsch* for 1501 by identifying the most significant astrological causes for the year—a conjunction of Mars and Jupiter, a lunar eclipse, and the ruler and coruler of the year.⁵⁵ According to Stabius, the favorable rays of the sun and Jupiter indicated a pleasant and peaceful year. The bulk of his *Practica Teutsch* was divided into two sections in which he offered detailed and specific predictions for the coming year. The first section included predictions about war and peace, diseases, and harvests as well as predictions about different classes of people and various cities and kingdoms. The second section was devoted to weather predictions. Stabius's general predictions for different religions and groups of people—the Christians would fare better than either the Jews or the Turks—contrasted with his more favorable prediction for Maximilian. Although he shied away from saying anything too specific, claiming that the King of the Romans or emperor was not subject to influence from the stars, Stabius assured Maximilian that he would be loved by his subjects and would have a successful year. Stabius read out of the stars similarly positive predictions for Vienna and Austria in general—both would be peaceful and prosperous.⁵⁶

Stabius's favorable predictions aligned with the emperor's own political goals and may have helped to cement a mutually beneficial relationship between them. Maximilian was eager to bolster his claims to the imperial title and to establish his authority in the empire. Stabius's astrological skills combined with popularity and accessibility of the *practica* literature offered another means of enlisting print in his political program. What made the *practica* such effective political tools was their broad appeal across multiple registers of society. Inexpensive and easy to read, *practica* reached a popular audience. At the same time, they were purchased, read, and analyzed by the more rarified and educated audiences. For example, one skilled astrologer covered a copy of Stabius's Latin *Practica Ingelstadiensis* for 1500 with horoscopes for eclipses, planetary conjunctions, and other celestial phenomena that occurred during that year (see figure 5.4).

Stabius sought to benefit from his relationship with Maximilian by gaining Habsburg favor and obtaining a position at court or the University of Vienna. During his visit to Vienna in 1497 he had succeeded only in securing a position for Conrad Celtis. Stabius's efforts in 1500 proved successful, for the following year he assumed one of the newly created chairs in mathematics at Celtis's *Collegium poetarum et mathematicorum*. Once in Vienna, he shifted his focus from popular texts to the elite Habsburg projects for Maximilian himself.[57] Alternatively, Maximilian did not give up his efforts to control the *practica* literature and turned to Georg Tannstetter, Stabius's most famous student, to produce the annual *practica*.

Tannstetter's Predictions

Tannstetter had been an attentive understudy. His first published work, the 1505 *Iudicium Viennense*, both resembled the form and content of Stabius's *Practica Teutsch* and reflected Tannstetter's attempt to establish himself in a new intellectual community. Three years after moving to Vienna Tannstetter was still a young academic struggling to demonstrate his erudition and to establish his authority. On the one hand, he larded his text with references to Ptolemy, Haly Abenragel, and Albumasar, rarely missing an opportunity to cite some authority. On the other hand, he explicitly linked his *Iudicium Viennense* to his position at the university: "I, Master Georg Tannstetter Collimitius, decided to write this annual prognosticon for the year 1505 in a simple and plain style: to honor the omnipotent God, to distinguish the flourishing University of Vienna, and as a service to the common man."[58] This rhetoric aligned Tannstetter's *Iudicium Viennense* with Maximilian's broad revitalization project. Maximilian had established the *Collegium poetarum et mathematicorum* expressly for the benefit of Vienna, Austria, and the German nation more broadly.

Joannes Stabius philosophus ac mathemati
cus beniuolis et ingenuis lectoribus felicitatem

Rognosticon anno a domico natali millesimiquingētesimi
currentis quo varios ac multiplices celestiū corporū in banc
inferiorem mortaliū regionē influxus et effectus expmunt cū
carmine seculari eo cp pfatus annus nouū seculū est inchoa
aturus ad honorem optimi maximi christi iesu eiuscp intemerate ac illi
bate matris sempercp virginis Marie necnon ad gloriam ac ornamētū
florentissimme Ingelstadensis achademie in publicū prodire iussimus
quo beniuolis ac meliori fidere natis lectoribus: rem grata efficeremus
Inuidis aūt indoctis et erroris nebula excecatis ad calcem buius progp
nostici epigramma lusimus quo rubigine attritos exacuere dētes et Let
beream valeant eraturare famem Uos igif bumanissimi lectores pro
solita vestra benignitate qua bonarū artium studiosos mira semp feli
citate pfequimini boc nostrū prognosticon vestris auspicijs editum leta
fronte suscipiatis.

Joannis stabij phi ac mathematici bimnus se
cularis.

Christe celorum decus atcp rector
Mater t christi veneranda virgo
Uota germani populi sereno
          Sumite vultu
Uos sacerdotum pietate clarus
Laudet et cetus precibus benignis
Uirgines lecte pueriq casti
          Carmina pangant
Sceptra qui summe moderaris aule
Et vagas mundi retines babenas
Orbe germano poteris nibil iam
          Uilere maius
Uirgo tu matres grauibus periclis
Libera in partu dominam polorum
Siue reginam vocitent olimpi
          Mitis adesto
Sancta germane sobolem pudicam
Procrees, terre populof a prisco

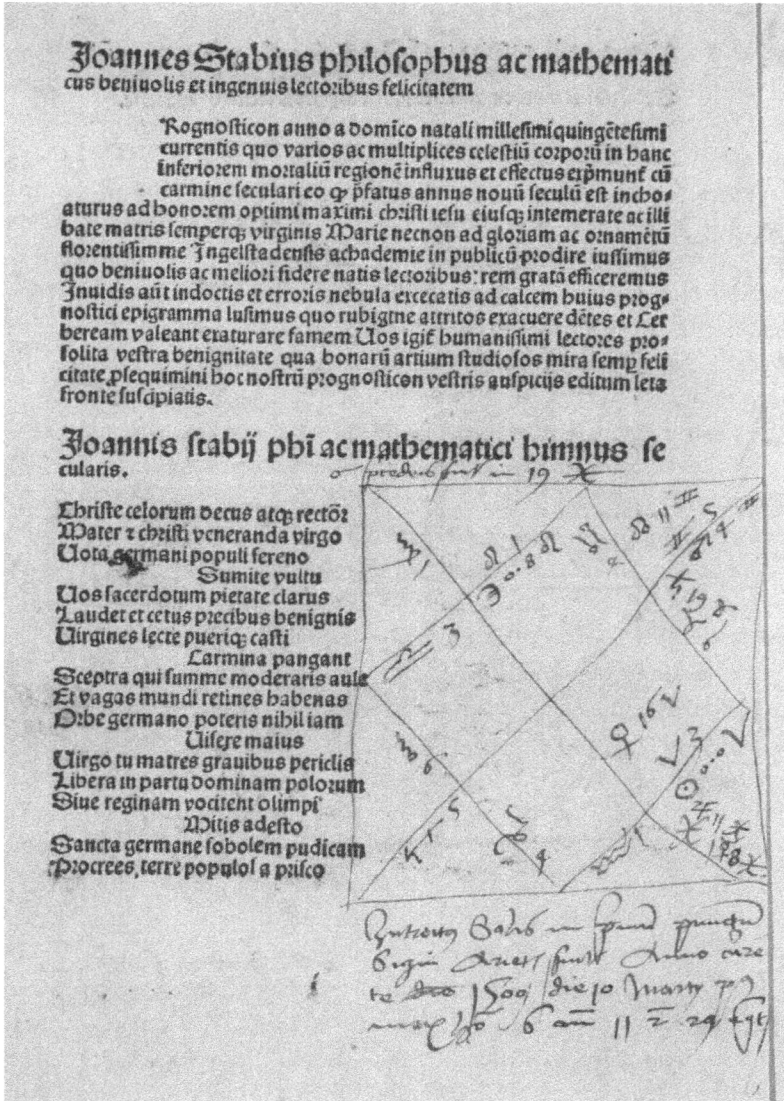

FIGURE 5.4. An astrologer worked through Stabius's *Practica Ingelstandiensis*. On the opening page he drew the horoscope for the entry of the sun into Aries. On subsequent pages he added horoscopes for the sun's entry into the other seasons of the year as well as horoscopes for conjunctions and eclipses. Johannes Stabius, *Practica Ingelstadiensis joannis Sabij phi ac mathematici benivolis et ingenuis lectroibus felicitatem* (n.p., 1499), © Österreichische Nationalbibliothek, Vienna Ink 8.H.85, 1v.

Following Stabius and other similar models, Tannstetter divided his *Iudicium Viennense* into two parts. The first treated specific predictions of peace and war, health, harvests, and predictions for specific groups of people and cities. The second offered detailed weather predictions for the coming year. Tannstetter opened with a discussion of the Lord of the Year.[59] He emphasized his own erudition by explaining the mechanism of celestial influence. Tannstetter claimed that the Lord of the Year and the planets in general affected terrestrial matter through a combination of their own direct influence and by altering the air in which all terrestrial things moved. He then constructed the horoscope for 1505—the *revolutio anni*—in order to determine the positions of the planets and the zodiacal houses at the moment the Sun entered Aries. From this horoscope Tannstetter determined that Mercury would be the Lord of the Year and that the Sun would be the coruler.[60] Tannstetter used the *revolutio anni* as the basis for all his predictions in his *Iudicium Viennense*.

Mercury, Mars, and Saturn played important roles in Tannstetter's predictions. Their locations in the horoscope and motions presaged war and destruction in the coming year: Mercury was the Lord of the Year, Mars was on the descendant, and Saturn was in retrograde.[61] Tannstetter brought this knowledge to bear on the most immediate of Maximilian's military and political concerns: the Landshut War of Succession in Bavaria.[62] According to Tannstetter the influences of Mars and Saturn indicated that Bavaria would continue to be beset by war in the coming year.[63] Since early 1504 Duke Albert of Bavaria-Munich had been at war with Duke Rupert of the Palatinate over control of Landshut. Maximilian's efforts to broker an agreement that divided the holdings were rejected by Rupert when he invaded Landshut and a number of other important Bavarian cities. Sensing a challenge to his authority, the emperor reacted immediately, gathering forces typically hostile to the Palatine and besieging various strongholds in Bavaria. Although by late 1504 the emperor's troops had made considerable progress, the war continued to occupy much of Maximilian's energy and resources. Not only did the war threaten Maximilian's own dominions in Tirol, it also forced him to delay his travel to Rome to receive the imperial crown. Finally, in the summer of 1505 the war concluded when all parties accepted a settlement proposed by Maximilian, one that allowed Maximilian to expand his Tirolean holdings.

Tannstetter then turned his attention to the diseases likely to affect his audience. Using the *revolutio anni*, he predicted a number of diseases and plagues for 1505. Mercury was in the sixth house, Saturn was in retrograde in the house of Mars, and both were affected by unfavorable rays from the sun. In a detailed analysis of this celestial configuration, Tannstetter warned of manias, melancholy, confusion, coughing, and gout. June, October, and November looked particularly bad, especially for people whose ascendant was the first part of Pisces, Virgo, or Gemini.[64] On the authority of Haly Abenragel, Tannstetter claimed that in addition to the general plagues and pestilences, countries under Cancer (France, Spain, and Portugal), as well as cities under Libra (Nuremberg and Ulm), would be particularly affected. Finally, invoking Albumasar's *Liber introductorii maioris*, Tannstetter claimed that because Mercury was the Lord of the Year and was subjected to the unfavorable rays of Saturn many destructive winds would ruin some crops in the coming year. Despite this warning, Tannstetter predicted relatively good harvests for wheat, rye, barley, and grapes.[65]

Turning from these broad predictions, Tannstetter focused first on the fortunes of different groups of people, arranged by religion, social status, or activity. He warned that many Christians and Jews would perish in the coming year. The Turks, by contrast, received a rather favorable report. After praising their attire and behavior, Tannstetter predicted that they would enjoy a calm year and would not conspire against the Christians. In July and August their fortunes would decline, though Tannstetter did not elaborate on why or what they should fear.[66] Tannstetter devoted the next few chapters to the fates of individuals and groups of people. He singled out Maximilian for a particularly encouraging prediction: "The invincible and most serene and always august King of the Romans will enjoy good fortune this year; he will obtain victory in wars and the people from other princes will hasten to his sacred majesty and will honor him. He will imprison many, however he will show them extreme grace."[67] Here Tannstetter seemed to allude to Maximilian's military engagement in Bavaria and the alliances he had formed to wage that war. At the same time, Tannstetter invoked the ideal image of Maximilian as strict but just, the same image that Maximilian portrayed in his own autobiographical writings and that Regiomontanus had found in the emperor's horoscope. Maximilian's good fortune contrasted with the varied predictions for the Christian princes, who would suffer loss and wars in the coming year.[68]

Tannstetter then turned his attention to general classes of people—surgeons and swordsmen, university masters and students, mothers and children, saturnine people, and the masses. In a combination of wishful thinking and hopeful suggestion, Tannstetter predicted that the university masters would be honored by kings and princes. Tannstetter concluded this first section with chapters on provinces, kingdoms, and various cities. He focused his predictions on locations both familiar to his local audience and important in Habsburg politics: Hungary, Bohemia, and the Austrian lands, Venice, Vienna, Nuremberg, and Buda.[69] The second half of Tannstetter's *Iudicium Viennense* dealt with weather prediction. He began with the general trends for each of the seasons before focusing on the full and new moons for each month and the weather associated with each.

The Later *Practica*

In 1511 Tannstetter published his own *Practica Teutsch*. In the intervening years since his *Iudicium Viennense* Tannstetter had established his reputation and grown comfortable in his position at the University of Vienna. He introduced a number of changes in his new *practica* that reflect his own growing familiarity with the genre and changes in his own understanding of astrology. His *Practica Teutsch* opened with a description of the most significant celestial causes for the coming year: an eclipse and some unfavorable aspects between the outer planets. This he followed with a determination of the Lord of the Year and the coruler—Mars and Mercury, respectively.[70] Malefic Mars was the Lord of the Year, prompting Tannstetter to predict destructive wars and extensive poisoning of the air, which would in turn bring about various pestilences.[71] The harvest, however, would be relatively prosperous, despite the fact that Mars would approach the Sun in the summer and cause hail and rain storms that would damage the vineyards. Given the importance of viniculture in Vienna, this was certainly a disturbing prediction. Tannstetter next added specific predictions about the fates of kingdoms and cities, focusing on those kingdoms most important to his audience, Hungary, Bohemia, and Austria.[72] Whereas previously Tannstetter had grouped people according to their social status or occupation, by 1511 he had adopted a more astrological approach that divided people into groups according to the planet that ruled them. This organization by planets was common in both contemporary medical practice and in society in general. Each planet ruled over people born at certain times and on certain days and influenced their development, personality, and profession. In the late fifteenth and early sixteenth centuries pictorial representations of *Planetenkinder*—the type of person born under the influence of a certain planet—became popular, appearing on ceilings and gables, in frescoes, and in book and manuscript illustrations. These images depicted the planets and the professions, activities, and temperaments they were supposed to rule.

By 1511 Tannstetter had also changed his approach to predicting the weather. In his 1505 *Iudicium Viennense* he had restricted himself to the full and new moons for each month and the weather for those days. In 1511 he expanded his predictions to include predictions for the days immediately before and after all four lunar phases. He drew attention to his expanded weather predictions in the opening lines of the second section of his *Practica Teutsch*.[73] Tannstetter's new approach aligned his *practica* much more closely with his wall calendars, the *practica* published by other astrologers, and with the predictive models outlined in Ptolemy's *Quadripartitum*. Ptolemy had provided guidelines for making both seasonal weather prediction as well as predictions for up to three days before and after each phase of the moon. Adopting the Ptolemaic model, Tannstetter outlined the general weather for the four seasons and then proceeded to offer detailed predictions for the four phases of the moon. These detailed predictions consumed more than half of Tannstetter's 1511 *practica* and continued to occupy a significant portion of all his subsequent *practica*. Tannstetter was not alone in offering such detail. Contemporary astrologers devoted as much effort and space to weather predictions in their own *practica*.[74] Tannstetter's decision to include more detailed weather predictions reflected his recognition that his audience expected such detail in a *practica*. For Maximilian, it was important to be seen as supportive of detail and precision—a key component of his political rhetoric and representation emphasized his commitment to precision. In that way, Tannstetter's new attention to detail and precision reinforced Maximilian's self-presentation. Tannstetter's *practica*, which displayed his skills through more and more detailed astrological predictions, were associated with the University of Vienna and the Habsburg court.[75] They both reflected and reinforced Habsburg authority. Maximilian's own reputation was enhanced by the skill of his astrologers, who in turn were known to benefit from Maximilian's patronage, thus carrying the official endorsement of the emperor.

Tannstetter's *Judicium Viennense* for 1512 and his wall calendar for that year were initially published by Wolffgang Huber in Nuremberg. Tannstetter himself understood his *practica* as complementing his wall calendars, as a dedicatory poem by his friend and colleague Ulrich Hutten made explicit. Hutten praised both Tannstetter's *practica* and his wall calendar, highlighting the astrologer's advice about propitious times for planting, when to take medicines or to let blood, to bathe, and when and where to expect war and peace, and his weather predictions.[76] When Huber printed the *practica* and the calendar, he underscored the connection between the texts by using the same woodcut on both (see figure 5.5).[77] The woodcut visually linked the *practica* to the calendar and reflected important astrological information: Jupiter, the Lord of the Year, stood on the right and spoke to Mercury, the coruler.

FIGURE 5.5. Title-page woodcut from the Wolffgang Huber edition of Tannstetter's *Judicium Viennense*. This same woodcut appeared on Tannstetter's 1512 calendar. It shows the Lord of the Year, Jupiter, and the coruler, Mercury. © Bayerische Staatsbibliothek, Munich, Res/4° Astr.p. 510, 22.

Huber might have prepared the woodcut specifically to illustrate Tannstetter's text.[78] However, Tannstetter's growing fame was such that his *practica* became immensely popular, with the consequence that they were soon pirated and surreptitiously printed in various cities throughout southern Germany.[79]

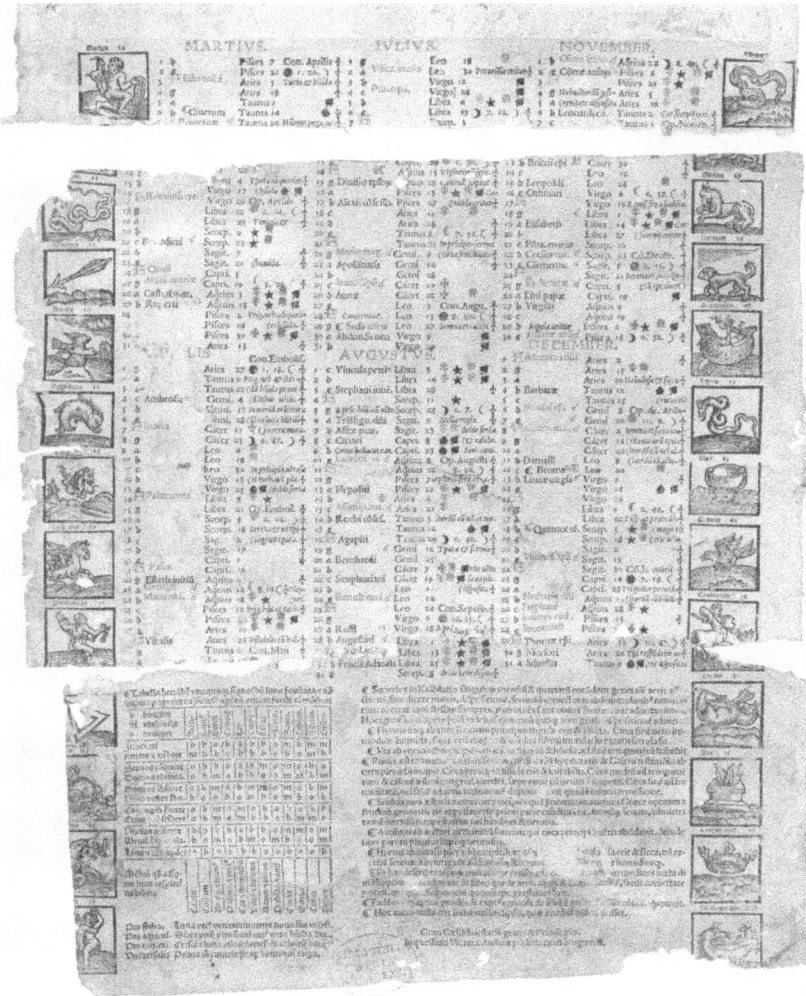

FIGURE 5.6. Bottom half of Tannstetter's 1527 calendar. The typical symbols are included in the calendar, along with specific weather predictions in the body of the calendar and general ones on the bottom. © Bayerische Staatsbibliothek, Munich, Einbl. Kal. 1527.

Henricus de Nussia, a printer in Cologne, quickly obtained a copy of Huber's edition so that he could print a version in his shop.[80] Nussia used a generic astrological woodcut that bore no relationship to Tannstetter's text. Hoping to capitalize on the popularity of Tannstetter's *practica*, Nussia was less concerned with the integrity of the text and the illustration than he was with having a copy to sell to his local audience.[81]

In 1512 Tannstetter was still adjusting his *practica* to reflect his changing understanding of these pamphlets. His *Judicium Viennense* for that year introduced a different way of correlating places with signs of the zodiac. In his earlier *practica*, Tannstetter had concentrated on kingdoms and cities and predicted their fates for the coming year. Now he shifted his focus from earthly locations to zodiacal signs, organizing terrestrial geography by the constellations. Ptolemy had first laid out a detailed set of rules for relating the zodiacal signs to geographic locations in his *Quadripartitum*. In the early sixteenth century astrologers modified his approach, which divided the zodiac into four groups of three signs each.[82] The new approach divided the zodiac into three groups of four signs each. Each group contained one sign from each season of the year. Each sign in a group was then related to a country or city on earth. By 1512 Tannstetter had begun to adopt this new system of astrological geography.[83] Tannstetter's pairing of zodiacal signs reflected a way of articulating the relationship between the constellations and geographic locations. Over the next few years Tannstetter would adopt completely the new division of the zodiac and organize the kingdoms and cities of Europe under their ruling signs.[84]

Tannstetter continued to refine his *practica* in order to base his predictions on more certain astrological foundations and more precise celestial causes, and to make his predictions more accurate. He was aware of contemporary debates and changing astrological practices. Just as his calendars reflected his broader intellectual efforts at the university and the court, Tannstetter's *practica* reflect his changing understanding of proper astrology. In 1517 he was prompted to reevaluate his astrological assumptions in light of Pico della Mirandola's attack on astrology. Tannstetter had encountered Pico's attack more than a decade earlier, but did not refer to it until now.[85] Tannstetter granted that Pico had refuted astrologers' reliance on the "ten revolutions of Saturn," but he did not consider Pico's attack particularly devastating, making only a dismissive reference to it.[86] Nonetheless, he used his characterization of Pico's attack to be more precise in his account of celestial causes. Previously he had determined only the Lord of the Year and the coruler. In 1517 he began to calculate the multiple corulers for different parts of the year and placed greater emphasis on the effects of eclipses.[87] In 1519 he gave up entirely on the Lord of the Year and replaced it with the rulers for each season, while maintaining the importance of eclipses. Just as he had relied on Ptolemy for determining the Lord of the Year, here again Tannstetter invoked another passage in Ptolemy to justify finding rulers for each season. The rulers for each season had a stronger influence on terrestrial events and provided, he implied, a better foundation for his predictions.[88] Tannstetter based his subsequent *practica* on this approach, at times finding both the ruler and the coruler for the seasons.[89] Tannstetter had exchanged a single Lord of the Year for eight planetary rulers for the year. The greater precision in identifying celestial causes reflected an evolving understanding of how celestial influences continually shaped the events throughout the year. Its closer attention to detail implied a concern for greater accuracy in his predictions. A few years later, Andreas Perlach used the same approach in his argument with Johannes Carion, who had based his predictions solely on the Lord of the Year.[90]

Tannstetter consistently linked his *practica* to immediate political events. Since 1508 Maximilian had been involved in a war with Venice. A year later, much of Europe was involved in this war, which by the end of 1511 showed no signs of ending.[91] Tannstetter combined his astrological analysis—in 1512 Mars was in the beginning of Scorpio—with his historical knowledge to predict that the war in Venice would continue through 1512 and probably grow worse.[92] When the war was still raging at the end of 1512, Tannstetter once again saw Mars as the cause of these wars and predicted that they would be more severe than before.[93] Diseases were no less disruptive or important to Maximilian's political agenda than were his many wars. Tannstetter recognized this fact and spent considerable effort to predict diseases for various parts of the empire and Europe. Most commonly, Tannstetter predicted an outbreak of the French Disease, reflecting the fear it instilled in the populace and its connection to imperial politics, recalling Maximilian's efforts to control it through a system of fines.[94] Seemingly every planetary configuration and *revolutio anni* threatened yet another outbreak of the French Disease.[95]

In other years the most threatening political or social event was related to the weather, insofar as it had, in ordinary times, the potential to harm crops, impede travel, and otherwise affect a prince's subjects. In extreme cases, it was an acute and terrifying social-political event in that perceived impending weather disasters threatened society and kingdoms. Such was the case in February 1524, when all the planets would conjoin in the watery sign of Pisces. As early as 1499 Johannes Stöffler and Jakob Pflaum had predicted this series of conjunctions. The dire predictions reached a fever pitch in the years leading up to the conjunctions as astrologers all over Europe produced inexpensive little tracts predicting the deluges and disastrous floods caused by the conjunctions.[96] Tannstetter joined this cacophony of voices in 1523, when he rejected the dire predictions in his lengthy *Libellus consolatorius*, which appeared in both Latin and German.[97] In order to reach a broader audience, he then condensed and simplified his argument in his *practica* for 1524. He repeatedly referred his readers to his *Libellus consolatorius* if they wanted further or more detailed reassurance. There is little new in this *practica*. He once again adopted the standard format and content. Predictably, the most significant causes of change in the coming year he traced to the ominous conjunction of all the planets in February in the sign of Pisces.[98] He discussed the rulers of the year, the likelihood of war and peace, impending diseases, the fate of the harvest, and the fortunes of various cities and people. Astrologers had predicted that most flooding from these conjunctions would happen in March. Consequently, Tannstetter offered his most detailed predictions for this month. Although he forecast heavy rain and storms, he stressed that there would be no significant flooding. The anxiety shared by Tannstetter's readers was reflected in both Tannstetter's precise predictions for the month and the efforts of at least one reader to observe and record the weather. Tannstetter had forecast a particularly severe rainstorm for the Friday following the full moon due in large part to Mercury's motion near the star Vega.[99] Tannstetter's reader confirmed this prediction, scrawling in the margin: "On Friday of the full moon there was an amazing rainstorm, unusual for our weather."[100] Tannstetter's predictions were not always so accurate. This reader noted that Tannstetter's forecast for clouds and rain had been incorrect: "it was a truly beautiful day."[101]

Tannstetter's *practica* for 1524 allowed him to disseminate his weather forecasts through a broad audience. As in his longer *Libellus consolatorius*, his goals were to calm the fears of people in Vienna and the environs and to dispel the notion that the conjunctions in February would bring about anything more than some severe rainstorms.[102] The threat of flood had taken such a strong hold on the populace that Tannstetter considered it worthwhile to produce a summary of his earlier works: "In this *practica* I will repeat everything I have said previously in my large *judicium* for the aforementioned year, 1524."[103] By producing a *practica*, Tannstetter was able to capitalize on a format that was familiar to many, easily understood, and, because of its brevity, cheaper than his longer texts. His efforts to reassure the Viennese colored the tone of his predictions. Unlike previous years, 1524 would see little in the way of war. Even the standard list of diseases, including leprosy, fistulas, and the pox, would not be as severe as usual.[104] Should his *practica* not convince people that there was little to fear, Tannstetter encouraged his reader to consult his longer work where he had explained all the same issues in greater detail.

Tannstetter's public-spirited motivation existed alongside his more self-serving interests. His *Libellus consolatorius* had been closely read, extensively quoted, and thoroughly criticized by Johannes Stöffler.[105] Tannstetter used his 1524 *Practica gemacht zu Wienn* to defend himself. He opened his work by claiming: "Here in this *practica*, you have once again my judgment and opinion regarding the pending consequences in the year 1524. From this you might finally understand, with what inequity and slander my name has been smeared."[106] He repeated this sentiment throughout his preface. Tannstetter viewed Stöffler's text as both an attack on his reputation and an infringement of his right to publish official astrological predictions. When Tannstetter published his *Practica gemacht zu Wienn*, he appended a letter from Emperor Charles V asserting Tannstetter's special privilege to compose and publish prognostications, *practica*, astrological booklets, and related texts. Anyone violating this privilege, which included reprinting his texts, and selling or purchasing reprinted texts, was liable for a fine of ten gold marks, half payable to the emperor and half to Tannstetter.[107] Tellingly, this letter was dated May 15, 1523, a couple of months after Tannstetter's longer texts, the *Zu eren und gefallen* and *Libellus consolatorius*, were published and then mined by Stöffler. Unfortunately for Tannstetter, his attempt to control his works failed miserably. His *Practica gemacht zu Wienn* for 1524 was printed by Johannes Singrenium in Vienna, with the letter of privilege, and shortly thereafter in Augsburg, by an unnamed printer and without the letter of privilege.[108]

* * *

Tannstetter continued to teach at the university for another decade, but he did not produce any more wall calendars or *practica*. For Tannstetter, they had served complementary but independent functions—each in its own way was a means of disseminating astrological knowledge and guiding people's lives. But by the late 1520s the functions these had played in his own work and in the astrological literature in general had changed substantially. By this time Tannstetter's students had largely assumed responsibility for producing wall calendars and *practica*, while he increasingly concentrated on medicine and producing medical texts.[109] As a new generation of astrologers incorporated these genres into their own work, they adapted them further to suit their own needs.

Tannstetter's *practica* for 1525, his *Juditium Astronomicum Viennense*, was one of his last.[110] When Tannstetter had began lecturing at the university, annual *practica* played an important role in daily life and were, consequently, excellent vehicles for establishing his own expertise as well as useful political instruments. By the time he wrote his last *practica*, authors composing eph-emerides and other texts that combined the aspects of a calendar with those of the *practica* were slowly changing the genre.[111] Tannstetter had firsthand knowledge of this evolution through his student, Andreas Perlach, whose work reflected the changes occurring in this literature.[112] When Tannstetter produced his last wall calendar, he seemed to adapt its content to compen-sate for the absence of a corresponding *practica*—his wall calendar for 1527 included additional information he had previously published in his *practica*, notably weather forecasts for various days throughout the year as well as for each of the seasons (see figure 5.6).[113]

Chapter Six

Ephemerides and Their Uses

On the first Thursday in March 1519, a number of students at the University of Vienna gathered to hear Andreas Perlach, then a young master in mathematics and astrology, deliver the first in a series of lectures on ephemerides. Perlach began with the basics. "The Arabic 'Almanac,' Latin 'Diale' or 'Diurnale,' Greek 'Ephemerides,' is a book in which the planets from day to day are presented. Take note: Each planet and each zodiacal sign has a certain resemblance or likeness to its character, which denotes that very sign or planet, therefore those images are not assigned by chance and accident. Accordingly the sign of Aries is such ♈, which seems to represent two horns like those of a ram."[1] Although basic, these lectures gave Perlach the opportunity to exhibit his knowledge. He used his ephemerides and his lectures on them to establish his expertise within his intellectual community; simultaneously demonstrating his expert knowledge within the confines of the university and disseminating that knowledge to the broader audience. They also linked Perlach to the local Habsburg court—German princes often relied on local university masters to produce ephemerides along with wall calendars and annual *practica*.[2] Ephemerides alone, however, offered little room to establish a broader astrological expertise. To demonstrate his command of astrological methods and uses, Perlach composed a short usage manual drawn from his university lectures. His manual complemented his ephemerides and together they enabled Perlach to display his technical and mathematical skills. Once he was secure in his position at the university and in Vienna more broadly, Perlach used his texts to address more explicitly political and social issues. At the local level, Perlach's astrological pamphlets were efforts to exercise social control by managing popular fear and expectations. At the same time, his pamphlets functioned on a broader, imperial level, supporting Ferdinand's political programs and efforts to be elected King of the Romans.

Ephemerides and Everyday Life in Early Modern Europe

Perlach began publishing astrological texts in 1518 when he produced his first almanac. In his early ephemerides Perlach did little to experiment with either format or content, instead conforming to the traditional format and content that Johannes Stöffler and Jakob Pflaum had established more than two decades earlier. Since 1499 Stöffler, at first alone and later with Pflaum, had produced ephemerides that contained a variety of astronomical and astrological data. Laid out in two tables on facing pages, this data included simple calendrical information such as the saints days and the dates of movable feasts, as well as more esoteric information that was needed by the astrologer to carry out his calculations. Stöffler and Pflaum simply carried over into print a model that had enjoyed a long manuscript tradition. The format remained common throughout the sixteenth century. Although Stöffler and Pflaum's ephemerides were calculated for Ulm, they enjoyed a wide circulation. Correcting the data from one ephemerides so that it applied to a different location was not a complicated process, but was time consuming. Perlach's *Almanach novum* offered the Viennese audience a solution to this laborious task.

Perlach announced on his title page that he had drawn his data "from the tables of the most learned man, master Johannes de Gmunden, formerly a studious alumnus at Vienna."[3] This accomplished two goals: it connected his work to the local intellectual tradition, and it signaled that his work was intended for the reader in Vienna. Both were important in ensuring that his ephemerides would be a success. Perlach indicated that his ephemerides was part of the tradition of Viennese scholarship that stretched back nearly a century. Perlach also connected his work to his contemporary context by claiming to have completed his ephemerides under the guidance of his mentor Georg Tannstetter.[4] By 1518 Tannstetter was something of a local celebrity, both a respected member of the university and *Leibartz* to Emperor Maximilian I. He had a reputation for publishing both learned texts, such as his 1514 edition of Peurbach's *Tabulae eclypsum* together with Regiomontanus's *Primum mobile*, and more

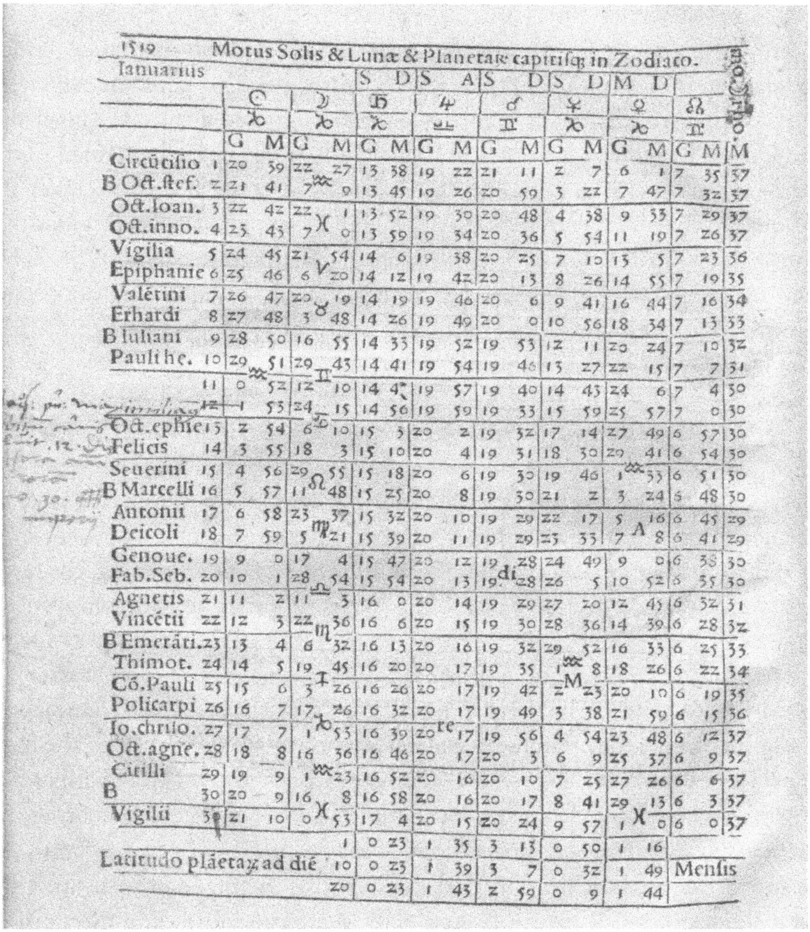

1519 Motus Solis & Lunæ & Planetarũ capitisq; in Zodiaco.

Ianuarius		☉		☽		♄ S	D	♃ S	A	♂ S	D	☿ S	D	♀ M	D	☊		
		♑		♑		♑		♎		♊		♑		♑		♊		
		G	M	G	M	G	M	G	M	G	M	G	M	G	M	G	M	M
Circũcisio	1	20	59	22	27	13	38	19	22	21	11	2	7	6	1	7	35	37
B Oct.stef.	2	21	41	7 ♒	9	13	45	19	26	20	59	3	22	7	47	7	32	37
Oct.Ioan.	3	22	42	22	1	13	52	19	30	20	48	4	38	9	33	7	29	37
Oct.inno.	4	23	43	7 ♓	0	13	59	19	34	20	36	5	54	11	19	7	26	37
Vigilia	5	24	45	21	54	14	6	19	38	20	25	7	10	13	5	7	23	36
Epiphanie	6	25	46	6 ♈	20	14	12	19	42	20	13	8	26	14	55	7	19	35
Valẽtini	7	26	47	20	19	14	19	19	46	20	6	9	41	16	44	7	16	34
Erhardi	8	27	48	3 ♉	48	14	26	19	49	20	0	10	56	18	34	7	13	33
B Iuliani	9	28	50	16	55	14	33	19	52	19	53	12	11	20	24	7	10	32
Pauli he.	10	29	51	29	43	14	41	19	54	19	46	13	27	22	15	7	7	31
	11	0 ♒	52	12 ♊	10	14	4[illegible]	19	57	19	40	14	43	24	6	7	4	30
	12	1	53	24	15	14	56	19	59	19	33	15	59	25	57	7	0	30
Oct.ephie	13	2	54	6 ♋	10	15	3	20	2	19	32	17	14	27	49	6	57	30
Felicis	14	3	55	18	3	15	10	20	4	19	31	18	30	29	41	6	54	30
Seuerini	15	4	56	29	55	15	18	20	6	19	30	19	46	1 ♒	33	6	51	30
B Marcelli	16	5	57	11 ♌	48	15	25	20	8	19	30	21	2	3	24	6	48	30
Antonii	17	6	58	23	37	15	32	20	10	19	29	22	17	5	16	6	45	29
Deicoli	18	7	59	5 ♍	21	15	39	20	11	19	29	23	33	7 A	8	6	41	29
Genoue.	19	9	0	17	4	15	47	20	12	19	28	24	49	9	0	6	38	30
Fab.Seb.	20	10	1	28	54	15	54	20	13	19 di	28	26	5	10	52	6	35	30
Agnetis	21	11	2	11 ♎	3	16	0	20	14	19	29	27	20	12	45	6	32	31
Vincẽtii	22	12	3	22	36	16	6	20	15	19	30	28	36	14	39	6	28	32
B Emerãti.	23	13	4	6 ♏	32	16	13	20	16	19	32	29	52	16	33	6	25	33
Thimot.	24	14	5	19	45	16	20	20	17	19	35	1 ♒	8	18	26	6	22	34
Cõ.Pauli	25	15	6	3 ♐	26	16	26	20	17	19	42	2 M	23	20	10	6	19	35
Policarpi	26	16	7	17	26	16	32	20	17	19	49	3	38	21	59	6	15	36
Io.chriso.	27	17	7	1 ♑	53	16	39	20 re	17	19	56	4	54	23	48	6	12	37
Oct.agne.	28	18	8	16	36	16	46	20	17	20	3	6	9	25	37	6	9	37
Cirilli	29	19	9	1 ♒	23	16	52	20	16	20	10	7	25	27	26	6	6	37
B	30	20	9	16	8	16	58	20	16	20	17	8	41	29	13	6	3	37
Vigilii	31	21	10	0 ♓	53	17	4	20	15	20	24	9	57	1 ♓	0	6	0	37
					1	0	23	1	35	3	13	0	50	1	16			
Latitudo plãetarũ ad diẽ					10	0	23	1	39	3	7	0	32	1	49	Mensis		
					20	0	23	1	43	2	59	0	9	1	44			

Figure 6.1. The January table from Andreas Perlach's 1519 *Almanach novum*. An anonymous reader noted Emperor Maximilian I's death. © Bayerische Staatsbibliothek, Munich, Res/4° Eph. Astr. 39, A2v.

popular texts, including yearly wall calendars and *practica*.[5] Perlach portrayed his work, and by extension himself, as part of this rich intellectual tradition.

Readers commonly used ephemerides as a diary in which they recorded significant events in the margins next to the days when they occurred.[6] These marginal notes ranged from the personal and familial events, to celestial and weather phenomena, to major political events (see figure 6.1).[7] By recording these events in the margins of an ephemerides the reader related those event to

specific celestial configurations and gestured to their astrological causes. Ephemerides were in the first instance astrological reference books that contained
the data needed to calculate planetary positions and to construct different
types of horoscopes. As a reference book it presupposed a body of knowledge
and a certain facility with that knowledge. The information contained in the
almanac's tables was inaccessible unless the reader could at least decipher the
various symbols and numbers in the tables. Correspondingly, this body of
knowledge informed the layout and content that an author used in his ephemerides. Like other ephemerides, Perlach's *Almanach novum* was just a series
of tables that contained little or no prose interpretation.

In 1518 Perlach also published *Usus almanach seu Ephemeridum* as an
introduction to his *Almanach novum* as well as an instruction manual for
all ephemerides. Both texts were printed by Hieronymus Vietor in Vienna,
in the same small size and typeface. Moreover, Perlach's *Usus* came off the
presses less than a month after his 1518 almanac, and the two were available by
mid-January, probably selling next to each other in Viennese bookshops. As
he did with his *Almanach novum*, Perlach linked his *Usus* to Viennese intellectual circles, in this case by claiming in his title that he had compiled his
text out of Tannstetter's commentaries. In his preface Perlach stated explicitly
the relationship between his two works, and at the same time emphasized his
debt to Tannstetter: "Since I have produced ephemerides for the coming year
that were composed through diligent calculation by me, urged on by the hope
of public benefit, I thought it worthy for the avid readers, if I would affix some
general canons drawn from the most erudite and copious commentaries, no
less derived from the 200 propositions of my teacher Georg Tannstetter Collimitius from Licoripensis."[8] Perlach used similar language in dedicating his
ephemerides to the Viennese Bishop Slatkonia.[9] In addition to linking these
two works together, the prefatory letters hinted at Perlach's motivation and
attempted to situate his works within a hierarchy of authority.[10] As a young
master at the university who hoped to secure a position at the imperial court,
Perlach could not dedicate his works directly to the emperor. Instead, he had
to choose patrons closer to his standing in the court hierarchy while, at the
same time, elevating his status. By dedicating his works to Tannstetter and
Slatkonia, Perlach suggested a patron-client relationship that could confer
authority on Perlach and his work.[11]

As with his ephemerides, Perlach had numerous models to draw from
when he composed his *Usus*. Both Perlach and his audience would have
known Stöffler and Pflaum's successful ephemerides. Although initially they
had not included any canons, by 1506 Stöffler and Pflaum began to preface

their tables with a short introduction on their use. Offering only minimal detail, they condensed their explanation of the ephemerides into three pages. They then presented canons for finding the latitude of the moon and the rising and setting times of the sun. They also offered a brief introduction to judicial astrology—casting horoscopes was the main reasons for purchasing and using an almanac.[12] Stöffler and Pflaum gave a brief account of the nature of the planets, the different aspects, and the mansions of the moon, and they treated typical astrological topics, including changes in the weather, appropriate times to begin tasks, take medicines, and cultivate crops. Following a standard horoscopic chart, they concluded with a short explanation of how to use the data in the ephemerides to construct a horoscope.[13]

Stöffler and Pflaum's introductory guide was a recognition that some of their readers needed additional instruction in order to read, understand, and use their almanac. At least in principle, the introduction made their ephemerides accessible to a wider audience. It was, however, hardly sufficient for the novice, who would have found the short explanations and disordered presentation difficult to follow. The introduction was often reprinted in subsequent editions of Stöffler and Pflaum's ephemerides but was never expanded and rarely corrected.[14] Some readers worked carefully through the introduction trying to understand the canons and how to apply them while others corrected the printing errors that had been introduced.[15] Whatever the limitations of Stöffler and Pflaum's introduction, the fact that it was regularly reprinted and that at least some contemporary readers annotated their copies reflected the need for such a text.

Perlach also looked to Tannstetter's lectures as a model. In his preface, Perlach had admitted that he drew his work from his mentor's lectures.[16] Perlach adapted his models to serve his own goals. He extended the discussion of issues he considered relevant, presented the canons in a more systematic order, and included uses that did not appear in Stöffler and Pflaum's introduction. The *Usus* reflected his experiences as both a student and master at the University of Vienna. Unlike Stöffler and Pflaum, who added their introduction almost as an afterthought, Perlach set for himself the task of writing an instructional guide to teach his reader how to use an ephemerides to draw horoscopes, and how to apply those horoscopes in everyday life.

The Scope and Content of the *Usus*

The first half of Perlach's *Usus* laid the mathematical foundations required to construct a horoscope. The second half presented the rules for applying astrology to daily life. Perlach first established the basic vocabulary and the basic information found on the first page of an almanac, such as the golden number and the date of Easter.[17] He defined the symbols used in ephemerides and explained what the numbers in the tables represent. After these preliminaries, the *Usus* became increasingly complex, instructing the reader in a range of astrological calculations. Ephemerides were necessarily general—it would have been impossible to include in the tables the planetary positions for every second of the day and for all locations. To the astrologer, however, knowing the exact position of the planets at any given moment was important for casting horoscopes. Consequently, Perlach devoted considerable effort to explaining how to calculate planetary positions, risings and settings, and motions for specific times of the day.[18]

The astrologer also had to confront the problem that in the early sixteenth century there was no standard for reckoning time. Perlach alerted his reader to the fact that some cities numbered hours from sunset, others from sunrise, and still others from midnight or noon.[19] In order to calculate the planetary and zodiacal positions the astrologer needed to know what method was used in any given location and had to convert this to the standard used in the ephemerides at hand. Perlach recommended that all times be converted to astronomical hours that started at noon and numbered the hours from zero to twenty-four. Along with calculating the exact time of day, astrologers also needed to calculate planetary positions and attend to planetary aspects—the relative positions of the planets to one another. Further, the astrologer had to take note of the planet's motion, whether normal, retrograde, or stationary, because these motions affected the duration and intensity of the planet's influence. Another important variable was the planet's position relative to the fixed stars, which altered the character of the influence.[20]

The sixteenth-century astrologer had to keep track of an enormous number of variables and significant details. Perlach had no intention of teaching his reader every facet of astrology. Instead, he sought a balance between overwhelming his reader with the mass of possible details and withholding vital information. Perlach achieved this balance through the extensive use of tables so that his readers had to consider only the most immediate details. He also helped his readers distinguish between relevant information and details that could safely be ignored. In the canon that discussed the changes in the apparent motion of the moon arising from the inequality of the natural day, Perlach provided a table to assist in calculating the change in the moon's location but then explained that it was a detail that most readers could ignore.[21] When discussing the sun and planets, Perlach explained that their motions were given only in degrees and minutes because they move rather slowly. Perlach worried that by including excessive detail he would alienate readers who did not already possess sufficient knowledge.[22]

Perlach's goal was to lead the reader up to the point where he could construct a horoscope for any given time, date, and place, the starting point for any astrological question or interpretation. The first half of Perlach's *Usus* culminates with a canon explaining how "To construct skillfully a figure of the heavens at any given time."[23] This canon bridged the gap between the mathematical first half of his work and the applications in the second half. For the astrologer this was the key canon that correlated the heavens to the horoscopic figure. The astrologer first calculated the cusps, or boundaries, of the horoscopic houses and then located the planets within those houses. Calculating the cusps required finding the degrees of the zodiac that corresponded to the beginning of each house. Although it was possible to calculate the cusps directly from the data in the almanac, here again Perlach insulated his reader from the details by providing a set of tables that facilitated the operation. Having calculated the exact time of day, the reader could read the horoscopic cusps directly from a set of tables.

In the second half of his *Usus* Perlach provided canons and guidelines for interpreting a horoscope. In his discussion of astrological rules and celestial influences Perlach was not an innovator. Instead, he merely codified in a number of tables and brief explanations common astrological knowledge. Although Stöffler and Pflaum's early ephemerides and their introductions did not include any information about the characteristics of the planets and zodiacal signs, or their relationships to parts of the human body or specific illnesses, contemporary medical texts often included an illustrated astrological man, which correlated the parts of the body to the different planets. Perlach's treatment, however, was more extensive and not limited to medical issues. He almost certainly borrowed his tabular presentation from the anonymous *Canon Joannis de monte regio in Ephemerides*, which had been published in Vienna in 1512. Perlach occasionally reproduced nearly word for word tables found in this earlier text, including many of the same abbreviations. Whether or not Perlach wanted his audience to recognize his borrowing, the similarity was not lost on contemporary readers, at least one of whom bound the *Canon Joannis de monte regio* immediately in front of Perlach's *Usus*.[24] Placing these two texts side by side emphasized the degree to which Perlach had used the former text as his model. Perlach not only borrowed the tables from this text, he also divided his own work into the same sections and presented them in nearly the same order.

The reader in the early 1500s who had these two texts bound together certainly saw such similarities, but more important he saw numerous differences. Perlach's *Usus* offered a fuller picture of the role for astrology in everyday life. He offered greater detail for each canon he shared with the *Canon Joannis de monte regio*. He also included canons and transitions missing from his models. The *Canon Joannis de monte regio* and Stöffler and Pflaum's introduction applied astrology to the weather, medical care and hygiene, and cultivating crops. Perlach adopted the same structure, but emphasized more explicitly the horoscope as the bridge between the heavenly bodies and the individual human body. Although the basic figure was probably common knowledge in the early sixteenth century, Perlach insisted on its role when he placed a generic chart in its logical position between his discussion of weather change and medical questions.

In addition to such transitional steps, Perlach explicitly broadened the application of astrology in everyday life, which was consistent with Maximilian's interest in expanding his management of the empire. Traditional sources had long pointed to the use of astrology in medicine, bathing, cutting hair, and purchasing and making clothing. Accordingly, Perlach relied heavily on Ptolemy, Pliny, the Roman poet Varro, the Arab physician Avicenna, and other classical and medieval sources when he wrote the *Usus*. His most common source was Ptolemy, whose *Quadripartitum* and *Centiloquium* were the standard reference works in astrology at the time. Perlach quoted directly from the *Centiloquium* repeatedly, using these quotations as authoritative support for his own much more elaborate explanations.[25] Unlike Stöffler and Pflaum and the author of the *Canon Joannis de monte regio*, who were content to provide a short list of astrological applications and rarely cited sources, Perlach scoured his classical sources for many more uses. Ptolemy, he claimed, warned people against passing on family possessions when Mars was the lord of the ascendant or joined to the lord of the second house.[26] In matters of inheritance, Perlach recommended that "it is, however, best when Jupiter is in the ascendant and the sun is at the *medium coeli*."[27] Similarly, Perlach enlisted Ptolemy as a source in his account of the best times to lay the foundations for new buildings and houses. Perlach's extensive use of Ptolemy's *Centiloquium* reflected both its position of authority in the sixteenth century and its use as a teaching text at the University of Vienna.[28]

When Perlach pushed beyond these traditional uses—in areas of personal hygiene and medicine as well as broader domestic issues—he cited neither authoritative classical sources nor contemporary models. He could draw on Varro and Pliny as sources on when to cut hair, but he was on his own when he gave instructions on when to cut finger- and toenails. He likewise found no support, or at least did not admit any, for his explanations of when to wean a child or when to hand a child over to a teacher. And all his classical sources were silent on the topic of when to purchase or sell goods, and when to give and receive loans.[29]

Perlach's *Usus* defined the body of knowledge in which his *Almanach novum* would be read and used. Here he recognized and addressed the uses of most interest to his audience. At the same time he identified a sphere of personal astrology that borrowed heavily from the more learned forms of astrology at the university, but was concerned with questions bearing directly on individual lives. This was a distinction that Ptolemy had established in his *Quadripartitum* when he divided astrological prognostication into general predictions, which pertain to kingdoms, cities, and races, and particular predictions grounded in nativities, which pertain to individuals.[30] Perlach provided his reader with an introduction to particular prognostications. Nowhere does he explain how to use astrology to address large social, religious, or political questions, which were the mainstay of elite astrology.[31] Although the functioning of astrology presumes a particular view of nature, nowhere did Perlach propound a natural philosophy or discuss issues of how astrology worked. His *Usus* was, after all, a usage manual and not a theoretical treatise. For similar reasons, Perlach avoided all discussion of casting and interpreting nativities. Such an activity required a much deeper understanding of astrology and a thorough discussion of theoretical issues that were absent from Perlach's introductory text.

Perlach used the *Usus* to assert his own expertise and authority and to establish himself within the Viennese intellectual community. He had previously published a single laudatory poem in Tannstetter's 1514 edition of Peurbach's *Tabulae eclypsum* and had only recently begun teaching at the university. Hoping to elevate his own standing at the university and to bring his expertise to the attention of the local Habsburg court, Perlach followed Tannstetter's model and entered the publishing world with two relatively simple astrological works, his *Almanach novum* and his *Usus alamach seu Ephemeridum*. Perlach's *Almanach novum* conformed to the expectations of this common genre and included little beyond tables of planetary data. Perlach had more freedom in his *Usus* to express himself and develop his own ideas. He skillfully rearranged and expanded instructional materials in circulation at the time. He combined the two models prevalent in the early 1500s into a single work, joining the two sections—the mathematics and the practice. Moreover, Perlach organized both sections in a more logical manner than he had found in his models, so that each half progressed stepwise from the simplest concepts to the more complicated calculations and interpretations. The details of Perlach's *Usus* interest us because those details mattered to Maximilian, who rewarded technical and mathematical expertise, particularly when it could be applied to social and political issues. Perlach recognized the emperor's intellectual preferences and acted in a way that advanced his own position within that framework. Therefore, Perlach's choices about technical details are at the core of what Maximilian privileged as a system of knowledge and knowledge creation.

Perlach's junior status in the intellectual and political community at the time was reflected in the colophon to his early works. A new master at the university, Perlach relied on his personal relationship with Tannstetter to publish his early works and to establish his reputation in Vienna.[32] In the end, Perlach's right to publish his *Almanach novum* and *Usus* was granted to him by virtue of his relationship with Tannstetter. Both men viewed their relationship as one between a mentor and a protégé.[33] Tannstetter regularly praised Perlach's ability and remarked that he had relied on Perlach's calculations in his own works.[34] In this way Tannstetter drew Perlach to the attention of Tannstetter's own powerful patrons, especially Emperor Maximilian I. At the same time, Perlach rarely passed up an opportunity to praise this teacher. When Maximilian died in 1519 Perlach continued to look to the Habsburg court, specifically focusing his efforts on Archduke Ferdinand, Emperor Charles V's brother, who continued Maximilian's program of institutional patronage and enlisting university masters at the Habsburg court. Within a few years Perlach succeeded in gaining a position at court in Vienna.

Perlach's Readers

Perlach was successful in reaching a broad audience, who read and annotated the *Usus*. They left marginalia, interlinear notes, and underlining throughout his text that reflected the range of scholarly annotations available in the early sixteenth century—correcting, digesting, and elaborating the text.[35] These readers' comments and notes give us at least a glimpse of how consumers made sense of the material, and in doing so we can move our analysis beyond the producers of propaganda and astrology. Political rhetoric and the tools of public image construction are effective only insofar as they engage audiences, yet most studies of either propaganda or astrology stop with the motivations and intentions of the authors, simply assuming that printed texts had audiences. Here we can see Perlach's audience engaging directly with the propaganda he produced. In this case, we see the ways in which Perlach's readers were taking seriously his expertise—even when they did not submit to it.

Some readers worked through Perlach's *Usus* correcting the errors listed in the errata at the end of the work. Most readers, however, expended greater effort to correct printing errors that had escaped the careful eye of the proofreader or occasionally to polish Perlach's serviceable Latin. His readers appropriated the text in different ways by highlighting various passages, underscoring relevant lines of the text, and adding summaries and key words in the margins. Most readers who bothered to annotate Perlach's text added explanation and clarifications, sometimes lengthy marginal notes that clarified or added to the meaning of the text.

Perlach's readers possessed varying linguistic skills as well as astrological knowledge. One reader who turned to Perlach's book to learn astrology also had to struggle with the Latin. He added synonyms for many unfamiliar words and wrote antecedents above numerous pronouns (see figure 6.2). In his trail of synonyms and resolved pronouns this reader revealed both what parts of the text interested him and how he incorporated the text into his existing knowledge. Although he worked through the entire text, he concentrated on the second half. He was particularly interested in those canons that pertained to the care of the body, perhaps prompted by one of the many plague outbreaks in Vienna at this time.[36] The same reader labored to clarify Perlach's various astrological references to such things as ruminating signs, Aries and Taurus, and beneficial planets, Venus and Jupiter.[37] Struggling through the *Usus,* he assimilated the information it contained and personalized his copy of the text. He ensured that his next encounter would progress more smoothly than his first.

Another reader, a Wolfgang Francis, extensively annotated his rather dilapidated copy of the *Usus,* which through hard use lost a number of pages while others that had fallen out were pasted in out of sequence.[38] Francis sat down in early 1519 with the *Usus* and a copy of an ephemerides for 1519 and worked through the text, adding numerous marginal notes and examples. Next to Perlach's discussion of how to determine the duration of an eclipse, Francis added a brief note about when eclipses could occur and then how long the eclipse that year would last.[39] Francis used Perlach's *Usus* just as it was intended, as a manual alongside an almanac. In the margins he worked out the operations Perlach outlined in the body of the text. He located relevant information from the ephemerides and used the instructions in the *Usus* to determine the length of the eclipse, its beginning and end, and how long the moon would be completely obscured by the shadow of the earth. But Francis did not confine himself to working out mathematical examples. He

Vtilitas Octaua.

Tempora Flebothomiæ Ventofifq; apta eligere.

Pro huius vtilitatis vfu, primū necefle eft ut Afpectus, Signa idonea, ipfa hominis cōplexio, ætafq; confideretur. Nam Luna ab im-
pedimentis quorū fupra mentionem fecimus libera, Ioui, Veneri,
Mercurioq; coniuncta, aut fextili, trino, quarto, oppofitoue radio
configurata in figno idoneo optimū dicimus, pro fanguinis emif-
fione. Signa aūt fanguineis congrua funt ea quibus terrefiris natura eft, ut ♉ ♍ & ♑ . Flegmaticis uero ignea, ut ♈ & ♌ . exceptis p-
tibus uie cōbufte. Colericis ea quibus aquea natura eft, ut ♋ & ♓
Melancolicis figna aeria, ut ♊ & ♒ . ♍ aūt ♌ & ♏ . figna non idonea ab aftrologis reputantur. Quod uero hactenus de Flebotomia
relatum eft, & ad ventofas pari modo referendū. Nam nihil inter
hęc difcriminis facimus, nifi ꝗ ventofe poft oppofitionem. Flebothomia uero ante eam rectius comodiufq; locum habet. Pręterea
& hora ipfa obferuāda eft diligentiffime, ita ꝗ afcendens fit ex fi-
gnis idoneis, dominufq; afcendentis & lunæ nō fint in quarta, aut
in octaua, neq; ullū cū dīio octauę domus afpectū habeant. Sint
& maleficae ab angulis remote, Septima aūte pręcipue & eius dīis
non impediantur, alioquin in flebothomatore facile cadet error.
Super omnia uero cauendum ne in eo mēbro flebothomia fiat, cū
luna id fignū tenuerit, quod mēbro illi dīiatur, ex Ptho. fententia
in uerbo 20. centiloquij fui. Addunt alij ea quoq; mēbra flebotho-
mo tangenda non effe, quę uel a figno afcendente, uel dīi afcendentis figno refpiciuntur, Addo ad hęc quod obferuatione dignū eft,
Lunam quattuor habere quadras, quarum prima a coniunctione
in quarta primam, iuuenili ętati fanguinis miffio prodeft. Secūda
iuuenili & virili. Tertia virili & fenili. Vltima fenij tantij idonea,
Aliis deinde omnibus locus eft, cū neceffitas aliud nō poftulat, fępe
eni pro morborum aut corporum qualitate, nihil horū aniaduer-
tere debet medicus. Poterit tamē fapiens medicus fi dies ipfa apta
non fuerit, horam faltem conuenienté ex iam dictis eligere.

Vtilitas Nona.

Tempora pro Pharmaco fumenda falutaria eligere.

Cum reliquorum cœleftium natura, influxufq; pernofcendus eft,
tū luna maxime in corpora noftra uis atq; poteftas, ftudiofo medico perquirēda, ut faluds gracia medicina exhibita cū ftellarum ad
fe habitudine naturaq; cōueniat, ifq; qui corporū cœleftiū eft influxus, cū rerū in terra mariq; nafcentiū, uirtuti & actioni confentiat,
In fignifero igitur tria figna funt humentia, a quibus humidę tri-
plicitatis appellatio orta eft, eā funt ♋ ♏ ♓. Quia vero, fimilia

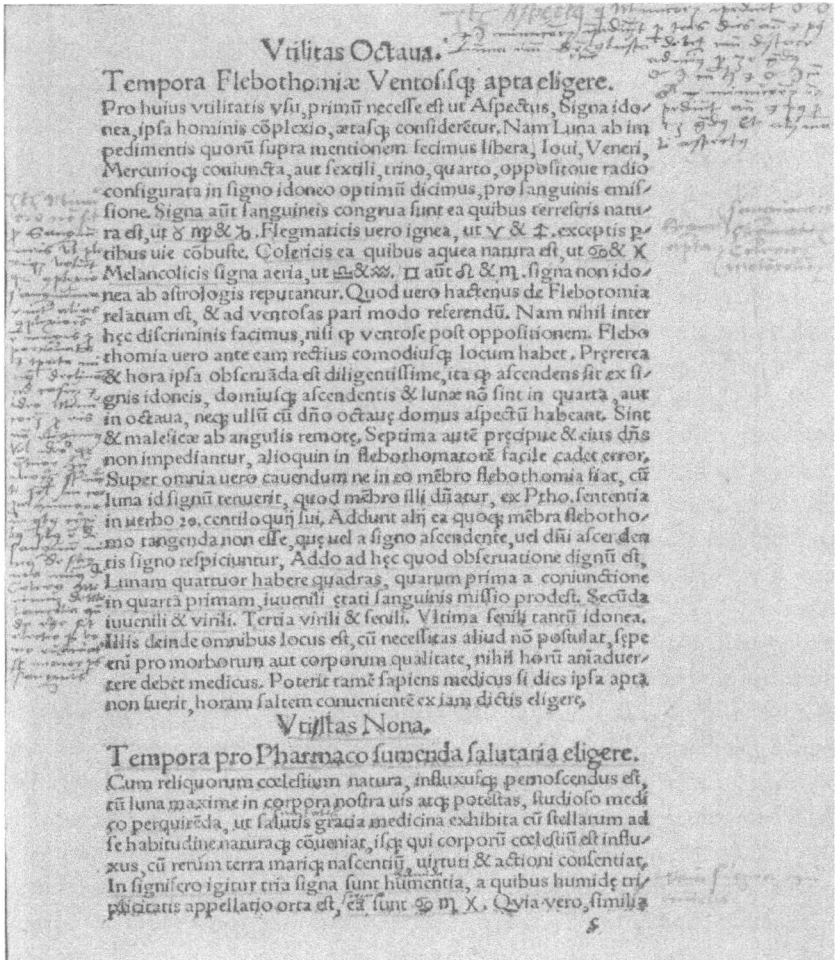

FIGURE 6.2. A page from Perlach's *Usus almanach* showing the different types of marginalia. The faint, red notes were probably written first, while the darker ones are
later. Perlach, *Usus almanach*, © Österreichische Nationalbibliothek, Vienna, 72.v.77,
e1r.

also added concrete examples to the text. Next to Perlach's tenth canon, on
the various ways of reckoning time and how to convert them to into astronomical time, Francis noted that in Bohemia they reckon time from sunset,
in Nuremberg from sunrise, and in Vienna they begin the day at midnight
and divide it into two twelve-hour halves.[40]

Despite Perlach's efforts to be thorough and logical, Francis found a few utilitas wanting and added further explanation to make them clearer. Some of his explanations were short, such as the definition he added for "conjunction": "when two bodies come into the same position."[41] In other places Francis added extensive marginal notes to clarify terms and to illustrate calculations. Perlach's instructions on how to determine the latitude of the five planets was anything but simple for Francis. He filled the page with more text than contained in the canon itself. Francis thought Perlach had too quickly skipped over certain key terms.[42] He scrawled a number of such clarifications in the margins. He spent considerable effort calculating the latitude of a planet when its motion changed from ascending to descending or vice versa. Here Francis added a much longer note in which he corrected a typographical error in Perlach's text, extended the text to include a case that Perlach had not discussed, and concluded the annotation with a detailed, real example. Francis used information from an ephemerides for 1518, perhaps Perlach's, to calculate the latitude of a planet when it progressed from the southern half of the zodiac to the northern half.[43]

When Wolfgang Francis sat down with the *Usus* he had not only the time to work carefully through the text, he also had at least two ephemerides at hand. In his marginal notes, Francis revealed his familiarity with both Ptolemy and Haly, whom he cited as corroborating evidence for one of Perlach's canons. Both authors were commonly treated in the university curriculum and the contemporary edition of Ptolemy's *Quadripartitum* included Haly's commentary.[44] What was lacking from these authors was the actual mechanics of astrology. They concentrated instead on how to interpret celestial configurations and relate them to events on earth. Francis used Perlach's *Usus* to improve his skills in making actual astrological calculations. He added the bulk of his comments to the first half of the text. Francis's marginal annotations were probably meant for himself and not for other readers—his hasty scrawlings and the decrepit state of his copy speak against Francis having shared it with anyone.

In the early sixteenth century, reading was not necessarily a solitary activity.[45] One pair of readers cooperated in working through Perlach's text. The first reader, by no means a complete novice in astrology, extensively annotated his copy of Perlach's text, underlining those sections that interested him and adding marginal headings. Although he left marks throughout the text, the practical canons in the second half received most of his attention. Using red ink, he underlined large portions of those canons that applied to personal issues such as bloodletting, taking medicines, weaning children, cutting hair and nails, and bathing. Like other readers who annotated their copies of the *Usus*, this reader transformed the text into an embodied memorial of his reading, taking control of Perlach's creation and making it into something that answered to his own needs. In reading and marking up the text, this reader used a system of annotations that distinguished between two types of marginal note. One type of note—keywords in the margins next to the underlined text—served as a simple reference device, highlighting the important passages and recalling their main points. The appropriate time to cut hair caught his attention. He underlined Perlach's recommendation—"if we want hair to regrow quickly and attractively, we will cut it during a waxing moon in a hair sign, but especially with the moon in Taurus, Virgo, and Libra"—and added in the margin "Hairy signs."[46] He also showed an interest in agricultural topics. Next to another canon, on times to plant and sow seeds, he highlighted the line that indicated propitious signs for planting and noted in the margin—"signs for planting and for sowing seeds."[47] This reader used a second type of note to pose a question or to indicate that he did not understand a particular point. He began these notes with the word "item." As with his other type of annotation, he underlined the relevant section of the text and entered a few key words in the margins.

A second reader engaged not only with the text, adding numerous marginal glosses, but also with the annotations left by the first reader. His marginal glosses served to explain sections of the text or tables in the text, and showed his mastery of the subject and the broader astrological literature. On the table that related the planets to parts of the body and to the conditions and illnesses they caused he inserted a number of cross references to the first book of Virgil's *Georgics*, to Pliny's *Historia naturalis*, and to Ptolemy's *Centiloquium*. On the same page he added explanations of the various illnesses listed in the table. Notably, he often appended to his explanations the German equivalents for Latin terms. In his description of mania he wrote: "Mania is called insanity or *morbus furoris*. Our *unsinnig*." Similarly, he glossed the Latin *apoplexia* by adding the German term "der Schlag."[48] This second reader also clarified the questions left by the first reader. When he encountered a marginal note preceded by "item," the second reader explained the confusing term or concept (see figure 6.2). Where the first reader had noted that he did not understand the term "cusp," the second reader added "the beginning of any house is called the cusp. And a planet is said to be at the cusp when it is within 5 degrees of the beginning of any house." He likewise explained the *medium coeli*, when to avoid bloodletting, the details of planetary aspects and other topics.[49]

Another reader brought considerable knowledge to bear on the first half of Perlach's *Usus*. He covered the title page with theoretical notes and comments. He recorded when the extra day in a leap year actually occurred—it was not simply that every fourth-year February had a twenty-ninth day, but that the "24th day of February is the place of the bisextus day; they say bisextus because twice in that month the 6th calens of March occurs."[50] According to contemporary theory this was correct, but because it did not bear on any of the calculations in the *Usus*, Perlach had omitted it. He also added notes about the *aureum numerum*, used to calculate when full moons would occur, and described the *ciclum solarem*, used in calculating which days of the year were Sundays.[51] Finding no appropriate margin in which to add these theoretical points, he annotated both sides of the title page.

He also added marginal and interlinear corrections or elaborations throughout the text, all of which point to thorough understanding of the subject. His corrections often required a knowledge of geography and an understanding of how celestial observations varied from place to place.[52] Next to Perlach's discussion of lunar eclipses, the reader added a diagram that represented five stages of the moon during an eclipse. He labeled each stage and added a brief description.[53] The mechanics of how and why the moon was obstructed by the earth were unimportant for Perlach's efforts, but this reader thought it worth his time to add the diagram and description to the text. This reader had a sophisticated understanding of celestial mechanics, how to manipulate the data in ephemerides, and how to calculate astrologically significant information.

The readers who left traces of their efforts in Perlach's text reflect the rich contours of his audience, an audience that ranged from novices, who had had little or no previous astrological experience, to experts who added substantive marginalia. These readers engaged with the text in ways that suited their own needs and particular interests. Common to them all, however, was an interest in praxis. They considered Perlach's *Usus* and *Almanach novum* practical tools. Their reading of these texts was oriented toward mastering them. They approached the *Usus* with a set of practical goals in mind—to learn the techniques contained in the text and to apply them to everyday life, especially personal and domestic topics.[54] This coincided with Perlach's own understanding of his work. Nowhere did Perlach's readers suggest that they used his texts for large-scale social or historical interpretations. In fact, Perlach himself had avoided any discussion of societal phenomena, including religious and political issues. In these early works, Perlach offered neither his own interpretations of contemporary political situations nor the tools that would enable his readers to carry out such interpretations. Instead of being explicit vehicles for a political agenda, Perlach's *Usus* and his early ephemerides served to establish his reputation and expertise in Vienna, both within the halls of the university and beyond its walls in the city itself.

Ephemerides and Politics

Perlach's early works had cemented his relationship with the local Habsburg court, which came to rely on his astrological expertise. Whatever had prompted Perlach to avoid political and social issues in his 1518 *Usus* and early ephemerides, by the late 1520s he was using his astrological works to engage with contemporary political issues. Throughout the decade he had continued to assist Tannstetter in carrying out calculations for important celestial phenomena such as the 1524 conjunction of all the planets in Pisces, and at the end of the decade composed two more ephemerides.[55] In his later works Perlach began to expand the content of his ephemerides to include interpretations that he had previously excluded from his texts. Perlach's early ephemerides provided primarily astrological data without any interpretation. They were reference texts that required a substantial body of knowledge. Now Perlach combined astrological data and astrological interpretations, reminiscent of the popular *practica* that continued to come off the presses.[56] Two important characteristics of Perlach's later ephemerides set them apart from earlier annual *practica* and his own early texts. On the one hand, *practica* did not include the tables of astrological data, including daily planetary positions, interplanetary aspects, and eclipses. *Practica* offered the reader general interpretations, weather and political predictions, possible diseases, and advice on crops and livestock. As such, the reader did not need to bring any special skill or knowledge to the text. Perlach's later ephemerides combined aspects from these two types of texts.

In 1528 Perlach produced an ephemerides that resembled his earlier works to the extent that it included primarily, though not solely, tables of astrological data. In the following year members of the court persuaded him to publish a more elaborate astrological pamphlet. In the dedication to Archduke Ferdinand, Perlach explained that Ferdinand's illustrious advisers had urged him to "publish in this year 1529 a more substantial Ephemerides, composed more cleverly and with greater effort, not only for the glory of your Majesty but also for the University of Vienna, which some time ago your Majesty began to restore to the students of the liberal arts and to the most famous lecturers of individual faculties."[57] Like Maximilian, Ferdinand had cultivated a relationship with the university through curricular reform and increased support for the faculty. And like his predecessor, he looked to the masters as a corporate body of experts who could be consulted on in matters of state. Perlach celebrated the close relationship between the archduke and the university.[58] His letter of dedication and his *Ephemerides* itself show that Perlach understood his work to contribute to the Habsburg political rhetoric. In the dedication Perlach implied that he was part of Ferdinand's extended circle of advisers or, at the very least, that they knew him and recognized his ability. Whereas Perlach's early letters of dedication had been aspirational, trying to establish or cement relationships with powerful patrons, his later letters of dedication reflected his enhanced authority and standing. No longer did he need to go through Tannstetter or other intermediaries. Now he could dedicate his works to the archduke.

An important aspect of Perlach's ephemerides were his astrological predictions regarding diseases and health, and war and peace. The broad predictions that Perlach had excluded from his earlier works played a significant role in his 1529 *Ephemerides*, which engaged explicitly in political issues. Perlach was by this time a respected professor at the university and a familiar figure in important political circles in Vienna. He understood the political advantages that he accrued by serving Archduke Ferdinand. Now, he drew attention to this fact in his preface when he indicated that members of Ferdinand's court had urged him to publish his ephemerides. Perlach thus articulated a new political and social function for his ephemerides and located himself and his work within a contemporary political and social crisis. In particular, he sought to refute the dire predictions published by Johannes Carion, astrologer at the court of Brandenburg. Carion had "in a recent publication foretold and prophesied many threats and tears for Austria." By showing that Carion's predictions were not accurate because he was ignorant of astrology, Perlach intended to assuage the fears of the citizens of Vienna.[59] By 1529 Perlach and his works had become part of Habsburg politics. His reputation as an important master at the University of Vienna and as author of both ephemerides and instruction manuals conferred authority on his *Ephemerides* and its political message.

Perlach used the first section of his text "to explain briefly the use of the ephemerides." Although his ephemerides still contained detailed astrological data, Perlach's new text presented the results of Perlach's astrological analysis and was intended for a general reader. If a reader wanted to learn how to carry out his own calculations, Perlach referred him to his *Usus*.[60] In the remaining pages of the introduction Perlach explained how to unravel the symbols and information in this text and then offered general guidelines about how to use these symbols and the information in his *Ephemerides*.

Perlach was particularly interested in weather predictions. He had calculated the positions of the planets relative to various fixed stars so that "any moderately ingenious reader" would be able to foretell the changes in the weather.[61] He explained how to interpret planetary qualities to predict the type of weather—hot, cold, rainy, or dry—and how to use planetary motions and positions to predict the duration and severity of the weather. Conjunctions between the superior planets—Mars, Jupiter, and Saturn—influenced large-scale weather patterns, whereas the motions of the moon and sun caused sudden changes in the weather.[62] Despite giving his reader rules of thumb for predicting the weather, however, Perlach often included his own predictions. Traditionally, prognostications that included weather predictions were the realm of the expert astrologer. Weather predictions could be highly significant, particularly in times of crises—for example, the conjunctions of 1524 that people feared would cause a deluge—and the success of princely control during times of crises depended on establishing and exercising authority over this domain during normal periods. Thus, Perlach's attention to the weather and his reluctance to leave such predictions entirely to his reader retained for himself—and by extension Ferdinand—authority over identifying and interpreting the celestial phenomena that influenced the weather.

Perlach did not confine his astrological predictions to the weather. He distilled from the celestial information various predictions about the coming year. He opened the third section with a discussion of the general and most powerful astrological phenomena for 1529, two eclipses that would occur in that year.[63] Perlach dismissed the first of these, a lunar eclipse, because it would occur below the horizon, and therefore have little or no effect on Vienna. A second lunar eclipse, however, would be just visible on the western horizon and together with subsequent interplanetary aspects, would cause a number of terrestrial events important to his Viennese readers. The effects of the eclipse were influenced by the malefic configuration of the planets and the unfavorable planetary aspects.[64] Perlach's long list of threatening features included many references to Mars and Saturn, traditional harbingers of doom. Perlach's predictions for the coming year were grounded in specific celestial causes. He arranged his interpretation of this eclipse into categories that most concerned his reader, issues such as broad weather predictions, peace and war, impending sicknesses and epidemics, crops, and general fortunes of cities and countries.

In general, he predicted a year marked by looting and killings due to the malefic effects of Mars in the house of Saturn. Although Perlach said little about farming, he offered an extensive list of diseases and health problems threatening the Viennese, including ulcers, hernias, bladder and kidney problems, and the pox. More generally they could expect breathing difficulties, fevers, and decay. Perlach's picture of 1529 was not reassuring. Perlach spent most of his effort trying to interpret the looming Turkish threat. For three years the Viennese has been living in fear of an Ottoman invasion. Perlach could not avoid this most pressing social and political issue. Winter would be a time of armed conflicts followed by a spring and summer marked by peace. Noting the retrograde of Mars, Perlach predicted unfortunate events for the fall, which began on September 13.[65] Despite Perlach's dire predictions, he concluded on a positive note by refuting Johannes Carion's disastrous forecasts for Austria.[66]

Educated in astrology and mathematics, Carion was the court astrologer for the elector Joachim I of Brandenburg. By 1524 he was producing yearly *practica* that included general weather and medical forecasts.[67] Throughout 1520s he also produced prognostications in which he warned of the political effects caused by various celestial events.[68] Although Perlach did not identify which of Carion's work he found problematic, his audience probably recognized Carion's popular *Bedeutnuss und Offenbarung*. In his pamphlet Carion outlined the political fortunes for various countries from 1527 to 1540. A lunar eclipse in Taurus in 1529 threatened suffering and loss throughout Europe and indicated that the German lands would be beset by massive bloodshed and wars. Austria, Carion predicted, would be bogged down by wars in Hungary and risked complete destruction.[69]

Perlach's contest with Carion reveals the importance of astrological pamphlets as political rhetoric. Threatening astrological predictions such as Carion's could incite public unrest and disobedience. One of the most threatening aspects of Carion's predictions was their popularity. Even the simple folk were familiar with his text and its predictions, making it all the more important for Ferdinand that Perlach discredit them.[70] In Vienna, with the Turkish armies advancing through Hungary and threatening to conquer the city, anxieties were high. Citizens were prepared to fear the worst. When Carion grounded his dire predictions in astrology, Perlach had to respond.

Perlach concluded his *Ephemerides* by explaining in detail Carion's errors and where he, Carion, had misunderstood astrology. Perlach would sharpen this charge a few years later when he accused Carion of practicing unnatural astrology.[71] For the moment, Perlach rejected Carion's use of Ptolemy, his interpretation of a recent lunar eclipse, and the ruler of the year for 1529. Perlach attacked Carion's predictions in the last chapter.[72] To correlate celestial phenomena to particular parts of the earth Perlach relied on a common sixteenth-century adaptation of standard Ptolemaic theory. Ptolemy had divided the zodiacal signs into four groups, each of which influenced a portion of the earth.[73] By the sixteenth century, astrologers had modified the technique, preferring to arrange the signs into three quadrangles of four signs that affected certain countries and cities.[74] Using this approach, Perlach claimed that, in general, lands under the influence of Scorpio would escape the worst effect. Perlach also drew on the theory of great conjunctions developed by Albumasar to predict that a conjunction of the superior planets in Leo would adversely affect lands under that sign.[75]

According to Perlach, Carion had based his predictions on two causes: a lunar eclipse that would occur late in the year and the *revolutio anni* for 1529. Perlach did not dispute the importance of these celestial phenomena for the coming year. Instead, he rejected Carion's interpretation of them, the events that they would cause, and the timing of those events. Perlach's goal in this was to allay the fears of Austrians and "especially the more simple folk" by showing that this eclipse portended nothing bad for his Austrian audience. Perlach turned his attention first to the lunar eclipse. Drawing on Ptolemy's *Quadripartitum*, he disputed Carion's dire predictions. The eclipse would cause little harm because, as Ptolemy had explained and recent experience had confirmed, only eclipses ruled by malefic planets caused ruin and destruction. Perlach cited two recent lunar eclipses to demonstrate his point: neither the eclipse in 1519 nor the one in 1526 had produced any harmful effects.[76] Nowhere did Perlach mention which planets would rule this eclipse. The reader had to infer that the rulers of the 1529 eclipse would be beneficent. Perlach offered the further reassurance. Because the eclipse would occur below the horizon, invisible to Austrians and the Viennese, it would not cause any harm to Vienna.[77]

In a similar manner, Perlach dismissed the other characteristics of Carion's prediction—the regions affected by the eclipse and when those effects would occur. According to Perlach, all astrologers agreed that eclipses do not affect all regions of the earth equally, but only those regions under the influence of quadrangle in which the eclipse occurs. The impending eclipse would occur in Taurus. The most learned contemporary astrologers, he told his reader, located Austria under Libra, which was not part of the quartet of signs that included Taurus. Whatever effects the eclipse would cause, they did not threaten Austria. For the same reason the eclipse did not threaten

Hungary. Not only had Carion misunderstood the relationship between the eclipse and the regions affected, he had failed to understand the timing of the effects caused by eclipses. Perlach pointed out that the effects of an eclipse in 1529 would be felt in 1530, not in 1529.[78] Perlach's readers had nothing to fear from this eclipse, at least nothing in 1529.

Perlach next attacked Carion's misunderstanding of the Lord of the Year, which Carion had claimed was the second cause of impending doom for Austria in 1529. Yearly *practica* often included a horoscope for the year calculated for the moment the sun entered Aries, identified the Lord of the Year, and from this information predicted the general outlook for the year. Perlach had included a sample horoscope for the year in his *Usus* and explained how to interpret it. Ten years later and in a different context, he suggested that the Lord of the Year was only one indicator, which was no more important than others. Drawing on another passage in Ptolemy, Perlach claimed that the year had not one but four horoscopes, one for each season. Of these four horoscopes, only one had Libra in a threatening position. He admitted that this happened to be the horoscope for spring, the same one used to determine the Lord of the Year, but pointed out that this configuration was tempered by Venus's favorable influences. Therefore Austria had nothing to fear in the coming year.[79] Perlach tried to convince his readers once again that Carion had failed to understand the finer points of astrological predictions when he based his predictions solely on the Lord of the Year.

This was not the last time Perlach and Carion debated proper astrological technique and interpretation.

The *Ephemerides* of 1531

Two years later Perlach produced another *Ephemerides*. This time he digested the astrological information even further, foregrounding a mix of general predictions and weather forecasts for the coming year. Unlike his earlier ephemerides and previous *Ephemerides*, his latest book was not intended for practicing astrologers. Perlach rearranged his text and altered its content to reflect his new understanding of his *Ephemerides* and his goal of reaching a broad audience who had little interest in practicing astrology. When composing his *Ephemerides* for 1531 Perlach assumed his readers had no astrological expertise or skill. His shift in style reflected a more robust understanding of the political uses of his work and his efforts to influence as broad an audience as possible with his message. Perlach turned his attention once again to the Turkish army in Hungary. His astrological analysis attempted to reassure his

Viennese audience, to bolster Ferdinand's position in Vienna, and to advance the archduke's political agenda at the imperial diet.

Following the siege of Vienna in the fall of 1529, the Turkish army had retreated back through Hungary toward Constantinople. The failure of the siege owed more to the damp Viennese climate and the long supply lines the Sultan Suleiman the Magnificent was trying to maintain than to Ferdinand's military prowess. Nevertheless, Ferdinand pursued the Turkish army through Hungary in the hopes of delivering the decisive defeat that would rid his lands of the Turkish threat once and for all. In November 1529 Ferdinand enjoyed considerable success on the battlefields in Hungary and fought all the way to Buda, but by January bad weather and lack of pay for his troops had brought his campaign to a halt. He was forced to put his plans for complete reconquest of Hungary on hold, though he had not given up this goal. His successes against the Turkish and Hungarian enemies had improved his image among his contemporaries, who began to take more seriously his efforts to drum up support for his campaign against the Turks.[80]

At the same time Ferdinand and his brother, Emperor Charles V, were trying to secure Ferdinand's election as King of the Romans. Ferdinand often stressed to the electors as well as to Charles that his success in repelling the Turkish threat was bound up with a secure position of power in the empire and the support of both the emperor and the German princes. One clear indication of this would be his election as King of the Romans. Charles arrived at the Augsburg Diet in June 1530 and asked that the electors recognize Ferdinand as King of the Romans.[81] Before the electors had decided whether or not they would elect Ferdinand, he had renewed his campaign against the Turks, led his troops back into Hungary, and successfully reached Buda. Although he once again failed to capture the city, his recent successes in Hungary further buoyed his position in the empire and reinforced his claim that with the necessary support he would be able to finally reconquer Hungary and expel the Turkish threat.

Such was the political context in which Perlach wrote and published his *Ephemerides* for 1531, which came off the presses in late December 1530. In his introduction Perlach pointed out his own debt to Ferdinand "who with a certain singular fondness embraces all students of learning, in particular those who apply themselves in astronomical matters, and especially me."[82] Perlach's relationship with the court had apparently grown stronger since he published his previous ephemerides in 1529. When he interpreted the solar eclipse in 1530 he drew numerous reassuring conclusions. Most important for his Viennese readers, Perlach argued that the Ottoman forces would suffer greatly from this eclipse. He grounded his bold claim in a detailed astrological analysis of the many malefic influences that promised to bring ruin to the Turkish army.[83] Perlach assured his readers that although 1531 would witness no eclipses, at least none that would be visible in Vienna, the effects from two eclipses in 1530 would be felt most strongly in the current year. Perlach began with a detailed description of the two eclipses, providing not only the location of the eclipses, their position with respect to the fixed stars, but also the qualities of those stars.[84] Perlach offered a detailed analysis of the harmful effects caused by the malefic influences of the stars and interplanetary aspects. He then launched into a long discussion of the timing of these harmful events, which would extend through the year. As he had argued in 1529 against Carion, here again Perlach admitted that not all eclipses produced the same types of effects, and that the effects themselves depended largely on the ruler of the eclipse. Leaving nothing to the reader's imagination, Perlach divided the eclipses into three sections—beginning, middle, and end—and analyzed the planetary rulers for each section as well as the mediating influences of the planets that participated in ruling the eclipse. Perlach's then listed the specific ills that his reader could expect from these eclipses. Here Perlach treated all the standard topics, including weather, livestock, farming, navigation, wars, and diseases. Finally, Perlach turned his attention to the regions and cities most likely to be affected by these eclipses. Again, he explained in great detail why certain countries could expect to suffer while others would have little to fear.[85] As before the effects of the eclipses would have the strongest influence in their own quadrangles and their most dire effects in the countries situated in those quadrangles.

Perlach made it clear that Vienna would soon be free of the looming Turkish threat. After living for years in the shadows of the Ottoman army, his readers must have found his prediction comforting. Perlach's predictions also aligned with Ferdinand's local political goals. Fear and the unrest it caused were real threats to public order and political authority. Managing those fears was an exercise in political management and power.[86]

In addition to assuaging the fears of local readers, Perlach's text attempted to persuade the emperor and the German princes that this series of celestial events presented an opportunity to vanquish the Turkish threat once and for all. Perlach drove this point home in his analysis of these eclipses, which functioned as a timely piece of propaganda in the contests between Ferdinand and Charles, on the one hand, and Ferdinand and the imperial electors and princes, on the other. Alluding to Ferdinand's recent successes against the Turks, Perlach claimed:

A certain person from the family of Emperor Maximilian, who is worthy of perpetual memory, came forward a few years ago. This person has the aforementioned planet Mars in his horoscope, and in the very place of the star, under which the Mahumet sect invaded, a fact that indeed I would consider, like a miracle, another misfortune for the Turks. He, if God should prolong his life, will use his own resources and those of his allies to root out, overthrow, and drive out this sect from those places, which previously through force and violence it had taken away from his predecessors. And he will be defender and avenger of all the evils that were brought about in former times.[87]

Here Perlach associated Ferdinand with Maximilian I, who was still remembered fondly by his subjects, even if the electors were more hesitant. Maximilian enjoyed the reputation of having ousted the Hungarian army that had defeated his father, Frederick III. Although Maximilian had relied largely on his own resources in this conflict, he had continually sought military and monetary support from the German princes for his campaign against the Turks. Like Maximilian, Ferdinand too depended on the financial and military support of the princes in the empire. Perlach reminded the princes of their role in Maximilian's military successes and pointed to their responsibility in vanquishing the Turks. Without their support at this opportune moment, when the stars were aligned for the defeat of the Turkish threat, the imperial forces would be unable to reconquer Hungary, and Europe would continue to live in fear of the next Turkish invasion. Whether it was successful in swaying any of the princes to vote for Ferdinand remains unknown; no doubt the approximately 350,000 guldens that Ferdinand paid to the electors were also influential in his election to King of the Romans in January 1531.[88]

* * *

Perlach began like so many other academic astrologers who hoped to secure a position at the local court, by publishing ephemerides. This literature allowed him to establish his reputation in the Viennese intellectual community and to begin climbing the social ladder. As Perlach became more secure in his position at the University of Vienna and in the community more broadly, he increasingly engaged in the day's pressing political issues. In his later ephemerides Perlach adopted aspects of both the yearly wall calendars and the *practica* literature. These later texts were accessible to a wider audience, one that was literate but not necessarily skilled in astrology. Their accessibility made Perlach's *Ephemerides* particularly useful in spreading a pro-Habsburg message that supported Archduke Ferdinand's interests. For Ferdinand, Perlach's astrological pamphlets lent intellectual and social authority to his political program. Astrology offered a persuasive and pervasive systematic body of knowledge for understanding political and social events. The popularity of Carion's *Bedeutnuss und Offenbarung*, which made threatening predictions for Austria, illustrates the power of astrology as a political instrument, making a response that refuted his predictions not only urgent but necessary as a political move for maintaining a citizenry who remained calm and confident in their ruler. In Perlach's expert hands, astrology offered a compelling set of mechanisms for combating and easing public anxieties. Moreover, the ubiquity and authority of astrology throughout all levels of society made it an especially effective tool for communicating political programs to the broadest possible audiences. In many ways Perlach's efforts to assuage the fears of the Viennese reflected Maximilian's goal to enlist all levels of society in his political program, a goal that both allowed for and obligated the Habsburgs to consider seriously their various publics.

Chapter Seven

Prognostications

Martin Luther had little patience for astrologers and their prognostications. Not only were astrologers continually changing his birthday, often to coincide with the Saturn-Jupiter conjunction of 1484, but for decades they had also predicted that the conjunctions of nearly all the planets in February 1524, in the sign of Pisces, would produce floods of biblical proportions.[1] Writing to Georg Spalatin in late March 1524—after the predicted flood had not materialized—Luther attacked the astrologers on both issues: "I previously saw that horoscope of mine that was sent here from Italy. But since the astrologers were so erroneous this year, it is not surprising if someone would even lie about my geniture."[2] Luther was on safe ground here. March had witnessed some heavy rains, but nothing too remarkable. His disapproval was, however, a lonely voice drowned out by the chorus of astrologers producing prognostications in the early sixteenth century. Both long-foreseen events, such as planetary conjunctions, as well as unpredictable phenomena, such as comets and meteorites, were prodigious events that provided astrologers and their patrons a useful resource for exercising political and social authority. Astrologers claimed to have a unique expertise that enabled them to analyze and interpret such events. They alone possessed the academic credentials, the theoretical knowledge, and the experience to understand celestial phenomena and to reveal their effects on the political and social world. While astrologers concentrated on urgent political and social issues, usually of most concern to their patrons, their printed prognostications were incredibly popular, finding readers across all levels of society. When a meteorite fell to earth or a comet streaked across the sky, audiences as well as patrons looked to astrologers for expert advice and, often, reassurances.

Prodigious celestial phenomena were not merely opportunities for interpretation; they required it. The rarity, in the case of portentous planetary conjunctions, or the unpredictability, in the case of comets or meteorites, made such events preternatural and pregnant with meaning and significance. However, interpretations of comets and planetary conjunctions were not universal or neutral. Instead they were hotly contested, and astrologers struggled to offer the most persuasive explanation. Through their astrologers, princes sought to shape and control the public understanding of and reaction to such events. Not only did such control reflect the prince's authority, calming public anxiety and fear was itself an exercise in political management. Consequently, when astrologers at rival courts published competing interpretations of prodigious celestial phenomena they never lost sight of their patrons' political or social agenda. Habsburg princes and the astrologers who gathered around them were no exception this rule. Their prognostications represented yet another facet of Emperor Maximilian I's and later Archduke Ferdinand's efforts to use astrology, complementing the other forms of popular astrological literature that the Habsburgs deployed.

The three of these Habsburg court astrologers discussed here—Johannes Stabius, Georg Tannstetter, and Andreas Perlach—composed prognostications at different times and in different social and intellectual contexts, but all did so with a keen eye trained on the interests of the ruling Habsburgs, at first Maximilian I and later Archduke Ferdinand. All three drew on astrology to interpret their specific social, intellectual, and political circumstances. At the turn of the sixteenth century, Stabius was a young academic hoping to secure a position at the imperial court. He drew on astrology to make sense of the rising Turkish threat, to justify his prophetic interpretation of history, and to advance his own status at the Holy Roman Court. By contrast, Tannstetter was well established when he offered his interpretation of the 1524 conjunctions. Responding to Archduke Ferdinand's expectations, Tannstetter leveled a scathing critique of those astrologers who had misunderstood the significance of the impending conjunctions. He then offered his own predictions. Finally, Perlach used the appearance of a comet in 1531 as an opportunity to continue his contest with Johannes Carion, astrologer to the Elector of Brandenburg and to urge German princes to support Archduke Ferdinand's efforts to rid Europe of the Turkish threat. In each case astrology provided insight into a pressing contemporary situation and a means of exercising some control over it.

Johannes Stabius, Great Conjunctions, and Prophecy

In 1502 Johannes Stabius accepted one of the chairs in mathematics in Conrad Celtis's newly established *Collegium poetarum et mathematicorum*. Later that same year Stabius was crowned poet laureate. Although Maximilian had given Celtis and his *Collegium* the right to recognize certain members of the college poet laureate and to bestow on them all the rights and privileges associated with that title, the emperor himself crowned the poet laureate. The position was an appointment that played an important role in Maximilian's larger political efforts, both within the Holy Roman Empire and in Europe more broadly.[3]

By the time Stabius assumed a chair at the *Collegium*, he had developed a reputation as both a skilled astrologer and a poet. While at the University of Ingolstadt he had produced *practica* that enlisted the planets and stars in support of Maximilian's political goals. Stabius had also composed poems that aligned with Maximilian's broader project. Although his celebratory poem of St. Koloman was not published for another decade, Stabius referred to it in his crowning ceremony.[4] Koloman was one of the Habsburg family saints that Maximilian enlisted to demonstrate his own sanctity. Stabius's poem and its accompanying woodcut by Hans Springinklee was one of Maximilian's many efforts to portray his saintly ancestors.[5]

Stabius had combined his astrological and prophetic skills with poetry in "Carmen Saphicum: ad Max. Ro. re. Trenice," which Celtis had published in 1501 along with his own *Ludus Diane*.[6] In the poem Stabius lamented the degeneration of society and urged Maximilian to remedy the situation. Fitting the celebratory nature of both Celtis's performance of the *Ludus Diane* and the publication of that work, Stabius placed his poem squarely in support of imperial propaganda. He saw Maximilian as intimately connected to the restoration of health and prosperity, which would return when the planets and constellations were once again favorably aligned. In "Carmen Saphicum" Stabius elevated Maximilian to a participatory role in realizing celestial influences. If the emperor acted at the appropriate time he could take advantage of the favorable planetary positions to reduce the conflicts between himself and the German princes and to generate support among the German princes and citizens for his imperial projects.[7] By the time Stabius moved to Vienna, he had already demonstrated how in his hands astrology could be used to support Maximilian's political goals.

Stabius used his chair at the *Collegium* as a transitional position before moving to Maximilian's court. To celebrate being crowned poet laureate, Stabius composed the *Pronosticon Johannes Stabii ad annos domini:M.D.III. & IIII*, dedicated to Maximilian. Stabius warned that a great conjunction between Saturn and Jupiter and later Mars would cause widespread unrest, destruction, and conflict in 1503 and 1504. The upheaval could be prevented if the princes stopped opposing Maximilian's reforms and instead supported his military and political goals. Once again, but this time on a larger stage, Stabius placed his astrological expertise in the service of the Habsburg political program. The poem was printed as both a broadsheet and a small pamphlet. The broadsheet could have served as decorations, adorning the walls of the university, while the pamphlets might have been sold as souvenirs or to a broader audience beyond Vienna.[8]

For the *Pronosticon* Stabius adapted the style and content of popular prognostication literature to fit the celebratory and political nature of his crowning as poet laureate. Stabius intended to impress his elite audience by displaying his erudition and eloquence. He wrote in hexameter verse that required substantial effort to untangle the syntax and meaning. To fit the meter Stabius employed a diverse and sometimes obscure vocabulary. At the same time, he made frequent allusions to the classical poets, especially Vergil and Ovid, in an effort to appeal to his learned and powerful audience.[9]

Stabius's *Pronosticon* combined astrological and prophetic approaches to understanding future social and political events.[10] He opened by recounting the important celestial causes, the conjunction of the three superior planets: "Icy Saturn is slowly moving out of Cancer, while violent Mars begins to approach it, mixing incongruous influences with the old man [Saturn]."[11] Stabius's contemporaries understood that great conjunctions between Saturn and Jupiter were portentous events that signaled significant changes on earth.[12] The conjunctions that would begin in 1503 and continue into 1504, however, promised to be worse than normal in large part because the two planets would conjoin three times as they progressed, regressed, and progressed once again through their orbits. In addition to the repeated and malefic conjunctions of Saturn, Jupiter, and Mars, the unfavorable influences of the sun and Mercury would worsen the effects.[13] For Stabius, these astrological causes reinforced a prophetic interpretation of history.

In the mid-thirteenth century, millennial speculations had revolved around Emperor Frederick II, casting him as the prophetic Last World

Emperor. He died before he could lead an army against the Antichrist and inaugurate the millennium. His premature death did little to quell speculations, which seemed to increase over the next two centuries in anticipation of the next Frederick. In 1452, Frederick of Habsburg became Emperor Frederick III, immediately prompting speculation that he was the long-expected Last World Emperor. Two years later the Ottomans conquered Constantinople and seemed poised to invade Europe. An explosion of prophetic literature in the latter half of the fifteenth century tried to make sense of the Turkish threat.[14] However, few people were less suited to be the prophetic emperor than Frederick, and toward the end of his reign authors began looking to other princes as the Last World Emperor.[15]

Johannes Lichtenberger was one of the more important of these fifteenth-century authors. In 1488 he published his *Prognosticatio*, an eclectic mix of prophetic traditions and rather simple astrological analysis.[16] Although he claimed to base his text on the Saturn-Jupiter conjunction in 1484, these astrological events played only a supporting role in his predictions.[17] Instead he grounded his interpretation of current and future events in a combination of the Methodian and Joachim prophetic traditions, both of which foretold of a Last World Emperor who would vanquish the great evil.[18] Combining the two, Lichtenberger prophesied that Maximilian would lead a unified Christian army against the infidels, vanquish the Ottoman threat, and usher in the Golden Age before handing the imperial insignia back to God in Jerusalem.[19]

In his *Pronosticon* Stabius seamlessly wove astrology and prophecy together to support his specific political commitments.[20] Like Lichtenberger, Stabius used a great conjunction as the starting point for his analysis and wedded his astrology to a prophetic interpretation of history and prediction for future. Stabius, however, drew on his more sophisticated understanding of astrology and gave it a more important role in his poem. That the planets and stars were responsible for terrestrial events is made clear in the woodcut that illustrated his poem. Influential rays stream from the planets to earth (see figure 7.1). Stabius drew on the qualities and characteristics of each planet to explain the rise of the Ottoman forces currently threatening the empire and all of Christendom. Stabius was at his best when he described Mars's role in these events: savage Mars would call forth from the deepest Hell the Furies, vengeance, hatred, and wrath. Mars would assail mankind and destroy kings, popes, and paupers alike. Most important, Mars would drive a fierce army down from the north to attack the borders of Christendom.[21] In a similar fashion Stabius rooted his ominous predictions in the planets.

FIGURE 7.1. A detail of the woodcut from Stabius's *Pronosticon* showing the planets in the zodiac and their rays streaming toward earth. © Bayerische Staatsbibliothek, Munich, 2 Einblat. IV, 7.

Stabius knew the prophetic traditions well and expected his audience to recognize his allusions to them. He borrowed repeatedly from the prophetic literature but always understood it within his astrological framework. He claimed that the prophesied judge of the Christian people had arrived, ushered in by deadly Saturn.[22] Invasion by the forces of evil as a prelude to the advent of the Antichrist was a standard theme in the prophetic traditions. The image of the Antichrist as coming from a "people engorged on human flesh," as Stabius colorfully put it, came directly from the Methodian proph-

fe qð vocaf regio folis.vbi cõfpexit gētes inmũdas et afpe-
ctu borribiles.fūt autē ex filijs Japhet nepotes.quoʒ imun
dicie videns exborruit.Comedebāt enim hij omnē cãticoruʒ
fpeciē oĩe coinqͥnabile.i.canes:mures:ferpētes:morticinoʒ
carnes:abortiua infirmabilia corpa.et ea que in aluo nõdum
p lineamēta coagulata funt.vel exaliqͥ parte mēbrorum pro-
ducta.compago formã figmenti poffit perficere formã vel fi-
gurã expͥmere iumētoʒ:necnõ etiã ʒ omnē fpēm feraʒ imun
darũ.mortuos aũt neqᷓꝗᷓ fepeleũt ß fepe comedũt illos.

¶ Quõ alexãder magnͧ. Gog ʒ Magog ppter
eoʒ turpitudinem in cafpys mõtibus incluferit.

Pbool er ꝗ
natͧ eft Alex
ander. Ifte te
xt'ipuꝗt vcm
Orofij ꝗ dicit
philippũ regē
macedonũ fu-
iffe przes puta
tiuũ Alexãdri
magni Alex-
ander magnͧ
occidit Dariũ
regē medoruʒ
ꝗ et pfaʒ vlti
mͧfuit qm in
ipo ceffauit re
gnum Perfa
rum et fuit fi-
lius Arfamis
ifto Dario br
Dañ.xj.ca.

FIGURE 7.2. Alexander's Gates holding back the forces of Gog and Magog. This woodcut comes from the 1498 edition of the Methodian prophecies. Herzog August Bibliothek, Wolfenbüttel, Inkunabeln/30-4-poet-1, b4r.

ecies.[23] In another set piece from prophetic literature, Stabius referred to the Caspian Gates that had until recently imprisoned the armies of the Antichrist. According to the prophecies, Alexander had waged war on Gog and Magog and their armies. In the end, he had confined them to the northern lands, trapped behind Alexander's Gates in the Caspian mountains.[24] Gog and Magog and their minions represented the Antichrist and his forces of evil. The image of the pious emperor before Alexander's Gates was common. Sebastian Brant's edition of the Methodian prophecies, his *De revelatione facta ab anelo beato methodio in carcere detento*, included a woodcut of a pious knight, perhaps meant to represent Maximilian, kneeling before the Caspian Gates while Gog and Magog along with their armies struggled to break through (see figure 7.2).[25] By the late fifteenth century, as the Ottoman forces marched ever closer to Europe, Europeans were increasingly convinced that the Antichrist had returned and had breached Alexander's Gates.

Despite these ominous images, Stabius concluded on a hopeful note. Like Brant before him, Stabius intended his dire interpretations of the planetary conjunction to incite the Germans to unite behind Maximilian to defeat the forces of evil. As heir to the Roman Imperium, Maximilian deserved the honor and allegiance of the empire's citizens, who must recognize their obligation to the emperor if they hoped to defeat the forces of evil.[26] Although Stabius foresaw a difficult and costly struggle—recent comets and the malefic influence of Mars threatened to cause wars, death, pestilences, and famine across the empire—if the princes supported the emperor they would in the end be victorious.[27]

When Stabius addressed Emperor Maximilian directly, he showered him with praise, holding him up as the Last World Emperor. Out of this period of chaos and destruction, Maximilian alone was able to usher in a period of peace and prosperity.[28] Stabius's confidence aligned with Maximilian's other efforts to gain support for his military and dynastic programs. During much of 1502, Maximilian had sought financial support from the German princes to travel to Rome to receive the imperial crown and then lead an army against the Turks.[29] The German princes had all but thwarted his efforts. To generate support for his cause, he printed numerous mandates and pamphlets to sway public opinion.[30] Stabius's poem conveyed a pro-Habsburg message to an elite audience of university officials, court secretaries, regional princes, and the emperor himself, who was in Vienna at the time, and thus complemented the emperor's other propaganda. Stabius warned the assembled dignitaries of the impending destruction that would follow from the conjunctions in 1503 and 1504 if they did not act to avert it. Stabius's prophetic interpretation of history, reading imperial history through the lens of biblical prophecy, complemented his astrological understanding of the present and his predictions for the future. If the princes offered the emperor their allegiance and military and financial support, Germany could vanquish the Turkish forces from the continent. Whether or not the emperor or his advisers were persuaded by the *Pronosticon*, Stabius anticipated that the emperor at least would appreciate his efforts and reward him accordingly.

With the *Pronosticon* Stabius reached an audience well beyond Vienna and the university. Both the pamphlet and the broadsheet editions were taken away as souvenirs and read and annotated after the celebration of Stabius's crowning as poet laureate. One reader underlined and annotated his copy of the pamphlet, drawing attention to the astrological causes as well as the results.[31] The broadsheet also attracted careful readers. One reader closely trimmed his copy of the broadsheet so that he could paste the text into the front of one of his books.[32] Another reader worked assiduously through his copy of the broadsheet, annotating nearly every line of the poem and adding marginal notes and diagrams. He found the *Pronosticon* a difficult text but clearly thought it was worth the effort to understand. He concentrated on clarifying the astrological terms and doctrines, glossed unfamiliar terms with simple explanations, and added marginal diagrams to illustrate the text. He paid equal attention to the prophetic sections of Stabius's poem, struggling to unpack the erudite work. These readers took Stabius's text, planetary conjunctions, and prophetic interpretations of those conjunctions seriously. Working carefully through the poem, they could not have missed Stabius's use of astrology and prophecy to analyze the present and the past, and to draw from that analysis a political conclusion, one that happened to support Maximilian (see figure 7.3).

Great conjunctions precipitated numerous prognostications that linked the ambitions of their authors as well as the political agendas of local princes. Stabius's *Pronosticon* was not unusual in its combination of politics and astrology. It demonstrated his commitment to the emperor and facilitated his move from the university to the court. Maximilian in turn validated Stabius's astrological expertise and recognized his political value. Shortly after crowning him poet laureate, Maximilian appointed Stabius to the position of court historian, where he was responsible for various imperial projects, including Maximilian's most famous pieces of imperial propaganda, his *Ehrenpforte* and his *Triumphzug*.[33] Stabius also produced complex, printed astrological instruments and celestial maps for the emperor.[34] Maximilian appreciated Stabius's efforts, heaping upon him the rewards that came with imperial patronage. Stabius received a noble title and a substantial sum of money from the imperial coffers.[35]

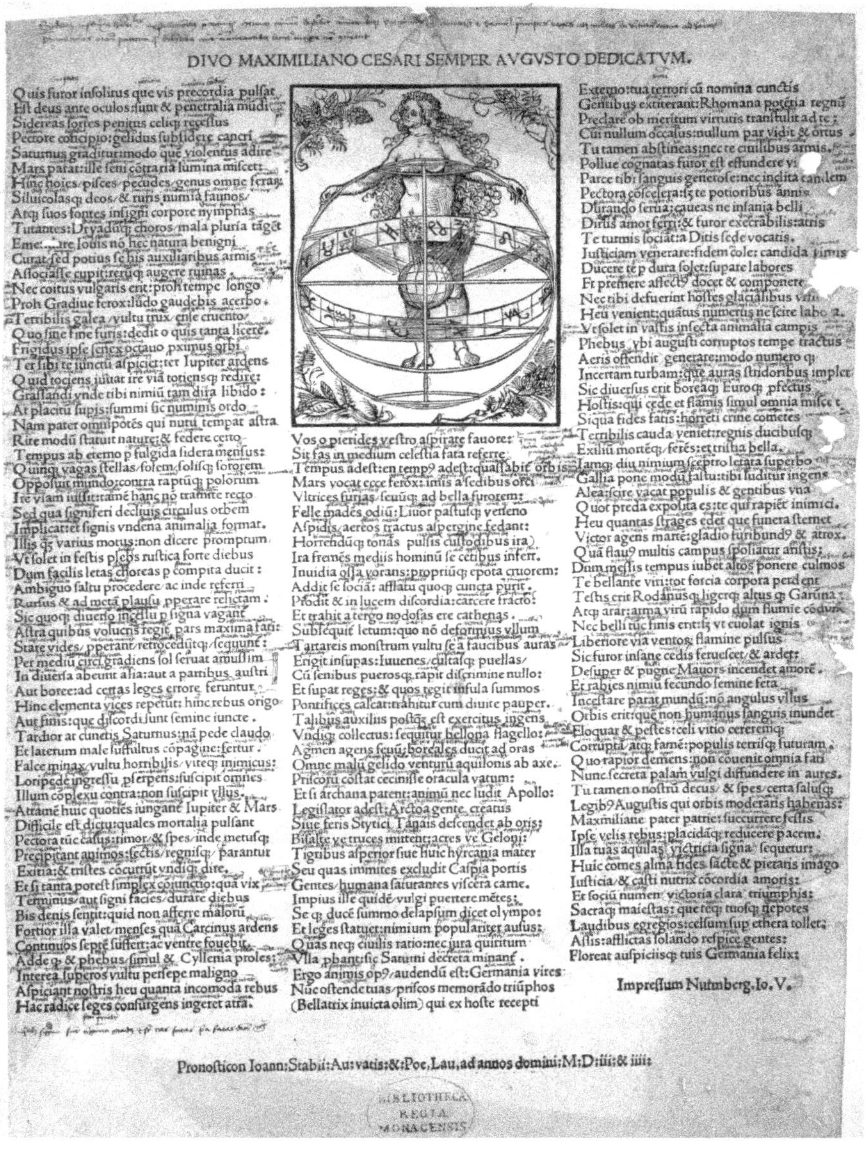

FIGURE 7.3. An annotated copy of Johannes Stabius's *Pronosticon* (1503). Stabius's text and the accompanying woodcut were printed simultaneously in a broadsheet format (shown here) and a small quarto pamphlet. The woodcut illustrates the position of the planets when Saturn and Jupiter were closest, in late May 1504. Bayerische Staatsbibliothek 2 Einblat. IV, 7.

Georg Tannstetter's Conciliatory Astrology

When Stabius composed his *Pronosticon* he was ascending the social and political ladder. Although he had influential friends and was aiming for a position at the court, he was still a relative newcomer in Vienna. Two decades later, Georg Tannstetter composed his prognostication when he was already an established astrologer in Vienna and familiar to the Habsburg court. Like Stabius, Tannstetter used a series of threatening planetary conjunctions as an opportunity to publish a pamphlet interpreting their effects. And like Stabius, Tannstetter was aware of the needs of his patron, Archduke Ferdinand. His prognostication reflected yet another example of a pro-Habsburg astrologer deploying astrology in support of the local Habsburg court.

Great conjunctions often sparked interest among astrologers and the public in general. The great conjunction that occurred in 1524 was, however, more significant than most. This time, Saturn and Jupiter would conjoin in a new triplicity, shifting from an air sign into a water sign. Moreover, this was only one of a series of conjunctions to occur that year. Along with Saturn and Jupiter, every other planet and the two luminaries would conjoin with at least three other planets in Pisces, a total of sixteen conjunctions in all. This series of conjunctions had long attracted attention of astrologers throughout Europe. As early as 1499 Johannes Stöffler and Jakob Pflaum had first predicted these conjunctions in their *Almanach nova*. In the 1510s the conjunctions began to attract considerable attention from astrologers and, consequently, to generate widespread fear.[36] The many works produced by astrologers between 1512 and 1524 adopted either an alarmist or consolatory tone, depending on the astrologer's religious and political situation.[37] In 1523, Georg Tannstetter used his astrological expertise and authority to combat the sensationalist pamphlets and to offer a reassuring interpretation of these conjunctions in his *Libellus consolatorius*.[38]

When Tannstetter wrote his *Libellus consolatorius*, he was an important figure at the University of Vienna and *Leibartz* to Archduke Ferdinand. He was also widely recognized for his wall calendars and yearly *practica*.[39] In the preface to his *Libellus consolatorius* Tannstetter suggested that Ferdinand's support for the university in general and astronomy in particular had compelled him to express his learned opinion about the impending planetary conjunctions.[40] Tannstetter implied that Ferdinand recognized his role as a master at the university to include serving the Viennese population, especially in times of crisis. As a member of the medical faculty, Tannstetter assisted during the many outbreaks of plague; just a year earlier, in 1521, Ferdinand had relieved him of his teaching responsibilities so that he could commit more time to serving as a physician. Tannstetter was later commended for his efforts.[41] Beyond providing medical assistance, he also published texts that were useful to the community. In 1521 Tannstetter had published his *Regiment für den Lauff der Pestilenz*, a short plague tract that offered readers advice on avoiding and treating the plague.[42] In 1523 the most pressing issue was the series of conjunctions that would occur the following February. Hoping to calm the fears of the Viennese, Tannstetter attempted to refute the common predictions of apocalyptic floods. He signaled his reassuring and hopeful predictions in the title of his work, *Libellus consolatorius*, and the woodcut that adorned the title page (see figure 7.4). Comforting verses from the Old and New Testaments frame the woodcut, reminding readers that God commanded the stars and protected the just. God's hand reaches down from the clouds as if guiding the personifications of the seven planets mingling together in the sign of Pisces. Farmers labor in the lower half of the image. An encouraging quotation from the Prophecy of Jeremiah divides the image: "Be not afraid of the signs of heaven." In the text itself, Tannstetter relied on three arguments to refute the dire predictions surrounding the 1524 conjunctions: authority from biblical and classical sources, astrological causation, and historical evidence.

In the opening chapter, Tannstetter mined the Bible for evidence that the world would not experience a second deluge in 1524. According to Genesis, God had promised Noah that he would not again use a flood to punish mankind. There is no cause, Tannstetter urged, to doubt that God would keep his word. Moreover, he argued that the astrologers predicting a deluge failed to realize that they could not predict when the Day of Judgment would arrive. Only God had foreknowledge of this event. Furthermore, those same astrologers misunderstood what sort of calamity would accompany the Day of Judgment. Citing Saint Peter, Tannstetter argued that the end of the world would come about through a conflagration and not through a flood. He found support for his conclusion in Seneca. There was no reason, he assured his reader, to fear a universal flood or to think these conjunctions portended the Day of Judgment.[43]

Tannstetter grounded his second argument in his understanding of astrological causation. He concentrated on the different effects caused by eclipses and conjunctions. Borrowing from Ptolemy's *Quadripartitum*, Tannstetter claimed that for every eclipse or conjunction there was a ruling planet, but that no single planet influenced the entire earth equally. Instead, the different planets hold sway over different regions. The ruler of any conjunction affects particular lands, those with which it shares a certain sympathy. Given that

FIGURE 7.4. The title page from Georg Tannstetter's *Libellus consolatorius* (1524). His title signaled his favorable interpretation. The woodcut reinforced his optimism. Encouraging quotations from the Old and New Testament as well as the guiding hand of God stay the harmful effects of the planets gathered in Pisces. Prominently displayed is a quotation from the Prophecy of Jeremiah, telling the reader "Be not afraid of the signs of heaven." © Österreichische Nationalbibliothek, Vienna, sig. *69.O.332.

different cities and countries have different characteristics, no single ruling planet could bring about the same effect in all kingdoms and countries. While some countries were damp and rainy, others were naturally arid and only rarely experienced rain or even clouds. Therefore, not every place would experience flooding or even increased rainfall as a result of this conjunction. The association between Pisces and rain was a local, northern association. North of the equator, the conjunctions in Pisces occurred in winter, while south of the equator they occurred in summer. While the Viennese associated Pisces with rain, snow, and cold weather, people south of the equator associated Pisces with hot and dry weather. Any rains in northern lands that might accompany the impending conjunctions owed more to the season and the character of Pisces and its controlling planets—Jupiter and Venus in the north—than to the effects of the conjunctions themselves. People south of the equator would expect warm, dry weather, as was typical for the sign of Pisces. According to Tannstetter's astrological analysis, the series of conjunctions not only would not but could not cause a universal deluge.[44]

In arguing against any universal deluge resulting from the conjunctions in Pisces, Tannstetter had implicitly confirmed the local nature of the effects. Each region could expect particular effects depending on the astrological relationship that existed between that locale and the zodiacal sign of Pisces and the planets that ruled the conjunctions. Tannstetter did not ignore these local effects but claimed that eclipses, not interplanetary conjunctions, were the main causes of local weather. He rejected an exclusive reliance on planetary conjunctions, at least when it came to making weather predictions.[45]

In rejecting the theory of conjunctions developed by Albumasar and his followers, Tannstetter found strange company in Pico della Mirandola. Pico had argued that Albumasar and his followers had wrongly used the mean conjunctions as if they were the true conjunctions when they tried to find causes for terrestrial events.[46] The fact that true and mean conjunctions often occurred at very different times and locations called the entire theory into question. Although Tannstetter generally dismissed Pico's criticism of astrology, he agreed with him on this point.[47] Tannstetter's argument here was the same one that he and Stiborius had put forward a decade earlier in their work on calendar reform. At that time they had chastised the Church for relying on mean motions of the sun and the moon in calculating the date of Easter.[48] In 1524 Tannstetter remained hostile to any theory based on mean motions of the planets. The difference between mean and true conjunctions was particularly significant when the conjunction occurred near the cusp of a zodiacal house. When astrologers commonly used mean motions of the planets to determine the location of a conjunction and to identify the causes of changes on earth, they risked incorrectly identifying both the sign in which the conjunction occurred and the ruling planet for that conjunction. Consequently, the terrestrial effects attributed to a conjunction or predicted to result from a conjunction were often inaccurate. Those effects could not be traced to the conjunctions themselves.[49] The French cardinal Pierre d'Ailly had employed Albumasar's theory of great conjunctions when in 1414 he composed his widely read and influential *Concordantia astronomie cum hystorica narratione*. This text was important for its attempt to account for all the significant changes on earth by referring to the great conjunctions and the ten revolutions of Saturn.[50] D'Ailly had argued that this theory provided the framework for understanding and healing the Great Schism. Tannstetter could not let such a work escape criticism.[51] In addition to his rejection of the general thrust of d'Ailly's work, two specific aspects attracted Tannstetter's attention. First, d'Ailly, like many astrologers, had explained the initial deluge by the third greatest conjunction. Tannstetter argued that such a universal effect from a single conjunction was astrologically impossible. Second, d'Ailly had relied on the ten revolutions of Saturn in his Christian-era chronology.[52] As he had done seven years earlier in his *practica* for 1517, and now once again with the support of Pico, Tannstetter rejected this basis.[53]

In his third argument, Tannstetter invoked the historical record.[54] He analyzed previous conjunctions and found no reason to fear that any conjunction, let alone those that were impending, could cause a deluge. In 1503 and 1504 Mars, Jupiter, and Saturn had repeatedly conjoined in the water sign of Cancer. It was thought that a similar set of conjunctions during Noah's lifetime had caused the first deluge. But, Tannstetter pointed out, the conjunctions in 1503–4 had produced no flooding. February 1465 and February 789 had witnessed great conjunctions between Saturn and Jupiter in Pisces, but again no flooding.[55] Tannstetter's most telling example was the series of conjunctions that had occurred in 670. In that year the same number of conjunctions had occurred in Pisces as would occur in 1524. He listed them for easy comparison. The large number of conjunctions in the watery sign of Pisces had caused no flooding.[56]

Only after Tannstetter had discredited the apocalyptic claims of other astrologers who predicted that the conjunctions would cause a second flood did he offer predictions of what might result from those conjunctions. In the second half of his book, Tannstetter laid out a series of predictions for 1524 based on the effects of a lunar eclipse and its ruling planets, and to a lesser degree the conjunctions in 1524. His predictions in this section of the book closely resembled those in his many *practica*: fortunes of peoples, likely diseases, the fate of the harvest, and which countries had the most to fear. Tannstetter took seriously his charge to calm the anxieties of the populace because fear constituted a real threat to public order. In order to reach a broad audience, he produced his *Libellus consolatorius* both in German and Latin.[57] Many other authors who composed both Latin and vernacular treatises on the 1524 conjunctions conveyed different messages in each type of work, adopting a reassuring stance in their vernacular works and a more alarmist position in their Latin tracts. Those authors hoped to advise the learned of the impending danger while calming the fears of the broader public that purchased the vernacular prognostications.[58] By contrast, Tannstetter used both his works to communicate the same reassuring message: there would be no flood in 1524.[59] He argued the same point in his *practica* for 1524. Throughout that work he referred to his larger *Libellus consolatorius*, encouraging the reader to consult the longer text.[60] With these three works, Tannstetter disseminated his reassuring political message across a broad and diverse audience.

Tannstetter's efforts to assuage the fears of the Viennese and the Austrians could only help Ferdinand's position in the city. Ferdinand had taken control of the five Austrian duchies in 1521, when he moved to Austria and took up residence in Vienna.[61] Since then he had struggled to assert his authority over his Austrian subjects. In 1522 he had quelled a rebellion of nobility from Lower Austria along with the Vienna city council. As a result the city lost many of its privileges, while Ferdinand failed to generate much support for his rule. By 1523 Ferdinand's domestic situation had not improved, and growing fears of a second deluge threatened to destabilize a populace that was already hostile to his rule. In this context, Tannstetter's *Libellus consolatorius* took on additional significance. His efforts to manage the fears of the Viennese had political implications. Moreover, Tannstetter's coherent message across his various texts, both the Latin and the vernacular texts as well as his *practica*, was calculated to calm the nobles—the audience for his Latin texts—and the broader public who would have typically purchased and read vernacular *practica*.[62] Finally, Tannstetter's support would have sent a useful political message. He was a respected master at the University of Vienna and associated with Archduke Ferdinand's court. By linking his texts to Ferdinand and Ferdinand to the university, Tannstetter implied that his texts were official predictions for the local Habsburg court. His support complemented other efforts to bolster Ferdinand's local reputation at time when the archduke was still struggling to assert his authority over his new subjects.

Andreas Perlach: Astrology, Prodigious Phenomena, and Academic Disputes

By the end of the decade, Ferdinand's local situation had improved while foreign threats to his reign had increased dramatically. Within the empire, the electors had yet to recognize his authority and elect him King of the Romans. At the same time, Turkish armies had defeated the Hungarian forces and were marching toward Vienna. Ferdinand was trapped between hostile German princes and an invading Turkish army. Not even his brother, Emperor Charles V, was particularly supportive of Ferdinand's efforts to establish his authority in the Austrian and German lands. Once again, Ferdinand found support in the local masters at the University of Vienna. In the late 1520s, Tannstetter's former student Andreas Perlach threw the authority of astrology behind the archduke's cause, applying his astrological expertise to interpreting a comet in such a way as to reassure the Viennese and to refute the dire predictions of a rival astrologer.

Along with planetary conjunctions, comets also offered astrologers opportunities to publish tracts that interpreted these phenomena with an eye on the immediate political or religious context. In the early sixteenth century comets were unpredictable, prodigious events. Astrologers often viewed them as standing outside of nature and as direct signs from God, an understanding that was reinforced and fueled by the religious tensions and conflicts of the Reformation.[63] Popular, vernacular literature treated comets either as signs of divine displeasure and future misfortune or, more radically, as indications of the Second Coming and Judgment Day. But astrologers also used comets for more worldly ends, drawing on these occasions to support political agendas.[64] The appearance of several comets between 1527 and 1539 heightened awareness of and raised concerns about the meaning of comets and their effects on earth.[65] Very bright comets in 1531, 1532, and 1533, and again at the end of the decade prompted a number of explanatory works.[66] Many astrologers composed vernacular pamphlets that combined political predictions with eschatological interpretations. In 1531 Andreas Perlach joined this chorus of voices with his vernacular *Des Cometen und ander erscheinung in den lüfften / Im XXXI. Jar gesehenn*, which supported Archduke Ferdinand's concrete political agenda and military goals. Perlach tried to reassure the Viennese, who were still living in fear of a renewed Turkish invasion, that the comet did not presage dire events. Finally, Perlach used his text to reassert his own astrological expertise and defend his academic reputation in his contest with Carion, the rival astrologer from Brandenburg.

The flurry of publishing activity around comets in the sixteenth century began in 1527 when Peter Creutzer published the first work on comets in two decades. Although a comet had been observed in 1506 throughout Europe and was recorded in numerous chronicles at the time, it had sparked little interest among astrologers.[67] This comet did, however, reappear in a supporting role in many of the cometary tracts produced in the 1530s and 1540s. The political, religious, and social atmosphere in the late 1520s and 1530s in Western Europe differed considerably from what it had been two decades earlier. During the same period, the audiences for pamphlet literature had expanded—millions of pamphlets had flowed from the presses between the appearance of the comet in 1506 and Creutzer's tract in 1527.[68] Authors and rulers had learned to exploit the printing press to their advantage. Cometary tracts quickly acquired a standard form and content that emphasized personal observation, the physical causes of comets, the type of comet, and concluded with various predictions.

Comets required astrologers to demonstrate a new form of expertise that was grounded not just in mathematics but also in observation. Predictable celestial phenomena like planetary conjunctions or eclipses did not require astrologers to make any observations. Instead, they relied on mathematical models and tables of planetary positions. Comets, by contrast, were seemingly random phenomena that appeared without warning. Even after they had been observed, their motion across the sky was not understood well enough to allow astrologers to make even short-term predictions. Consequently, astrologers grounded their prognostications in observations.[69] They reported detailed accounts of when and where they first spotted the comet and recounted its motion across the sky. They also described its shape and the direction of its tail. Johannes Schöner, who was renowned for his astronomical observations, described in detail his viewing of the comet of 1531: "On the 15th day of August at almost 9 PM I saw a comet at around 9 degrees above the horizon in the quarter of the sky between the western horizon and midnight, whose body and tail were directly in a straight line with the last two stars of Libra, or rather named Ursa maior."[70] The title-page woodcut depicts what Schöner had described, including Schöner observing the comet (see figure 7.5).

Johannes Virdung, the professor of astrology at the University of Heidelberg, was even more detailed than Schöner in reporting his observations of the comet.[71] But it was not sufficient to have observed a comet; the expert astrologer also had to know what caused it and had to demonstrate his command of the textual authorities, notably Aristotle. Virdung understood that his authority to interpret the comet of 1531 was grounded in this two-pronged

FIGURE 7.5. The title page from Johannes Schöner's *Coniectur odder abnemliche auszlegung Joannis Schöners vber den Cometen so jm Augstmonat des M.D.XXXj. jars erschinen ist* (Nuremberg: Peypus, 1531). © Bayerische Staatsbibliothek, Munich, sig. Res/4 Astr.p. 90 h#Beibd.14.

approach that combined observations with information from textual sources. After he detailed the path of the comet, he culled from Aristotle his rich discussion of the causes of this comet. Comets were warm exhalations from earth drawn up through the atmosphere by the attraction of the planets until they were ignited by elemental fire. Finally, he distinguished between types of comets by the shape and direction of the tail.[72] By recounting these three aspects—personal observation, causes, and types—the astrologer demonstrated his authority to interpret the effects of a comet. Virdung concluded his text by outlining the comet's effects and when these events would occur.[73]

Since antiquity comets had been the harbingers of war, disease, famine, and drought. Sixteenth-century astrologers continued to see comets as ominous signs. Like other astrologers in 1531, Perlach rushed to produce an interpretation of that year's comet, sending his work off to the printer by mid-November. Perlach's predictions in *Des Cometen und ander erscheinung in den lüfften* were intended to reassure the Viennese that they had little to fear from this comet, and certainly not another Turkish invasion (see figure 7.6). He also used *Des Cometen* to defend proper astrology and to attack Carion, accusing him of practicing magic rather than the "natural art of astrology."[74] Perlach opened his work with a long justification for both his efforts to interpret the comet and his practice of astrology. He claimed that many people, both nobles and common people, had asked him about the comet, its origin and significance.[75] Out of respect for their friendship he had been moved to write his current work. It was also more efficient and effective than replying to each person.[76] Although astrology provided only general knowledge of future events, Perlach thought even general knowledge helped people prepare for the future. Proper astrology could not offer precise predictions, a point he returned to throughout his preface. What separated correct, natural astrology from demonology or magic was this lack of precision and certainty. Perlach emphasized that his work was based on the natural art of astrology, which was not in conflict with God. In fact, Perlach claimed that his expertise in astrology was a gift from God.[77]

For Perlach there was no tension between a fatalist astrology and an omnipotent God. It was, instead, the proper method of predicting the future and the degree of certainty those predictions could carry. Moreover, Perlach was arguing about who had the right to make such predictions. In his preface to the reader, Perlach stated what was at stake:

> Dear friendly Reader, you should not assume that my Prognostication, which I have taken from the natural art called Astrology, must occur. Because, as Ptolemy, the pagan master and unbeliever, writes: The wise man rules over the stars. Much more so a godly Christian person, in his prayers to God, may rule over the stars and their influence. So I have always had this practice in my *Judicia*, to write or publish nothing contrary to natural causes. In that way the earnest man sees, that I utilize only the basis of the natural art named astrology, and no other. Thus none of my *Judicia* should compare to those of Carion, who used not the natural art of astrology, but rather some other, which he perhaps should not mention. . . . If Carion had [used] a proper basis of the

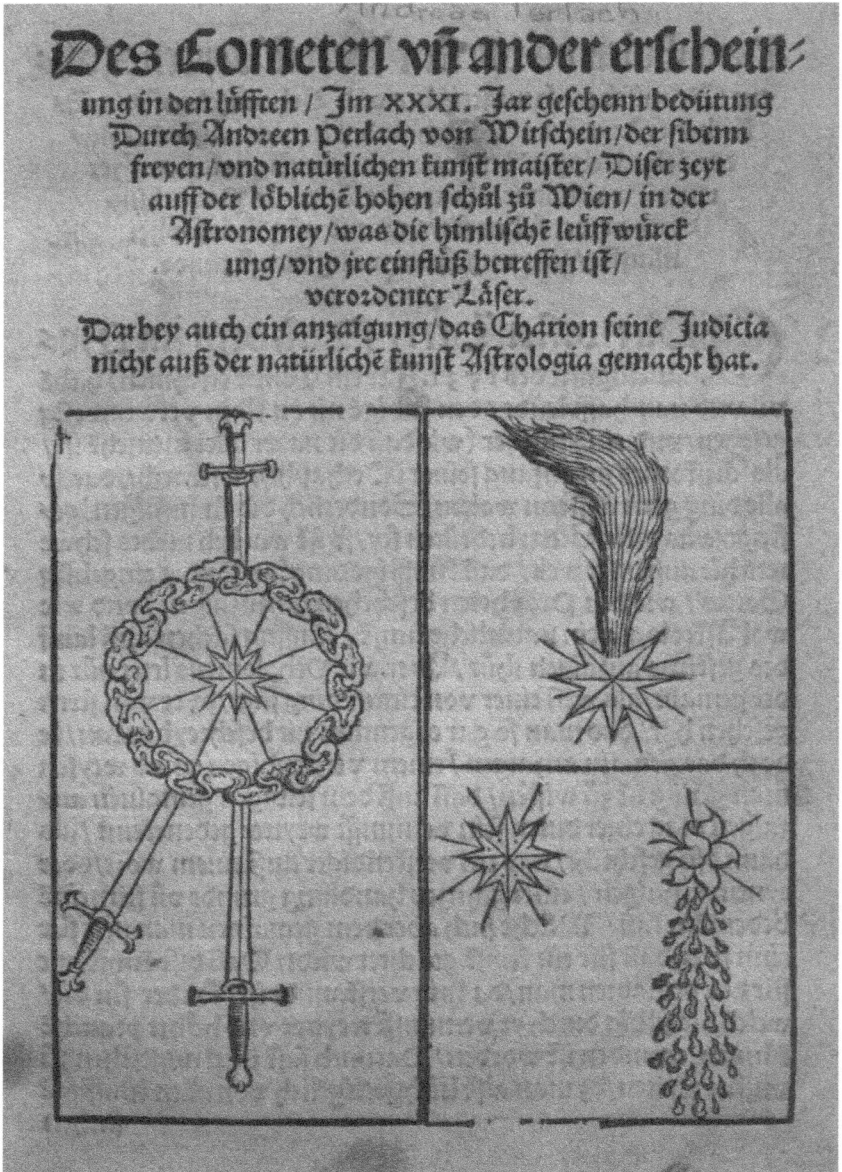

FIGURE 7.6. The title page from Andreas Perlach's *Des Cometen* (1531). Perlach indicated in the subtitle that Carion had not used the "natural art of astrology" in making his predictions. The right half of the woodcut depicts three types of comets. Augsburg, Stadst- und Staatsbibliothek, 4 Math 316/Beibd.

natural art of astrology, it would not have been possible for him to make so many shameful errors in his little book, which contrary to the basis of this method was published. And such shameful errors, which even a beginner would reprimand, should silence such a man, who intends to be seen the most learned Astrologer, and is mistaken by many for one (although by no truly educated and reasonable man).[78]

Perlach's diatribe was intended both to defend his own work and to once again discredit Carion and his predictions, which he had recently republished in a new, expanded edition of his *Bedeutnus und Offenbarung*. This pamphlet had provoked Perlach's fury two years earlier, but apparently, his first critique had done little to check its popularity. Whereas he had limited his earlier criticisms to Carion's understanding of astrology, this time Perlach expanded his attack to include both Carion's knowledge of astrology and his character in general.[79]

The core of Perlach's polemic focused on Carion's misuse of standard astrological concepts. Perlach sometimes explained how Carion failed to understand the point and other times simply listed his errors.[80] According to Perlach, Carion's ignorance of the *revolutio anni* prevented him from realizing that he had miscalculated a number of his annual horoscopes and in these cases had determined the wrong Lord of the Year. Consequently, all of Carion's predictions based on these horoscopes were wrong. Carion also failed to understand the basic astronomy that undergirded his astrological predictions. He did not know the distances of the planets from the sun or how to calculate the duration of an eclipse. He even failed to realize that he had to adjust the data from Stöffler and Pflaum's *Almanach novum* to reflect his location in Berlin.[81] Even when Carion calculated eclipses correctly he did not interpret them properly. As an example, Perlach returned to the topic that had occupied him in his previous attack on Carion, the lunar eclipse on October 17, 1529. Perlach and his audience knew that Carion had previously predicted that the Turkish army would conquer Vienna, a prediction that had been reprinted in Carion's latest *Bedeutnus und Offenbarung*.[82] Perlach was quick to point out that by October 14, 1529, the Turkish siege had been broken and the Turks driven back deep into Hungary.[83]

Although Perlach had no difficulty in enumerating Carion's various mistakes and errors, he feared that others would mistake Carion's predictions for valid astrology.[84] Perlach specified his criticism when he suggested that Carion had used some sort of magic to make his predictions. Although he stops short of accusing Carion of practicing demonic magic, his allusion to Pelagius the Hermit implied such illicit activities.[85] Pelagius owed much of his local fame to Johannes Trithemius, who claimed that Pelagius had taught Trithemius's teacher, Libanius Gallus.[86] Trithemius identified Pelagius with Fernando de Córdova, thereby linking himself through his education to one of the most famous fifteenth-century prodigies.[87] From Pelagius, Trithemius had learned both licit natural magic and illicit demonic magic. Trithemius often claimed that he and Pelagius practiced only licit magic. The reputations of the two men were, however, quite different. During a trip to Paris, Pelagius was suspected of practicing demonic magic, and some people even considered him the Antichrist.[88] Trithemius's claims to the contrary notwithstanding, he had a reputation as a demonologist, was said to have conjured spirits for Emperor Maximilian I, and might have been the "man especially learned in the black arts" who instructed the young Maximilian in these subjects.[89] Perlach's reference to Pelagius recalled for his readers the questionable activities of both Pelagius and Trithemius: "It also occurred to me, that he took his entire *judicium* from the books on conjuring spirits of that master Pelagius the Hermit from Majorca, since such a book has been copied in Berlin and not for the first time brought to Austria. I have seen it with my own eyes. Previously there have been such people, as Ptolemy showed, who use such arts and say that they produce their *judicia* from astrology, though they have no ability or training in it (as with this Carion)."[90] Perlach offered this as one reason for attacking Carion, but he had yet another reason. Carion had misrepresented Perlach's initial critique in the 1531 *Ephemerides*. Carion's response had provoked Perlach to retort that "this Carion either understands no Latin, or must be possessed by a malicious and evil spirit."[91] Perlach opted for the latter. Perlach also envied the respect afforded to Carion, particularly by those people who considered themselves well-educated. It was an insult to Perlach that even *lectores ordinares* in astronomy were among the people who hailed Carion the most learned astrologer. The only way to account for such collective error was to blame it on the meddling of Satan.[92]

Having committed the bulk of his energy to attacking Carion, Perlach finally turned to interpreting the comet of 1531. In order to stress his natural astrology, Perlach opened with a long discussion of the causes of change on earth, in general, and the specific causes of comets. All change on earth, he said, was caused by the motions of the planets and their positions relative to each other and the fixed stars. Comets were a particular type of atmospheric phenomenon, related to shooting stars, meteor storms, and similar spectacles. These atmospheric phenomena were also produced by the motions of the planets; namely, Mars attracted vapors out of the earth and into the atmosphere where they were ignited. What distinguished one apparition from another was simply its duration.⁹³ The comet in 1531 had been caused by two eclipses the previous year. Both had occurred in a fiery sign and had been ruled by Mars, a fiery planet. At this point Perlach turned his attention to explaining why a comet indicated general sickness. Each planet possessed certain characteristics, either favorable or unfavorable, and would draw out of the earth fluids with similar qualities. These fluids then affected the trees, plants, animals, and humans. When a planet approached and drew out the poisonous and harmful vapors, these vapors poisoned the air, which in turn caused destruction and decay among people, animals, and plants alike. The same noxious gasses that poisoned the air and harmed people became comets and other fiery appearances.⁹⁴ According to Perlach, then, comets were signs of general sickness and death because they shared the same cause, not because they were somehow a link in a chain of causes or because God used them as signs.⁹⁵ For Perlach, comets were preternatural but not supernatural. He summed up his position in the closing lines of the section: "from all of this one may assume that the cause that drew up the material of this comet, also caused the deaths in Vienna and other places."⁹⁶

In the remaining sections of Perlach's text he offered a lengthy discussion of when the comet appeared and its course. He recorded not only the zodiacal houses, but also which fixed stars and planets it conjoined. Perlach then turned to what dire events would follow the appearance of this comet. He predicted a heat wave, drought, and food shortages, as well as a general poisoning of the waters, particularly standing water, which is not purified by flowing through the sand and rocks.⁹⁷ He also warned that the poisoning of the air would likely cause infestations of poisonous worms, locusts, beetles, and toads that would destroy crops and livestock.⁹⁸ Despite these dire predictions, Perlach concluded his tract on an encouraging note.

Perlach's efforts to discredit Carion in an attempt to calm public concern were an exercise in political management, just as Tannstetter's had been. Unlike Tannstetter, however, Perlach intended to reach audiences beyond Vienna and Lower Austria. Addressing the Turkish threat required more than a calm populace; it required the military and financial support of the German princes. Although the Turks had recently been driven from Vienna, the Ottoman forces just to the east of Vienna remained the most urgent problem and greatest source of unease for the Viennese. The Viennese and people in Lower Austria lived in the shadow of the Ottoman threat. Perlach used the last section of his text to reassure his audience that the Turkish army would not return and capture the city but would instead be defeated by, he implied, Ferdinand's forces.

The Mahumet sect will have to fear downfall, so the moon (which is on
their shield and helmet) reveals, for in this year for almost an entire month
it approaches their vile star, in the eye of Taurus, and many times will eclipse
that same star. Because also the moon will become ruinous from the same star.
However, this Sect is not to be underestimated, since Jupiter comes into its
sign, that is Scorpio. At the end of the year, it comes to their star. Therefore,
this Sect will take many significant risks, through war and occasional attacks
against their enemy.[99]

Perlach lifted this prediction almost word for word from his *Ephemerides* for
1531. Now, however, the situation had changed slightly. Although by the fall
of 1531 Ferdinand had been elected King of the Romans, the German princes
were still withholding their financial and military support, and Europe was
still living in fear of further Turkish invasions. Once again Perlach reminded
his reader of Ferdinand's potential and the favorable astrological influences.
If at this key moment the German princes threw their support behind Ferdinand, he could finally drive the Ottomans out of Europe and recapture Hungary.[100] Perlach used the comet to advance Ferdinand's case, reminding his
readers that they had failed to offer the support Ferdinand needed. Although
they were "disobedient people," they had another opportunity to support the
king's efforts and to help drive the Turks out of Europe.[101]

* * *

In the three decades that passed between the time Johannes Stabius wrote
his *Pronosticon*, Georg Tannstetter composed his *Libellus consolatorius*, and
Andreas Perlach completed his *Des Cometen*, the general situation in Vienna
had changed little. In the details, however, the contexts for each work changed
significantly. All three drew on astrology to help make sense of their particular
political and social details. Moreover, they recognized that astrology was a
useful tool for expressing political agendas and exercising social control.

When Stabius wrote his *Pronosticon*, the Turkish threat was on the rise,
but had not yet reached Vienna, and Maximilian was viewed by many as the
prophetic Last World Emperor who would vanquish the infidel and inaugurate a period of peace and prosperity. Stabius employed astrology to show that
the advent of the Turk was related to the conjunctions of the superior planets
and that these same conjunctions and planetary motions were responsible for
the outbreak of plagues and wars. Astrology reinforced the predictions he
found in the prophetic literature and further cemented Maximilian's reputation as the Last World Emperor. The correspondence between astrology and

prophecy could then be used to support using this same prophetic literature to predict Maximilian's role as the Last World Emperor and final victory over the Turks.

Tannstetter used his expertise and status to assist Archduke Ferdinand in establishing his authority in Vienna. When the hysteria surrounding the 1524 conjunctions reached a fever pitch, Tannstetter produced a work, in both German and Latin, aimed at calming the fears of the people. He argued that astrology, understood properly and in harmony with other forms of knowledge, demonstrated that the conjunctions in Pisces in 1524 would not produce any ill effects. Tannstetter's texts and predictions were associated with the court. His effort to manage fear, both popular and elite, was an exercise of political authority.

When Perlach wrote his *Des Cometen*, Vienna was reeling from a recent Turkish siege. Ferdinand had succeeded in lifting the siege and driving the Turks back to Buda, or at least he claimed the credit. But he had failed to vanquish the Turkish forces. Vienna and Austria, and Europe as a whole were living in fear of a renewed invasion. Consequently, Perlach feared that Carion's dire predictions had struck a nerve in the Austrian population, adding to their already heightened sense of insecurity. Perlach attempted to manage the fears of the Viennese and people in Lower Austria by offering favorable interpretations of the comet. As with Stabius and Tannstetter, Perlach employed astrology to understand his particular political and social situation and to exercise some influence on that situation. And as with his two predecessors, Perlach aligned his political message with the interests of the local Habsburg court. Maximilian had envisioned a consistent and overt role for astrology in Habsburg politics and had worked to implement that vision. Ferdinand continued his predecessor's efforts and relied on astrologers and their predictions at key moments in his reign. By the late 1520s astrology had become an important part of Habsburg political rhetoric.

Conclusion

Astrology and Maximilian's Legacy

On January 12, 1519, Emperor Maximilian I of Habsburg died in Wels. He was not alone. His favorite astrologer and *Leibartz*, Georg Tannstetter, stood at his side while his secretary, astrologer, and propagandist, Joseph Grünpeck, waited in the antechamber. Maximilian left this world, just as he had entered it, surrounded by astrologers. Those who attended the emperor's birth and death reflected Maximilian's commitment to and reliance on astrology and astrologers throughout his reign. He not only celebrated his own expertise in the science of the stars but also regularly drew on the advice and support of the astrologers he attracted to his court. Twenty-five years after his death, Johannes Schöner praised the emperor for his skills and numbered him among the most important patrons of astrology.[1] Further, Schöner honored the emperor by using his geniture for nearly every example in his magisterial *Judicium nativitatum libri tres*. Schöner was not alone in his opinion. Forty years later in his history of great patrons of astrology, Heinrich Rantzau singled out Maximilian as a lover of the art.[2] The emperor's support of astrology was anything but accidental. He recognized that astrology provided a powerful political tool. As a systematic body of knowledge astrology was embedded in all registers of society, from the learned and political elite to popular culture. Astrology and the expert astrologers who deployed it purported to use nature as evidence, guide, and justification for political actions. Astrology also augmented other forms of knowledge, notably biblical and political prophecy. For Maximilian astrology played an important role in his efforts to assert and reinforce his authority against the resistance of rebellious princes, uncooperative popes, hostile neighbors, and truculent citizens.

Maximilian's use of astrology began with the emperor himself. Central to his self-understanding and the image he projected was his own expert knowledge of astrology. From his earliest efforts to craft his autobiography, Maximilian foregrounded the role of astrology in every aspect of his reign. He invoked his horoscope to explain key aspects of his personality and to justify his rulership and guarantee his future successes. Because it offered unique insight into the character and actions of political rivals and allies, Maximilian claimed to have learned astrology in order to facilitate and expand his diplomatic efforts. His autobiographical corpus constructed his memorial in both words and images, forming at once an idealized monument shaping how contemporaries viewed Maximilian and normative portraits of the ideal Habsburg prince. For Maximilian, the ideal prince exploited the science of astrology in all facets of politics through personal knowledge and expertise and through privileged access to sources of knowledge.

The imperial court was not, however, bounded by walls nor did it define the limits of Maximilian's world. Maximilian and his successor, Archduke Ferdinand, developed patronage practices that extended to include the university. Maximilian turned to the University of Vienna to draw astrologers into his court and as a body of experts who could be tapped for advice and intellectual support in his political endeavors, thus enhancing his reputation. Moreover, he recognized that their expertise could provide a framework for political decisions. Maximilian revitalized the university and funded a series of institutional developments intended to reestablish the University of Vienna as an important center for teaching astrology and astronomy. In this, Maximilian was part of a larger movement in the thirteenth through sixteenth centuries that saw princes rely increasingly on the local universities in their ideological projects and efforts to produce political propaganda.[3] His reliance on the university helped to foster a new type of patronage practice that flourished in the later sixteenth century. Princes began to supplement personal patronage, where individuals were brought into the confines of the court, with the patronage of extra court institutions such as universities and academies. Here, the university or the academy adopted a role in representing the will of the prince and legitimating his authority.[4]

Maximilian's efforts to portray himself as a prince who benefited from the advice of expert astrologers was not a fiction. The coherent system of predictive and explanatory knowledge proffered by astrologers became a cornerstone in Maximilian's courtly politics, from Brant's broadsheets supporting his war against the French and his efforts to establish a centralized military to Grünpeck's explanation of a series of planetary conjunctions that had occurred a decade earlier as the cause of the advent and spread of French Disease, which Maximilian used as an opportunity to enact a set of social reforms among the citizens in the empire.

Astrology offered more than persuasive propaganda; it also provided concrete mechanisms to guide political actions. For the emperor, astrological instruments transformed astrology from an abstract body of knowledge to a specific set of techniques and devices that he used in various political situations, bolstering his authority among the growing bureaucratic class and to disseminate his image as a patron and student of astrology to rival courts. They also played important symbolic roles in Maximilian's politics. In 1506 Maximilian employed the *Clipeus Austrie* designed by Andreas Stiborius to guide his peace negotiations with the Hungarian King Ladislaus. This instrument revealed Maximilian to be a competent astrologer and scholar. The imperial historian Johannes Stabius produced a number of ornate printed astrological instruments for the emperor. On the one hand, these were functional devices and could be used to tell time or to determine propitious moments to undertake different political actions such as triumphal entries. On the other hand, they complemented Maximilian's other refined propaganda, in that these elaborate prints were sent as gifts to princes and important members of the emerging administrative class of lower nobility, imperial free knights, and upper bourgeoisie.

Capitalizing on the ubiquity of astrology, the emperor exploited the power of print to publicize his political agenda to a broad audience that included every level of society. The emperor enlisted astrologers at his court and the university in his propaganda campaign to promulgate a pro-Habsburg agenda to audiences beyond the narrow confines of elite society through a variety of astrological genres. Astrological pamphlets, broadsheets, and advertisements were posted on notice boards, read out in town hall meetings, and sold in the markets, becoming indispensable vehicles for communicating Habsburg and imperial interests. Maximilian transformed cheap ephemeral print into an important instrument of governance and broadcast his political message through it.

Of the popular astrological literature, almanacs were the most sophisticated and required the most knowledge. Andreas Perlach produced almanacs as well as a manual detailing the proper astrological techniques and methods. Perlach's early almanacs and his *Usus almanach*, then, reflected the importance that the Habsburg rulers placed on controlling precise, technical knowledge.[5] As Perlach's authority and position improved, he adapted the format and contents of his almanacs, expanding the political reach and significance of his texts. Even the most ephemeral astrological texts, wall calendars and the *practica* that complemented them, became important instruments in Habsburg politics. These texts drew on a visual vocabulary and astrological imagery along with words to convey their content and were wildly popular in the late fifteenth and early sixteenth centuries. The Habsburgs recognized the political utility of controlling the production of these ephemeral texts both to assert their authority over yet another facet of astrological knowledge. Through Tannstetter's wall calendars and *practica*, Maximilian and later Ferdinand attempted to centralize the interpretations of celestial phenomena and to publicize those interpretations through all registers of society.[6] Tannstetter invested his wall calendars and *practica* with considerable rigor and precision, well beyond the needs of most of his audience. In addition to spreading his fame beyond Vienna, Tannstetter's wall calendars and *practica* helped to ensure his success at both the university and the court and to publicize the Habsburg political and social programs to a broad audience.

Prognostications composed in response to particular celestial phenomena were another popular astrological genre that Maximilian and Ferdinand appropriated for their political programs. Extraordinary events such as the appearance of a comet or rare planetary conjunctions attracted widespread public attention and offered astrologers further opportunities to interpret the natural world. Stabius, Tannstetter, and Perlach all seized on these prodigious events as evidence of Maximilian's preordained right to rule and justification for Habsburg authority within the empire. In each case, astrology provided the framework for understanding the effects of some spectacular celestial event. Stabius, Tannstetter, and Perlach offered expert interpretations of these events intended to mollify the fears of local audiences and to generate support for the Habsburg agenda.

Throughout his reign and at all levels of his dynastic program Maximilian continually enlisted astrology as a body of predictive knowledge about the natural world that had clear application in politics. His reliance on astrology and patronage of astrologers established practices that continued for more than a century both among his Habsburg successors and German princes across the empire. Maximilian's legacy had a lasting impact on both the

Habsburg dynasty and the broader relationship between science and politics in early modern Europe.

The importance Maximilian attached to scientific knowledge became a key component of Habsburg politics. It was at the imperial Habsburg courts in the later sixteenth century that Maximilian's interests in astrology as well as prodigies, portents, and their collection, found their greatest expressions.[7] Ferdinand of Tyrol amassed an enormous collection of curiosities and sports of nature at castle Ambras near Innsbruck.[8] Emperor Maximilian II arranged in Vienna an impressive collection of art and nature, including a unicorn's horn that is still on display in the *Schatzkammer*. The collections and passions of Rudolf II were, beyond question, the acme of the tradition begun by Maximilian I. Rudolf himself recognized his own debt to Maximilian in his efforts to enlist art and science in his political program.[9] Moreover, Rudolf's enthusiasm for and patronage of astrology, astrological instruments, alchemy, magic, art, and his collection of natural artifacts were extreme expressions of his predecessor's attempts to find divine legitimation in the person of the emperor himself. Maximilian's patronage of astrology along with the arts and literature, his attempts to locate his authority in the prodigies, monsters, and celestial phenomena can be seen as the origins of a tradition that became a Habsburg passion.[10]

Maximilian's efforts to establish patronage networks that linked individual experts as well as institutions to the court emerged as a central characteristic of early modern politics. By the end of the sixteenth century, especially in the Germanies, princes increasingly viewed local universities, alchemical laboratories, and observatories as bodies of experts to be consulted in political affairs. Princely patrons devoted considerable energy and resources to supporting various scientific projects that promised to give them privileged access to natural knowledge. They recognized that natural knowledge had political instrumentality, that it could be enlisted in their broad princely agendas. Princes drew on natural knowledge to impose religious conformity or promote religious tolerance, to ground their authority in some purported natural order that mirrored God's order, to assert their sovereignty over rebellious subjects, to extend state control throughout their territories and beyond, or to support entrepreneurial ventures and to generate the revenues that were increasingly important in European economies.[11]

Similarly, in his use of astrology as courtly science and form of propaganda, Maximilian's representation of himself as both a skilled practitioner and generous patron of the sciences prefigured the late sixteenth-century German prince practitioner. These princes not only supported alchemical and astrological projects, they participated in them. By the end of the sixteenth century German princes such as Emperor Rudolf II and Mortiz of Hessen celebrated their own technical expertise throughout their political, religious, and economic programs.[12] These later princes understood that their own expertise and skills would be an important part of their self-presentation, that the politics of representation drew on their technical, sometimes esoteric scientific knowledge. At the beginning of the century, however, such a role for scientific knowledge in the construction of a political identity was not yet common practice. For reasons that we may never fully understand, Maximilian realized that his own reputation was enhanced by both supporting the production of scientific knowledge and celebrating his own abilities in that body of knowledge. His success as a political actor, in establishing the House of Habsburg as one of the foremost dynasties in Europe, ensured that his model would be emulated and developed by future princes and heads of state.

Finally, in light of his efforts to enlist all levels of society in his political program, Maximilian's use of astrology reveals an expanding role for scientific knowledge in politics and in shaping public opinion. Astrology, which was a traditional and academically respected body of knowledge and was embedded in popular and elite culture, gave Maximilian a tool that purported to use nature as evidence, guide, and justification for political actions. Astrology complemented or often supplemented other forms of knowledge production, including prophetic and genealogical pursuits. Emperor Maximilian I developed patronage networks of scientific experts, exploited the reflected credibility of those experts, and used astrology's predictive and explanatory techniques to advance his political and dynastic programs. Within a century princes throughout the Holy Roman Empire and across Europe were using scientific knowledge and the expert purveyors of that knowledge to shape public opinion and advance their own political agendas. Emperor Maximilian I's reliance on astrology demonstrates that the intimate relationship between science and politics existed long before it became a defining characteristic of the modern world.

NOTES

Introduction

1. The letters are transcribed in Eugenio Albéri, ed., *Le Relazioni Degli Ambasciatori Veneti Al Senato Durante Il Secolo Decimosesto* (Florence, 1862), 59–66.

2. For a fuller account of this war, see Hermann Wiesflecker, *Kaiser Maximilian I. Das Reich, Österreich Und Europa an Der Wende Zur Neuzeit* (Munich: R. Oldenbourg Verlag, 1981), 1–153.

3. Marino Sanuto, *I Darii Di Marino Sanuto* (Venice, 1886), 291.

4. Monica Azzolini, *The Duke and the Stars. Astrology and Politics in Renaissance Milan* (Cambridge, MA: Harvard University Press, 2012), 205–6.

5. H. Darrel Rutkin, "Astrology," in *Early Modern Science*, ed. Katherine Park and Lorraine Daston (Cambridge: Cambridge University Press, 2006), 541–61. See also Azzolini, *The Duke and the Stars*, 26–27.

6. Robert Westman opted for the more nuanced "science of the stars." See *The Copernican Question*, 10 and ch. 1.

7. This was a pragmatic choice—it is easier and less confusing to write astrology and astrologer than try reflect the many different ways contemporaries referred to themselves and their discipline—and for this group of historical actors, the terms astrologer and astrology do not seem to do violence to how they would have described their own activity.

8. See Larry Silver's excellent study on Maximilian's artistic endeavors: *Marketing Maximilian. The Visual Ideology of a Holy Roman Emperor* (Princeton, NJ: Princeton University Press, 2008).

9. For example, see Karl Vocelka, *Die politische Propaganda Kaiser Rudolfs II. (1576–1612)* (Vienna: Verlag der Österreichischen Akademie der Wissenschaften, 1981); R. J. W. Evans, *Rudolf II and His World. A Study in Intellectual History, 1576–1612* (Oxford: Oxford University Press, 1973); Helen Watanabe-O'Kelly, *Court Culture in Dresden from Renaissance to Baroque* (Houndmills: Palgrave Macmillan, 2002); Tara Nummedal, *Alchemy and Authority in the Holy Roman Empire* (Chicago: University of Chicago Press, 2007); Eric H. Ash, *Power, Knowledge, and Expertise in Elizabethan England* (Baltimore, MD: Johns Hopkins University Press, 2004).

10. We typically associate propaganda with the deceitful underbelly of modern statecraft. In the best cases, it conjures up white lies and half-truths, while in the hands of totalitarian regimes it is a tool of deceit and oppression. Garth Jowett and Victoria O'Donnell consider manipulation and deception key components of propaganda; however, Nicholas Cull notes that World War I changed the nature of propaganda to emphasize the intentional misleading nature. Jacques Ellul has argued vigorously that propaganda has become a necessary aspect of our modern, technological society. Before

Ellul, Edward Bernays argued that controlling public opinion was a necessary part of any democratic society. Jowett and O'Donnell, *Propaganda & Persuasion* (London: Sage, 2011), 44; Cull, *The Cold War and the United States Information Agency. American Propaganda and Public Diplomacy, 1945–1989* (Cambridge: Cambridge University Press, 2008), 9; Ellul, *Propaganda. The Formation of Men's Attitudes* (New York: Alfred A. Knopf, 1969); Bernays, *Propaganda* (New York: H. Liveright, 1928).

11. Peter Burke, Kevin Sharpe, and Jason Peacey all make this point in their studies on Louis XIV, the Tudor monarchy, and the English Revolution. See the introductions to Burke's *The Fabrication of Louis XIV* (New Haven, CT: Yale University Press, 1992); Sharpe, *Selling the Tudor Monarchy. Authority and Image in Sixteenth-Century England* (New Haven, CT: Yale University Press, 2009); Jason Peacey, *Politicians and Pamphleteers. Propaganda during the English Civil Wars and Interregnum* (Aldershot: Ashgate, 2004).

12. Scholars have devoted considerable attention to political rhetoric and representation in early modern England. See, for example, Joan Coutu, *Persuasion and Propaganda. Monuments and the Eighteenth-Century British Empire* (Montreal: McGill-Queen's University Press, 2006); Roy Strong, *Splendor at Court. Renaissance Spectacle and the Theater of Power* (Boston: Houghton Mifflin, 1973); Roy Strong, *The Cult of Elizabeth: Elizabethan Portraiture and Pageantry* (London: Pimlico, 1999); Sharpe, *Selling the Tudor Monarchy*; Kevin Sharpe, *Image Wars. Promoting Kings and Commonwealths in England 1603–1660* (New Haven, CT: Yale University, 2010); Mary Hill Cole, *The Portable Queen. Elizabeth I and the Politics of Ceremony* (Amherst: University of Massachusetts, 1999); Sydney Anglo, *Images of Tudor Kingship* (Surrey: Seaby, 1992). Later sixteenth-century Germany has also received attention. See, for example, Thomas DaCosta Kaufmann, "From Mastery of the World to Mastery of Nature: The Kunstkammer, Politics, and Science," in *The Mastery of Nature. Aspects of Art, Science, and Humanism in the Renaissance*, ed. Thomas DaCosta Kaufmann (Princeton, NJ: Princeton University Press, 1993); Margit Altfahrt, "Die Politische Propaganda Für Maximilians II. (Erster Teil)," *Mitteilungen des Instituts für Österreichische Geschichtsforschung* 88 (1980); Margit Altfahrt, "Die politische Propaganda für Maximilians II. (Zweiter Teil)," *Mitteilungen des Instituts für Österreichische Geschichtsforschung* 89 (1981); Vocelka, *Die Politische Propaganda Kaiser Rudolfs II.*

13. We need not adopt a teleological understanding of print to recognize its impact. All studies of print owe a debt to Elizabeth L. Eisenstein, *The Printing Revolution in Early Modern Europe* (Cambridge: Cambridge University Press, 1983). See also Anthony Grafton's review and critique of Eisenstein's original work, "The Importance of Being Printed," *Journal of Interdisciplinary History* 11 (1980). Adrian Johns offers a different set of criticisms in *The Nature of the Book. Print and Knowledge in the Making* (Chicago: University of Chicago Press, 1998). Stephan Füssel offers a convenient overview of print's importance in *Gutenberg Und Seine Wirkung* (Frankfurt: Insel Verlag, 1999). Andrew Pettegree's more recent survey is also helpful: *The Book in the Renaissance* (New Haven, CT: Yale University Press, 2010). Historians of religion and of the Reformation have long recognized the importance of print in disseminating the Reformation's message. See Pettegree and Matthew Hall, "The Reformation and the Book: A Reconsideration," *Historical Journal* 47 (2004). Mark Edwards, *Printing, Propaganda, and Martin Luther* (Berkeley: University of California, 1994), connects print to religious propaganda. Hans-Joachim Kohler's

edited volume drew attention to the role of pamphlets in the Reformation: *Flugschriften Als Massenmedium Der Reformationszeit: Beiträge zum Tübinger Symposion 1980* (Stuttgart: Klett-Cotta, 1981). See also "The Flugschriften and Their Importance in Religious Debate: A Quantitative Approach," in *"Astrologi Hallucinati": Stars and the End of the World in Luther's Time,* ed. Paolo Zambelli (Berlin: Walter de Gruyter, 1986). Robert Scribner's work on illustrated pamphlets has connected print culture to popular and oral culture; see *For the Sake of Simple Folk. Popular Propaganda for the German Reformation* (Oxford: Oxford University Press, 1994). See also Marina Frasca-Spada and Nicholas Jardine, eds., *Books and the Sciences in History* (Cambridge: Cambridge University Press, 2000).

14. Peter Parshall, "Prints as Objects of Consumption in Early Modern Europe," *Journal of Medieval and Early Modern Studies* 28 (1998).

15. Sharpe, *Selling the Tudor Monarchy,* 29–34.

16. See, for example, Bertrand Taithe and Tim Thornton, "Propaganda: A Misnomer of Rhetoric and Persuasion?," in *Propaganda. Political Rhetoric and Identity 1300–2000,* ed. Bertrand Taithe and Tim Thornton (Gloucestershire: Sutton, 1999).

17. Mimicking Maximilian, Henry VII minted a large gold coin bearing his image. See P. Grierson, "Origins of the English Sovereign . . .," *British Numismatic Journal* 33 (1965): 118–34; "Propaganda: A Misnomer of Rhetoric and Persuasion," in *Propaganda, Political Rhetoric and Identity, 1300–2000,* 6–7; Thornton, "Propaganda, Political Communication and the Problem of English . . .," in *Propaganda, Political Rhetoric and Identity,* 41.

18. Roy Strong, *Holbein and Henry VIII* (London: Routledge & Kegan Paul, 1967).

19. Kevin Sharpe, *Selling the Tudor Monarchy,* 30.

20. Cole, *The Portable Queen.* The Tudors were not the first to use visual propaganda; see Sarah Gaunt, "Visual Propaganda in England in the Later Middle Ages," in Taithe and Thornton, *Propaganda.*

21. Taithe and Thornton, *Propaganda*; R. G. Cole, "The Reformation Pamphlet and Communication Processes," in *Flugschriften Als Massenmedium der Reformationszeit,* ed. Hans-Joachim Köhler (Stuttgart: Klett-Cotta, 1981).

22. On prophecies as a contested political space during Henry VIII's reign, see Sharon L. Jansen, *Political Protest and Prophecy under Henry VIII* (Woodbridge: Boydell Press, 1991). Suzanne Lewis has argued that one of the functions of the Bayeux Tapestry was to make the change to Norman authority seem natural. Suzanne Lewis, *The Rhetoric of Power in the Bayeux Tapestry* (Cambridge: Cambridge University Press, 1999).

23. Sydney Anglo has questioned the effectiveness of these elite forms of propaganda. He has argued that scholars read too much into erudite propaganda, claiming that most contemporaries would not have had the knowledge to unpack all the symbols and meanings bound up these images and pageants. See Anglo, *Images of Tudor Kingship.*

24. Evans, *Rudolf II and His World.*

25. Vocelka, *Die Politische Propaganda Kaiser Rudolfs II.*

26. Altfahrt, "Die Politische Propaganda Für Maximilians II. (Erster Teil)"; Altfahrt, "Die Politische Propaganda für Maximilians II. (Zweiter Teil)."

27. Robert Scribner's work is important for understanding the mechanisms by which printed pamphlets reinforced and spread popular unrest. See Robert W. Scribner, "Oral

Culture and the Diffusion of Reformation Ideas," in *Popular Culture and Popular Movements in Reformation Germany*, ed. Robert W. Scribner (London: Hambledon Press, 1988); Scribner, *Popular Culture and Popular Movements in Reformation Germany* (London: Hambledon Press, 1988); Scribner, *For the Sake of Simple Folk.*

28. See especially chapter 18, "How Rulers Should Keep Their Promises," in Quentin Skinner and Russell Price, eds., *The Prince* (Cambridge: Cambridge University Press, 1988), 61–63.

29. J. P. D. Cooper, *Propaganda and the Tudor State. Political Culture in the Westcountry* (Oxford: Oxford University Press, 2003), 12.

30. Manfred Hollegger, "Erwachen vnd Aufsten als ein starcker Stryter. Zu Formen und Inhalt der Propaganda Maximilians I.," *Propaganda, Kommunikation und Öffentlichkeit* 6 (2002): 223–24.

31. Hollegger, "Erwachen vnd Aufsten," 227–34. On Maximilian's patronage of music, see Theophil Antonicek, Elisabeth T. Hilscher, and Harmut Krones, eds., *Die Wiener Hofmusikkapelle. Georg Von Slatkonia Und Die Wiener Hofmusikkapelle* (Vienna: Böhlau, 1999).

32. Quoted in Stephan Füssel, *Gutenberg and the Impact of Printing* (Aldershot: Ashgate, 2005), 155.

33. On Maximilian's book collections, see Theodor Gottlieb, *Büchersammlung Kaiser Maximilians* (Leipzig, 1900). On the political uses of libraries, see Howard Louthan, *The Quest for Compromise. Peace Makers in Counter Reformation Vienna* (Cambridge: Cambridge University Press, 1997), 65–87; Martyn Rady, "The Corvina Library and the Lost Royal Hungarian Archive," in *Lost Libraries. The Destruction of Great Book Collections Since Antiquity*, ed. James Raven (New York: Palgrave, 2004); Csaba Csapodi, and Klára Csapodi-Gárdonyi, *Bibliotheca Corviniana. The Library of King Matthias Corvinus of Hungary* (New York: Frederick A. Praeger, 1969); Csaba Csapodi, *The Corvinian Library: History and Stock* (Budapest: Akademai Kiado, 1973). On Maximilian's early use of medals, see Gregory Todd Harwell, "Aurea Condet Saecula (Per Arva Saturno Quondam). Imperial Habsburg Medals from the Coronation of Frederick III (1452) until the Succession of Maximilian I (1494). Art and Legitimacy between Feudalism and Absolutism" (PhD diss., Princeton University, 2005).

34. Nearly a century ago Peter Diederichs's *Kaiser Maximilian I als politischer Publizist* (PhD diss., University of Heidelberg, 1931) laid the foundations for such work, which continues through the most recent studies. See also Manfred Hollegger's more recent work on Maximilian's printing efforts, "Erwachen Vnd Aufsten."

35. In addition to Silver's *Marketing Maximilian*, see also Heather Kathryn Suzanne Madar, "History Made Visible: Visual Strategies in the Memorial Project of Maximilian I" (PhD diss., University of California, 2003).

36. H. Th. Musper, ed., *Kaiser Maximilians Weisskunig*, 2 (Stuttgart: W. Kohlhammer Verlag, 1956), 226.

37. See "The Secretaries of Rulers," in Skinner and Price, *The Prince*, 80–81.

38. Hollegger, "Erwachen vnd Aufsten," 224–25 and Hermann Wiesflecker, *Kaiser Maximilian I. Das Reich, Österreich Und Europa an Der Wende Zur Neuzeit* (Munich: R. Oldenbourg Verlag, 1986), 452.

39. German historiography stretching back more than a century to Heinrich Ulmann's *Kaiser Maximilian I. Auf urkundlicher Grundlage dargestellt* (Stuttgart: Cotta, 1884) typically condemns Maximilian. More recent Austrian scholarship such as Hermann Wiesflecker's *Kaiser Maximilian I. Das Reich, Österreich und Europa an der Wende zur Neuzeit* (Munich: R. Oldenbourg Verlag, 1971–1986) portrays the emperor in a more favorable light. During his long career at the University of Graz Wiesflecker supervised numerous dissertations grounded in careful and detailed analysis of the political record. His work and that of his students have given us a much fuller picture of Maximilian's reign.

40. For example, in *Marketing Maximilian* Silver examines Maximilian's use of visual arts while Jan-Dirk Müller's *Gedechtnus. Literatur und Hofgesellschaft um Maximilian I.* (Munich: Wilhelm Fink Verlag, 1982) concentrates on the emperor's literary output.

41. Helmut Grössing and Franz-Graf Stuhlhofer have produced the most sustained work on science at Maximilian's court and the University of Vienna. See, for example, Helmuth Grössing, *Humanistische Naturwissenschaft: Zur Geschichte der Wiener Mathematischen Schulen des 15. und 16. Jahrhunderts* (Baden-Baden: Valentin Koerner, 1983) and Franz Graf-Stuhlhofer, *Humanismus zwischen Hof und Universität. Georg Tannstetter (Collimitius) und Sein Wissenschaftliches Umfeld im Wien des frühen 16. Jahrhunderts* (Vienna: WUV-Universitätsverlag, 1996).

42. Andrea Baltl's dissertation catalogs some of the scientific and artistic projects that Maximilian supported but provides little or no analysis of those projects and why the emperor cared about them. Andrea Baltl, "Maximilians I. Beziehungen zur Wissenschaft und Kunst" (PhD diss., Universität Graz, 1967). Stuhlhofer's book on Georg Tannstetter offers a useful biography of one astrologer-counsellor at Maximilian's court. See Graf-Stuhlhofer, *Humanismus zwischen Hof und Universität.*

43. See Pamela Smith's recent review of the history of early modern science, "Science on the Move: Recent Trends in the History of Early Modern Science," *Renaissance Quarterly* 62 (2009).

44. See, for example, Ash, *Power, Knowledge and Expertise*; Eric H. Ash, ed., *Expertise. Practical Knowledge and the Early Modern State* (Chicago: University of Chicago Press, 2010); Nummedal, *Alchemy and Authority*; Pamela H. Smith, *The Business of Alchemy: Science and Culture in the Holy Roman Empire* (Princeton, NJ: Princeton University Press, 1994).

45. Certainly the most influential of this early material was the work by Aby Warburg and Fritz Saxl. See, for example, "Italian Art and International Astrology in the Palazzo Schifanoia, Ferrara," "On Images of Planetary Deities in the Low German Almanac of 1519," and "Pagan-Antique Prophecy in Words and Images in the Age of Luther," all in Aby Warburg, *The Renewal of Pagan Antiquity*, trans. David Britt (Los Angeles: Getty Research Institute for the History of Art and the Humanities, 1999), 732–57, 758–59, 760–75. See also Saxl's magisterial *Verzeichnis astrologischer und mythologischer illustrierter Handschriften des lateinischen Mittelalters*, vols. 1 and 2 (Heidelberg: C. Winter, 1915, 1927) and *Catalogue of Astrological and Mythological Illuminated Manuscripts of the Latin Middle Ages*, vols. 3 and 4 (London: Warburg Institute, 1953–66) as well as his lectures, especially "The Revival of Late Antique Astrology" and "The Belief in Stars in the Twelfth Century," in Fritz Saxl, *Lectures* (London: Warburg Institute, 1957). Franz Cumont's and

Franz Boll's many studies of ancient astrology were also important early work. See Grafton's perceptive comments in his review; Anthony Grafton, "Starry Messengers: Recent Work in the History of Western Astrology," *Perspectives on Science* 8, no. 1 (2000).

46. Some book-length studies include Anthony Grafton, *Cardano's Cosmos. The Worlds and Works of a Renaissance Astrologer* (Cambridge, MA: Harvard University Press, 1999); Lauren Kassell, *Medicine & Magic in Elizabethan London. Simon Forman: Astrologer, Alchemist, & Physician* (Oxford: Clarendon Press, 2005); Steven Vanden Broecke, *The Limits of Influence: Pico, Louvain, and the Crisis of Renaissance Astrology* (Leiden: Brill, 2003); Günther Oestmann, *Heinrich Rantzau und die Astrologie* (Braunschweig: Braunschweigisches Landesmuseum, 2004); Hilary M. Carey, *Courting Disaster. Astrology at the English Court and University in the Later Middle Ages* (New York: St. Martin's Press, 1992); and Claudia Brosseder, *Im Bann der Sterne. Caspar Peucer, Philipp Melanchthon und andere Wittenberger Astrologen* (Berlin: Akademie Verlag, 2004).

47. I regret that I have not been able to incorporate all of Robert Westman's work into this book, as Westman offers much more than an explanation of how Copernicus developed his heliocentric theory in *The Copernican Question. Prognostication, Skepticism, and Celestial Order* (Berkeley: University of California, 2011), esp. chaps. 1–3.

48. Oestmann, *Heinrich Rantzau und die Astrologie*; Reisinger, *Historische Horoskopie*. For example, see the work by Helmuth Grössing and Franz Graf-Stuhlhofer on astrology in Vienna: "Versuch Einer Deutung Der Rolle Der Astrologie." See also Grössing, *Humanistische Naturwissenschaft*; Graf-Stuhlhofer, "Zu Den Hofastronomen"; Graf-Stuhlhofer, *Humanismus zwischen Hof und Universität*; and Brosseder's work on astrology in the Wittemberg context: Brosseder, *Im Bann der Sterne*; Claudia Brosseder, "The Writing in the Wittemberg Sky: Astrology in Sixteenth-Century Germany," *Journal of the History of Ideas* 66 (2005).

49. Azzolini takes political historians to task for avoiding astrology in their histories of politics. A reciprocal charge could be leveled at many historians of astrology who pay too little attention to politics in their histories of astrology. See Azzolini, *The Duke and the Stars*, 5–7.

50. Azzolini, *The Duke and the Stars*.

51. Ryan, *Kingdom of Stargazers*.

Chapter One. Astrology and Maximilian's Autobiography

1. Regiomontanus has received considerable scholarly attention. The standard biography is Ernst Zinner, *Leben und Werken des Joh. Müller von Königsberg gennant Regiomontanus* (Osnabrück: Otto Zeller, 1968). On Regiomontanus's astronomical calculations and work, see E. Glowatzki and Helmut Göttsche, *Die Tafeln des Regiomontanus: Ein Jahrhundertwerk* (Munich: Institut für Geschichte der Naturwissenschaften, 1990). See also James Steven Byrne, "A Humanist History of Mathematics? Regiomontanus's Padua Oration in Context," *Journal of the History of Ideas* 67, no. 1 (2006): 41–61 and James Steven Byrne, "The Stars, the Moon, and the Shadowed Earth: Viennese Astronomy in the Fifteenth Century" (PhD diss., Princeton University, 2007).

2. On the close connections between the Habsburg courts and astrologers, see Michael Shank, "Academic Consulting in 15th-Century Vienna: The Case of Astrology," in *Texts and Contexts in Ancient and Medieval Science. Studies on the Occasion of John E. Murdoch's Seventieth Birthday*, ed. Edith Sylla and Michael McVaugh (Leiden: Brill, 1997), 245–70.

3. On Peuerbach's authorship of this horoscope, see Helmuth Grössing, *Humanistische Naturwissenschaft: Zur Geschichte der Wiener mathematischen Schulen des 15. und 16. Jahrhunderts* (Baden-Baden: Valentin Koerner, 1983); Helmuth Grössing and Franz Stuhlhofer, "Versuch einer Deutung der Rolle der Astrologie in den persönlichen Entscheidungen einiger Habsburger des Spätmittelalters," *Anzeiger der phil.-hist. Klasse der Österreichischen Akademie der Wissenschaften* 117 (1980): 267–83.

4. Shank, "Academic Consulting," 261; Alphons Lhotsky, "Kaiser Friedrich III.: Sein Leben und seine Persönlichkeit," in *Aufsätze und Vorträge*, ed. Hans Wagner and Heinrich Koller, vol. 2: *Das Haus Habsburg* (Vienna: Verlag für Geschichte und Politik, 1971), 156.

5. On the similarities between the two genitures, see Grössing, *Humanistische Naturwissenschaft*, 90–91. One of the surviving copies, BSB Clm 453, was copied by Regiomontanus.

6. Regiomontanus, "Epistola ad quandam Imperatricem judicium astrologicum de ejusdem filio continens," ÖNB cod. lat. 5179. especially fol. 2r–4r.

7. Throughout the sixteenth and into the seventeenth century astrologers evaluated the various methods to rectify a geniture. See, for example, Johannes Schöner's comments in the opening pages of his *Nativitatum libri tres* or Melanchthon's introduction to Schöner's *Opera mathematica*. See Johannes Schöner, *De iudiciis nativitatum. Libri tres* (Nuremberg: Ioannem Montanum & Ulricum Neuberum, 1545), 1r–3v; Johannes Schöner, *Opera mathematica* (Nuremberg: Ioannem Montanum & Ulricum Neuberum, 1551), 57r; Oger Ferrier, *A Learned Astronomical Discourse, of the Judgement of Nativities*, trans. Thomas Kelway (London: Charlewoodhouse, 1593), fols. 1v–3r; Lilly, *Christian Astrology, Modestly Treated in Three Books* (London: John Partridge, 1647), 500–19.

8. "constat autem hunc natum esse Anno domini 1459 currente die 22 Marcii hora post meridiem 5ta precise completa diebus equatis." Regiomontanus, "Epistola ad quandam Imperatricem judicium astrologicum," 1v.

9. Regiomontanus, "Epistola ad quandam Imperatricem judicium astrologicum," 2v–3v, 5v–8v.

10. "Hec omnia praemissa sunt tamquam fundamenta atque radices huius nativitatis & revolutionum eius sequentium." Regiomontanus, "Epistola ad quandam Imperatricem judicium astrologicum," 9r.

11. Regiomontanus, "Epistola ad quandam Imperatricem judicium astrologicum," 9r–11v, 17r–18r.

12. "Significatur itaque ratione martis, natum futurum audacem, potentem, fortem, iracundum, armorum cupidum, bellorum auctorem, animosum, sine pavore mortis pericula aggredientem. Neminem sibi praeferet, nec cuiquam se humiliabit. plurimum de suis viribus confidet. impetum & violentiam in adversarios suos plerumque faciet. rebelles destruet. proeliis interesse cupiet. In omnibus his tamen honorem obtinebit. auxilia amicis praestabit. inimicus nocebit. legaliter in cunctis se geret." Regiomontanus, "Epistola ad quandam Imperatricem judicium astrologicum," 10v.

13. Regiomontanus, "Epistola ad quandam Imperatricem judicium astrologicum," 11r.

14. Regiomontanus, "Epistola ad quandam Imperatricem judicium astrologicum," 17v.

15. Tacitus, *Annals, Books IV–VI, XI–XII*, Loeb Classical Library (Cambridge, MA: Harvard University Press, 1937), Book 12, 68. See also Josèphe-Henriette Abry, "What Was Agrippina Waiting For? (Tacitus, Ann. XII, 68–69)," in *Horoscopes and Public Spheres. Essays on the History of Astrology*, ed. Günther Oestmann, H. Darrel Rutkin, and Kocku von Stuckrad, vol. 42, Religion and Society Series (Berlin: Walter de Gruyter, 2005), 37–48.

16. Shank, "Academic Consulting," 261; and Grössing and Stuhlhofer, "Versuch einer Deutung der Rolle der Astrologie," 280, both attribute the calculation to Nihili.

17. Lhotsky, "Kaiser Friedrich III," 156; Grössing, *Humanistische Naturwissenschaft*, 88.

18. Bylica's horoscope for Frederick survives in Martin Bylica, "Nativitates," BJ cod. 3225, 11r. See also Darin Hayton's "Expertise *Ex Stellis*: Comets, Horoscopes, and Politics in Renaissance Hungary," *Osiris* 25 (2010): 44; and "Martin Bylica at the Court of Matthias Corvinus: Astrology and Politics in Renaissance Hungary," *Centaurus* 49 (2007): 193.

19. See Monica Azzolini's *The Duke and the Stars. Astrology and Politics in Renaissance Milan* (Cambridge, MA: Harvard University Press, 2012).

20. Alfred Schmid, *Augustus und die Macht der Sterne. Antike Astrologie und die Etablierung der Monarchie in Rom* (Cologne: Böhlau, 2005), esp. 19–54; Tamsyn Barton, *Ancient Astrology.* (London: Routledge, 1994), 40–41.

21. See the excellent critical edition Abū Maʿšar, *On Historical Astrology. The Book of Religions and Dynasties (On the Great Conjunctions)*, trans. Keiji Yamamoto and Charles Burnett, vol. 2: *The Latin Versions* (Leiden: Brill, 2000).

22. For a general survey of the importance of Albumasar's *On the Great Conjunctions* for the history of the Catholic Church, see J. D. North, "Astrology and the Fortunes of Churches," in *Stars, Minds and Fate: Essays in Ancient and Medieval Cosmology*, ed. J. D. North (London: Hambledon, 1989), 59–89. A more detailed and specific study is Laura Ackerman Smoller, *History, Prophecy, and the Stars. The Christian Astrology of Pierre d'Ailly, 1350–1420* (Princeton, NJ: Princeton University Press, 1994).

23. For a good discussion of astrology's role in establishing the legitimacy of the early Roman emperors, see Schmid, *Augustus und die Macht der Sterne*, 245–77. See also Dimitri Gutas, *Greek Thought, Arabic Culture. The Graeco-Arabic Translation Movement in Baghdad and Early ʿAbbāsid Society (2nd–4th/8th–10th centuries)* (London: Routledge, 1998), esp. 45–53.

24. On Maximilian's use of visual strategies for establishing his authority, see Larry Silver, *Marketing Maximilian. The Visual Ideology of a Holy Roman Emperor* (Princeton, NJ: Princeton University Press, 2008); Heather Kathryn Suzanne Madar, "History Made Visible: Visual Strategies in the Memorial Project of Maximilian I" (PhD diss., University of California, 2003).

25. Jan-Dirk Müller, *Gedechtnus. Literatur und Hofgesellschaft um Maximilian I.* (Munich: Wilhelm Fink Verlag, 1982). See more recently Heather Kathryn Suzanne Madar, "History Made Visible."

26. Larry Silver's recent masterful study *Marketing Maximilian* puts Maximilian's efforts into historical perspective.

27. Christopher S. Wood, "Maximilian I as Archeologist," *Renaissance Quarterly* 58 (2005): 1128–74; Frank L. Borchardt, *German Antiquity in Renaissance Myth* (Baltimore, MD: Johns Hopkins University Press, 1971). See also Wood's wider ranging discussion in *Forgery, Replica, Fiction. Temporalities of German Renaissance Art* (Chicago: University of Chicago Press, 2008).

28. "wer ime in seinem leben kain gedachtnus macht, der hat nach deinem tod kain gedächtnus und desselben menschen wird mit dem glockendon vergessen, und darumb so wird des gelt, so ich auf die gedechtnus ausgib, nit verloren, aber das gelt, das erspart wird in meiner gedachtnues, das ist ain undertruckung meiner kunftigen gedächtnus, und was ich in meinem leben in meiner gedächtnus nit volbring, das wird nach meinem tod weder durch dich oder ander nit erstat." H. Th. Musper, ed., *Kaiser Maximilians Weisskunig*, 2 (Stuttgart: W. Kohlhammer Verlag, 1956), 226.

29. Astrology certainly was not the only mechanism for structuring the past or imposing a pattern on the present. It was, however, a particularly powerful system, as Johannes Lichtenberger recognized when he tried to justify his prophetic interpretations of the past and the present by grounding them in astrology. See Dietrich Kurze, "Johannes Lichtenberger (†1503): Eine Studie zur Geschichte der Prophetie und Astrologie," *Historische Studien* (1960).

30. Hermann Wiesflecker, "Joseph Grünpecks Redaktion der lateinischen Autobiographie Maximilians I," *Mitteilungen des Instituts für österreichische Geschichtsforschung* 78 (1970): 416.

31. Hermann Wiesflecker, *Joseph Grünpecks Commentaria und Gesta Maximiliani Romanorum Regis. Die Entdeckung eines verlornen Geschichtswerkes* (Graz: Jos. A. Kieneich, 1965), 15; Wiesflecker, "Joseph Grünpecks Redaktion."

32. Hans-Otto Burger, "Der *Weisskunig* als Literaturdenkmal," in *Kaiser Maximilians Weisskunig*, ed. H. Th. Musper (Stuttgart: W. Kohlhammer Verlag, 1956), 16.

33. For a recent analysis of the relationship between autobiography and astrology, see Anthony Grafton, *Cardano's Cosmos. The Worlds and Works of a Renaissance Astrologer* (Cambridge, MA: Harvard University Press, 1999), 178–98.

34. Maximilian, "Fragmente Einer Lateinischen Autobiographie Kaiser Maximilians I," *Jahrbuch der kunsthistorischen Sammlungen des allerhöchsten Kaiserhauses* 6 (1888): 423–24.

35. Maximilian, "Fragmente Einer Lateinischen Autobiographie," 426.

36. Maximilian, "Fragmente Einer Lateinischen Autobiographie," 425.

37. On the importance of the Burgundian models, see Wiesflecker, "Joseph Grünpecks Redaktion," 416. For Molinet's chronicles, see Georges Doutrepont and Omer Jodogne, eds., *Chroniques de Jean Molinet*, 3 (Brussels: Palais des Académies, 1935–37). On Olivier de La Marche, see Catherine Emerson, *Olivier de La Marche and the Rhetoric of 15th-Century Historiography* (Woodbridge: The Boydell Press, 2004).

38. On Maximilian's comment to Pirckheimer, see Burger, "Der *Weisskunig* als Literaturdenkmal," 16.

39. The French Disease, dubbed so because it first appeared in the French troops besieging Naples and spread from them up the Italian peninsula, was also known as Great Pox. In 1530 the Italian physician Girolamo Fracastoro first used the term syphilis to refer

to the disease. I refer to it as the French Disease to avoid labeling it with modern-day attributes. For Grünpeck's views on the French Disease, see chapter 2 of this volume.

40. See Darin Hayton, "Joseph Grünpeck's Astrological Explanation of the French Disease," in *Responding to Sexual Disease in Early-Modern Europe*, ed. Kevin Siena (Toronto: Centre for Renaissance and Reformation Studies, 2005), 241–74. See also Paul Albert Russel, "Syphilis, God's Scourge or Nature's Vengeance? The German Printed Response to a Public Problem in the Early Sixteenth Century," *Archiv für Reformationsgeschichte* 80 (1989): 286–306; Paul Albert Russel, "Astrology as Popular Propaganda: Expectations of the End in the German Pamphlets of Joseph Grünpeck (†1533?)," in *Forme e destinazione del messaggio religioso. Aspetti della propaganda religiosa nel cinquecento*, ed. Antonio Rotondò (Florence: Leo S. Olschki, 1991), 165–95.

41. On the relationship between these two versions and Grünpeck's final redaction, see Wiesflecker's "Joseph Grünpecks Redaktion" and *Joseph Grünpecks Commentaria*.

42. Silver, *Marketing Maximilian*, 2.

43. The manuscript is in the Haus-, Hof- und Staatsarchiv in Vienna, ms. Böhm no. 24. Many of the illustrations have been reproduced in Otto Benesch and Erwin M. Auer, *Die Historia Friderici et Maximiliani* (Berlin: Deutscher Verein für Kunstwissenschaft, 1957).

44. Silver, *Marketing Maximilian*, 2; Hans Mielke, *Albrecht Altdorfer. Zeichnungen. Deckfarbenmalerei. Druckgraphik* (Berlin: Reimer Verlag, 1988), 68–83.

45. "per ocium ac quietem mathematicis disciplinis operam dedit, siderum motus, terre et maris condiciones, et tocius mundi diversitatem ab eiusce artis magistris exactissime sciscitatus est, at tantamque coelestis discipline cognitionem pervenit, quod ex stellarum congressibus pleraque futura euidit, extant proprii chirographi vaticinia de Maximiliani filii tocius vite successibus et fine, . . . visuntur et monumenta manuum suarum regiis in biblithecis, quibus ex genituris quorundam regum naturas et mores definit." Joseph Grünpeck, *Historia Friderici IV. et Maximiliani I. ab Jos. Grünbeck*, ed. Joseph Chmel 1838), 72–73.

46. Grünpeck, *Historia Friderici IV. et Maximiliani I.*, 81.

47. Grünpeck, *Historia Friderici IV. et Maximiliani I.*, 81–82.

48. Joseph Grünpeck, "Lebensbeschreibung Kayser Friederichs des III. und Maximilians I," cod. lat. 7419, 37v–8r.

49. The competing models of interpretation were by no means easily distinguished. Nor did the prophetic model fade. For a discussion of different models as they relate to late antiquity, see David Potter, *Prophets and Emperors. Human and Divine Authority from Augustus to Theodosius* (Cambridge, MA: Harvard University Press, 1994), 15–17. Simon de Phares dismissed subjective prognostication in his *Recueil des plus célèbres astrologues*. Analyzing Simon's text, Jean-Patrice Boudet argues that prophecy was not respected at courts and among educated astrologers. Boudet recognizes, however, that Simon's rejection of popular prophecy might not have been universal. Gabriella Zarri finds prophets at a number of Italian courts. See Jean-Patrice Boudet, "Simon de Phares et les rapports entre astrologie et prophétie à la fin du Moyen Âge," in *Les Textes prophétiques et la prophétie en occident (XIIe–XVIe siècle)*, ed. André Vauchez (Rome: Ecole française de Rome, 1990), 327–52, esp. n. 39; Gabriella Zarri, "Les Prophètes de cour dans l'Italie de la Renais-

sance," in ibid., 359–85. Although Simon rejected one form of prophecy—subjective divination and its reliance on an inspired individual—he was a proponent of astrology and other forms of inductive divination. For his rejection of inspired individuals, see Boudet, "Simon de Phares," 336.

50. On Brant's political propaganda, see chapter 2 of this volume.

51. On the importance and significance of the Maximilian's white armor, see Burger, "Der *Weisskunig* als Literaturdenkmal," 19.

52. On the *Weisskunig* as a mirror of princes, see Georg Misch, "Die Stilisierung des eigenen Lebens in dem Ruhmeswerk Kaiser Maximilians, des letzten Ritters," *Nachrichten von der Gesellschaft der Wissenschaften zu Göttingen* (1930): 435–59; Marjorie Dale Wade, "The Education of the Prince: A Mirror of Reality and Romance in Maximilian's *Weisskunig*" (PhD diss., University of Michigan, 1974).

53. Maximilian even tried to communicate his mandates and other official proclamations to as broad an audience as possible, relying on *Volkslieder* to spread his message among illiterate citizens. See Manfred Holleger, *"Erwachen vnd Aufsten als ein starcker Stryter.* Zu Formen und Inhalt der Propaganda Maximilians I," *Propaganda, Kommunikation und Öffentlichkeit* 6 (2002), 226.

54. Maximilian had at least one copy of Hartlieb's translation of *Alexander* in his personal library. See Theodor Gottlieb, *Büchersammlung Kaiser Maximilians* (Leipzig, 1900), 103. For a discussion of Maximilian's intentions to produce a *Volksbuch* edition of the *Weisskunig*, see Burger, "Der *Weisskunig* als Literaturdenkmal," 19–20.

55. On Peutinger, see Josef Bellot, "Konrad Peutinger und die literarisch-künstlerischen Unternehmungen Kaiser Maximilians I," *Philobiblon* 11 (1967): 171–90. On Peutinger's role as supervisor and participant, see Silver, *Marketing Maximilian*, 4–6.

56. See Pia F. Cuneo, "Images of Warfare as Political Legitimization: Jörg Breu the Elder's Rondels for Maximlian I's Hunting Lodge at Lermos (ca. 1516)," in *Artful Armies, Beautiful Battles. Art and Warfare in Early Modern Europe*, ed. Pia F. Cuneo, History of Warfare Series (Leiden: Brill, 2002), 97.

57. Musper, *Kaiser Maximilians I. Weisskunig*, 219.

58. Musper, *Kaiser Maximilians I. Weisskunig*, 219.

59. "Nun was der alt weiß kunig gar kunstreich in dem erkennen des gestirns und erkennet durch den einflus und aus dem regirer des himelzirkls, darunder das kind geporen was, das dasselb kind in diser welt in die höchst regirung kumen, und durch ine vil wunderlich sachn und grosse streit beschehen sollen." Musper, *Kaiser Maximilians I. Weisskunig*, 220.

60. For numerous examples of comets presaging deaths, famines, and disasters, see Hartmann Schedel, "The Nuremberg Chronicle. A Facsimile of Hartmann Schedel's *Buch der Chroniken*," (1493), e.g., fol. 76r, 157r, 167v, 220r.

61. Johannes Hartlieb, "Alexander," Cod. Pal. germ. 88, 19v–20r.

62. Larry Silver, "Nature and Nature's God: Landscape and Cosmos of Albrecht Altdorfer," *Art Bulletin* 81 (1999): 194–214.

63. Musper, *Kaiser Maximilians I. Weisskunig*, 223.

64. Musper, *Kaiser Maximilians I. Weisskunig*, 223.

65. When Maximilian explains *"gehaim wissen und erfarung der welt"* he includes a

discussion of religious hierarchies, secular estates, and a summary of five important subjects he needs to master to be an effective ruler: "der erst, von der almechtigkeit gots; der ander, von dem einfluss der planeten; der drit, von der vernunft des menschn; der viert, von der zu vil senftmuetigkeit in der regierung; der funft artikl, zu streng in dem gewalt." Musper, *Kaiser Maximilians I. Weisskunig*, 223.

66. "Nachdem und der jung weiß kunig nun das haimlich wissen der erfarung der welt bewegt und zu gueter maß funden het, wie vor davon geschriben ist, da gedacht er in im selbs, wie ime kunftiglichen notthun wurde, die stern und einflus mit irer wurkung zu erkennen; sonst mocht er die natur der menschen nit volkumenlich erlernen, das im in dem haimlichen wissen der erfarung der welt ain mangl sein wurde." Musper, *Kaiser Maximilians I. Weisskunig*, 224. Maximilian's use of "haimlich" is related to his use of "gehaim" and similar terms he used in the previous chapter of *Weisskunig*, where context indicates that these terms pick out knowledge that Maximilian considers central to ruling. Hans-Otto Burger's modern German translations for these various terms underscore the political uses of this *gehaim wissen und erfarung der welt*. Burger translates *haimlich wissen* as "politisches Wissen" and "gehaim wissen und erfahrung der welt" as "politische Wissenschaft." See Musper, *Kaiser Maximilians I. Weisskunig*, 328–29.

67. See Steven J. Williams, *The Secret of Secrets: The Scholarly Career of a Pseudo-Aristotelian Text in the Latin Middle Ages* (Ann Arbor: University of Michigan Press, 2003); W. F. Ryan and Charles Schmitt, eds., *Pseudo-Aristotle, The Secret of Secrets: Sources and Influences*, vol. 9, Warburg Institute Surveys and Texts (London: Warburg Institute, 1982). Nine English versions of the *Secretum* have been published recently in M. A. Manzalaoui, ed., *Secretum secretorum: Nine English Versions*, Early English Text Society no. 276 (Oxford: Oxford University Press, 1977).

68. See the chapter titled "Of the Dyfference of Astronomy," in Robert Copland's English translation Manzalaoui, *Secretum secretorum*, 331–33. Maximilian owned at least one version of this text, "Liber thesaurorum ad regem Alexandrum seu Secretum secretorum," Österreichische Nationalbibliothek cod. lat. 2476. On Maximilian's ownership, see Gottlieb, *Büchersammlung Kaiser Maximilians*, 106.

69. Musper, *Kaiser Maximilians I. Weisskunig*, 224.

70. See Maximilian, *Die Abenteuer des Ritters Theuerdank von 1517* (Cologne: Taschen, 2003).

71. Silver, *Marketing Maximilian*, 7.

72. On these associations, see Burger, "Der *Weisskunig* als Literaturdenkmal," 17.

73. "Der Ernhold bedeut das gerucht und gezeügnus der warhait so einem yeden menschen bis in sein grüben nachvolgt. Sy sein güt / oder pösz / darumb wirdet. Er bemeltem Jungen Fürsten Tewerdank für / und für zügestellt / sein leben wesen unnd getaten züoffenwaren und zubezeuugen mit der warhait." Melchior Pfinzing, "Clavis," in Maximilian, *Die Abenteuer des Ritters Theuerdank von 1517*, facsimile ed. (Cologne: Taschen, 2003), A2r.

74. For the importance of Virgil for Maximilian's political propaganda, see, for example, Marie Tanner, *The Last Descendant of Aeneas. The Habsburgs and the Mythic Image of the Emperor* (New Haven, CT: Yale University Press, 1993); E. L. Harrison, "Virgil, Sebastian Brant, and Maximilian I," *Modern Language Review* 76 (1981): 99–115.

75. See, for example, Dieter Blume, *Regenten des Himmels. Astrologische Bilder in Mit-

telalter und Renaissance, vol. 3, Studien aus dem Warburg-Haus (Berlin: Akademie, 2000), 158–94.

76. For the use of *Planetenkinder* on churches, see Amelia J. Carr and Richard L. Kremer, "Child of Saturn. The Renaissance Church Tower at Nideraltaich," *Sixteenth Century Journal* 17 (1986): 401–34.

77. Maximilian, *Die Abenteuer des Ritters Theuerdank von 1517*, ch. 25.

78. Maximilian, *Die Abenteuer des Ritters Theuerdank*, ch. 52.

79. Maximilian, *Die Abenteuer des Ritters Theuerdank*, chs. 70, 83, and 92.

80. On the role of astrological pamphlets in Maximilian's political agenda, see chapters 5 and 6 of this volume.

Chapter Two. Astrology as Imperial Propaganda

1. "Der jung weiß kunig fraget in seiner jugent gar oft von den kuniglichn geschlechten, dann er het gern gewist, wie ain jedes kuniglich und furstlich geschlecht von anfang herkumen were, darinnen er in seiner jugent kain erkundigung erfragen möcht. Darab er dann oft ainen verdrieß trueg, das die menschen der gedächtnuss so wenig acht nämen. Und als er zu seinen jarn kam, sparet er kainen kosten, sonder er schicket aus gelert leut, die nichts anders teten, dann das sy sich in allen stiften, klostern, puechern und bey gelerten leutn erkundigeten alle geschlecht der kunig und fursten und ließ solichs alles in schrift bringen zu er und lob denen kuniglichn und furstlichn geschlechten. In sölicher erkundigung hat er erfundn sein mandlich geschlect von ainem vater auf den andern biß auf den Noe, das sonst ganz undertruckt und die alten schriften, darauf nichts mer geacht worden ist, verloren weren worden." Maximilian, *Kaiser Maximilians Weisskunig*, vol. 1: *Textband* (Stuttgart: W. Kohlhammer Verlag, 1956), 225.

2. Laschitzer provides a summary of twenty of these. See Simon Laschitzer, "Die Geneologie des Kaisers Maximilian I," *Jahrbuch der kunsthistorischen Sammlungen des allerhöchsten Kaiserhauses* 7 (1888), 31–39.

3. Glenn Elwood Waas, *The Legendary Character of Kaiser Maximilian I* (New York: Columbia University Press, 1941), 118; Joseph Ritter von Aschbach, *Die Wiener Universität und ihre Humanisten im Zeitalter Kaiser Maximilians I.* (Vienna: Wilhelm Braumüller, 1877), 369.

4. On Trithemius, see Noel L. Brann, *Trithemius and Magical Theology. A Chapter in the Controversies over Occult Studies in Early Modern Europe* (Albany: State University of New York Press, 1999); Michael Kuper, *Johannes Trithemius: Der schwarze Abt* (Berlin: Clemens Zerling, 1998).

5. "unius Abbatis Sponhaimensis fulatur, quem Ego non pro historico sed fabulatore omnium fabulosissimo reputo." Johannes Stabius, "Conclusiones super genealogiis Domus Austriacae," Vienna, Österreichische Nationalbibliothek cod. lat. 3327, 5v.

6. On Maximilian's genealogical projects and their political functions, see Larry Silver, *Marketing Maximilian. The Visual Ideology of a Holy Roman Emperor* (Princeton, NJ: Princeton University Press, 2008), esp. ch. 2.

7. E. L. Harrison, "Virgil, Sebastian Brant, and Maximilian I," *Modern Language Review* 76 (1981): 99–115, esp. 100.

8. Dieter Wuttke, "Sebastian Brant und Maximilian I. Eine Studie zu Brants Donnerstein-Flugblatt des Jarhes 1492," in *Die Humanisten in ihrer politischen und sozialen Umwelt*, ed. Otto Herding and Robert Stupperich (Soppard: Harald Boldt Verlag, 1976), 154–55; Harrison, "Virgil, Sebastian Brant, and Maximilian I," 103–5.

9. Wuttke, "Sebastian Brant und Maximilian I," 158–60.

10. Jan-Dirk Müller, "Poet, Prophet, Politiker: Sebastian Brant als Publizist und die Rolle der laikalen Intelligenz um 1500," *Zeitschrift für Literaturwissenschaft und Linguistik* 10 (1980): 108–11; Dieter Wuttke, "Erzaugur des heiligen römischen Reiches deutscher Nation: Sebastian Brant deutet siamesische Tiergeburten," *Humanistica Lovaniensia* 43 (1994): 110.

11. Maximilian convened the diet at Worms but later moved it to Freiburg. He complained that the princes were resisting his efforts to secure support and hoped that he would have greater influence if the diet was relocated to a city under his control. Hermann Wiesflecker, *Kaiser Maximilian I. Das Reich, Österreich und Europa an der Wende zur Neuzeit*, vol. 2: *Reichsreform und Kaiserpolitik. 1493–1500. Entmachtung des Königs im Reich und in Europa* (Munich: R. Oldenbourg Verlag, 1975), 271–301.

12. On pamphlets, see the work by Hans-Joachim Kohler, especially "Die Flugschriften. Versuch der Präzisierung eines geläufigen Begriffs," in *Festgabe für Ernst Walter Zeeden zum 60. Geburtstag*, ed. Horst Rabe, Hansgeorg Molitor, and Hans-Christoph Rublack (Munster: Aschendorff, 1976), 36–61; and "The Flugschriften and Their Importance in Religious Debate: A Quantitative Approach," in *"Astrologi hallucinati." Stars and the End of the World in Luther's Time*, ed. Paolo Zambelli (Berlin: Walter de Gruyter, 1986), 153–75. On Maximilian's use of pamphlets during the Hebrew debate about Reuchlin and Pfefferkorn, see Heiko A. Oberman, "Zwischen Agitation und Reformation: Die Flugschriften als 'Judenspiegel,'" in *Flugschriften als Massenmedium der Reformationszeit. Beiträge zum Tübinger Symposium 1980*, ed. Hans-Joachim Kohler, Spätmittelalter und Frühe Neuzeit, vol. 13 (Stuttgart: Klett-Cotta, 1981), 269–89; Johannes Schwitalla, "Deutsche Flugschriften im ersten Viertel des 16. Jahrhunderts," *Freiburger Universiätsblätter* 76 (1982): 37–58.

13. Ottavia Niccoli, *Prophecy and People in Renaissance Italy*, trans. Lydia G. Cochrane (Princeton, NJ: Princeton University Press, 1990); Richard L. Kagan, *Lucretia's Dreams. Politics and Prophecy in Sixteenth-Century Spain* (Berkeley: University of California, 1990); Germanna Ernst, "Astrology, Religion and Politics in Counter-Reformation Rome," in *Science, Culture and Popular Belief in Renaissance Europe*, ed. Stephen Pumphrey, Paolo L. Rossi, and Maurice Slawinski (Manchester: Manchester University Press, 1991), 249–73; Patrick Curry, "Astrology in Early Modern England: The Making of a Vulgar Knowledge," in *Astrology, Religion and Politics in Counter-Reformation Rome*, ed. Stephen Pumphrey, Paolo L. Rossi, and Maurice Slawinski (Manchester: Manchester University Press, 1991), 274–91; Patrick Curry, *Prophecy and Power. Astrology in Early Modern England* (Princeton, NJ: Princeton University Press, 1989); Bernard Capp, *English Almanacs, 1500–1800. Astrology and the Popular Press* (Ithaca, NY: Cornell University Press, 1979), esp. 67–101.

14. Jan R. Veenstra, *Magic and Divination at the Courts of Burgundy and France. Text and Context of Laurens Pignon's Contre les devineurs (1411)* (Leiden: Brill, 1998); Hilary M.

Carey, *Courting Disaster. Astrology at the English Court and University in the Later Middle Ages* (New York: St. Martin's Press, 1992); Laura Ackerman Smoller, *History, Prophecy, and the Stars. The Christian Astrology of Pierre d'Ailly, 1350–1420.* Girolamo Cardano illustrates how authors accommodated their works to different audiences. See Anthony Grafton, *Cardano's Cosmos. The Worlds and Works of a Renaissance Astrologer* (Cambridge, MA: Harvard University Press, 1999), esp. 109–27.

15. Sebastian Brant, "De corrupto ordine vivendi pereuntibus," ll. 29–194 in Sebastian Brant, *Stultifera navis* (Basel, 1498), 145v.

16. Brant, *Stultifera navis*, ll. 290–304.

17. "Credite Germani mox tempora plena periclis. Ventura; & magnis cuncta replenda malisHeu quantum vereor ne nos fata impia tangant. Et sceptrum a nobis imperiumque trahant. Aspicite hanc caeli quam cernitis oro figuram: Se cancro iungunt sydera saeva nimis Saturnum/Martem/atque Iovem variabile signum Coniungens cancri; friget & humet aquis, Indicat atque senes/iuvenesque & Martia corda. Et clerum/instabili mobilitate frui. Quaeque agere incipient: quicquid placet: ocius illud Prorepet cancri more/ modoque retro Inde graves clades: patriae & co[mmun]is Erynnis Et mala provenient: dii prohibete minas. Multa quidem nobis astra/&fera fata minantur: Cogitat at nulls tam prope adesse diem. Tempus erit: sceptum a nobis tolletur / & ibit Longius: ah saltem theuthona terra dole. Quis mihi / quis lachrymas dabit: ut deflere ruinam Communem possim: vel gemere interitum?" Brant, *Stultifera navis*, ll. 305–22.

18. Brant, *Stultifera navis*, ll. 325–40; ll. 349–98; ll. 449–78.

19. By the late fifteenth century images of the world turned upside down became a common tool used by social and religious reformers. Ernst Roberts Curtius, *European Literature and the Latin Middle Ages* (Princeton, NJ: Princeton University Press, 1988), 94–98; R. W. Scribner, *For the Sake of Simple Folk. Popular Propaganda for the German Reformation* (Oxford: Oxford University Press, 1994), 148–89.

20. There is no tension between Brant's use of astrology here and his apparent rejection of astrology and false prophets found elsewhere in his *Ship of Fools.* See Dieter Wuttke, "Sebastian Brants Verhältnis zu Wunderdeutung und Astrologie," in *Studien zur deutschen Literatur und Sprache des Mittelalters. Festschrift für Hugo Moser zum 65. Geburtstag,* ed. Werner Besch, Günther Jungbluth, Gerhard Meissburger, and Eberhard Nellmann (Berlin: Erich Schmidt Verlag, 1974), 272–86.

21. Modern critical editions and translations of Albumasar's works include editions of the medieval Latin translations. See Abū Maʿšar, *The Abbreviation of the Introduction to Astrology: Together with the Medieval Latin Translation of Adelard of Bath* (Leiden: Brill, 1994); Abū Maʿšar, *On Historical Astrology*; Abū Maʿšar, *Liber introductorii maioris ad scientiam judiciorum astrorum,* 9 vols. (Naples: Instituto universitario orientale, 1995–1996). On Alcabitius, see Alcabitius, *Al-Qabīsī (Alcabitius): The Introduction to Astrology,* vol. 2, Warburg Institute Studies and Texts (London: Warburg Institute, 2004).

22. Österreichische Nationalbibliothek cod. lat. 5318 contains Arnold of Friburg's translation that had been checked and corrected by Burkharten Kechk of Salzburg in 1474. Alcabitius, "Isagoge in astrorum iudicia ab Arnoldo de Friburgo in germanicam linguam translata," Vienna, Österreichische Nationalbibliothek cod. lat. 5318, 107r–28v.

23. For the manuscript tradition and printed editions of Albumasar's *Conjunctio*

magnis, see Abū Ma'šar, *On Historical Astrology. The Book of Religions and Dynasties (On the Great Conjunctions)*, trans. Keiji Yamamoto and Charles Burnett, vol. 2: *The Latin Versions* (Leiden: Brill, 2000), xi–xxx. For Alcabitius, see *Al-Qabīsī (Alcabitius)*, 156–98.

24. The best introduction to this theory is J. D. North, "Astrology and the Fortunes of Churches," in *Stars, Minds and Fate: Essays in Ancient and Medieval Cosmology*, ed. J. D. North (London: Hambledon, 1989), 59–89.

25. Alcabitius, *Liber isagogicus Alchabitii de planetarum conjunctionibus* (Venice, 1485), cc8r.

26. An excellent example of this eclectic tradition is found in an early sixteenth-century copy of Alcabitius. Throughout this text a reader has added marginalia to indicate where Alcabitius's text agrees with or is in tension with Albumasar, Haly Abenragel, Leopold of Austria, Abubacher, Ptolemy, and Almansor. The text itself cites Messahalah and Dorotheus of Sidon. See Alcabitius, "De iudiciis astrorum interprete Johanne Hispalensi," Vienna, Österreichische Nationalbibliothek cod. lat. 5275, 221r–54v.

27. On d'Ailly, see Smoller, *History, Prophecy, and the Stars*.

28. For example, d'Ailly's texts were bound between a copy of "Liber de locis stellarum fixarum cum ymaginibus" and Arnold of Friburg's translation of Alcabitius. See Österreichische Nationalbibliothek cod. lat. 5318, 38r–105v.

29. Dietrich Kurze, "Johannes Lichtenberger (†1503): Eine Studie zur Geschichte der Prophetie und Astrologie" (1960).

30. Steven Vanden Broecke has argued that Pico della Mirandola's famous critique of astrology, which included an attack on conjunctionist astrology, had a considerable impact by the early sixteenth century. Within the Viennese context, the first clear evidence of any direct engagement with Pico's critique does not occur until 1517 when Georg Tannstetter dismisses easily Pico's attack on astrology in his *judicium* for that year. See Steven Vanden Broecke, *The Limits of Influence: Pico, Louvain, and the Crisis of Renaissance Astrology* (Leiden: Brill, 2003). For Tannstetter's comment, see his *Judicium Astronomicum Viennense. anni M.CCCCC.xvij. Ad nobiles et providentissimos dños Magistrum civium et universum inclite urbis Vienne Senatum per Georgium Tannstetter Collimitium artium et medicine doctorem diligenter elaboratum* (Vienna: Johannes Singrenius, 1516), Vienna, Universitätsbibliothek I.545.631 ES., A2v.

31. Christine E. Ineichen-Eder, "A Computus Notebook by Sebastian Brant (CLM 26618)," *Scriptorium*. 35, no. 1 (1981): 91–95.

32. Johannes Hispalensis, "Tractatus de signis coelestibus eorumque effectibus," Österreichische Nationalbibliothek cod. lat. 5463, f. 155r–v, 157v. John of Seville's ambiguity is understandable, as the Arab sources themselves were not clear on which method to employ. For example, Latin translations of Albumasar often included variations in the chorography. See Albumasar, "Liber magnarum conjunctionum," Vienna, Österreichische Nationalbibliothek cod. lat. 2436, f. 250v–256r. The editors of the modern editions of Albumasar's *On Great Conjunctions* have highlighted the ambiguities in the chorographic tradition, see Abū Ma'šar, *On Historical Astrology. The Book of Religions and Dynasties (On the Great Conjunctions)*, trans. Keiji Yamamoto and Charles Burnett, vol. 1: *The Arabic Original* (Leiden: Brill, 2000), 513–19, 606.

33. See, for example, Claudius Ptolemy, *Quadripartitum in Julius Firmicus Maternus*,

Astronomicon Lib. VIII. per Nicolaum Prucknerum Astrologum nuper ab innumeris mendis vindicati. His accesserun. CLAVDII PTOLEMAEI ἀποτελεσμάτων, *quod Quadripartitum vocant, Lib. IIII* (Basel, 1533), 11. See also Alcabitius, "Isagoge," 108r.

34. Albumasar, *De magnis coniunctionibus* (Venice, 1515), C1r, C1v–C2r.

35. Brant, *Stultifera navis*, ll. 309–24, 419–48.

36. Brant, *Stultifera navis*, ll. 455–82, 505–24.

37. Brant, *Stultifera navis*, ll. 505–12.

38. Brant, *Stultifera navis*, l. 580. On the emendation to the text to read "Alcathoe," see Harrison, "Virgil, Sebastian Brant, and Maximilian I."; H. B. Gottschalk, "The Conclusion of Brant's 'De corrupto ordine vivendi pereuntibus,'" *Modern Language Review* 77 (1982): 348–50.

39. Wiesflecker, *Kaiser Maximilian I*, 217–49.

40. Hermann Wiesflecker, *Kaiser Maximilian I. Das Reich, Österreich und Europa an der Wende zur Neuzeit*, vol. 5: *Der Kaiser und seine Umwelt. Hof, Staat, Wirtschaft, Gesellschaft und Kultur* (Munich: R. Oldenbourg Verlag, 1986), 344; Dieter Mertens, "Maximilian I. und das Elsass," in *Die Humanisten in ihrer politischen und sozialen Umwelt*, ed. Otto Herding and Robert Stupperich (Boppard-am-Rhein: Harald Boldt Verlag, 1976), 177–201.

41. Wuttke, "Sebastian Brants Verhältnis," 281.

42. On Grünpeck, see Albin Czerny, "Der Humanist und Historiograph Kaiser Maximilians I. Joseph Grünpeck," *Archiv für österreichische Geschichte* 73 (1888): 315–64. See also Paul Albert Russel, "Astrology as Popular Propaganda: Expectations of the End in the German Pamphlets of Joseph Grünpeck (†1533?)," in *Forme e destinazione del messaggio religioso. Aspetti della propaganda religiosa nel cinquecento*, ed. Antonio Rotondò (Florence: Leo S. Olschki, 1991), 165–95.

43. On the popularity of Pavia for German students, see Paul Grendler, *The Universities of the Italian Renaissance* (Baltimore, MD: Johns Hopkins University Press, 2002), 82–92.

44. For example, the famous Polish astrologer Martin Bylica graduated in 1463 from Krakow before ending up at the Hungarian court of Mathias Corvinus. See Darin Hayton, "Expertise *Ex Stellis*: Comets, Horoscopes, and Politics in Renaissance Hungary," *Osiris* 25 (2010): 27–46. See Mieczyslaw Markowski, "Die Astrologie an der Krakauer Universität in den Jahren 1450–1550," in *Magia, Astrologia e Religione nel Rinascimento* (Wrocław, 1974), 83–89.

45. Grünpeck's letter is reprinted in Czerny, "Der Humanist und Historiograph Kaiser Maximilians I," 355–57.

46. Bernhard Waldkirch and Joseph Grünpeck to Conrad Celtis, 29 October 1496. Hans Rupprich, ed., *Der Briefwechsel des Konrad Celtis* (Munich: C. H. Beck, 1934), 224–25.

47. Wiesflecker, *Kaiser Maximilian I*, 241–56.

48. Joseph Grünpeck, "Prognostikon für 1496–1499," Munich, Bayerische Staatsbibliothek, cgm. 3042, 1v–3r.

49. On Lichtenberger, see Kurze, "Johannes Lichtenberger (†1503)"; Dietrich Kurze, "Prophecy and History: Lichtenberger's Forecasts of Events to Come (From the Fifteenth

to the Twentieth Century); Their Reception and Diffusion," *Journal of the Warburg and Courtauld Institutes* 21 (1958): 63–85; Dietrich Kurze, "Popular Astrology and Prophecy in the Fifteenth and Sixteenth Centuries: Johannes Lichtenberger," in *"Astrologi hallucinati,"* 177–93.

50. The best account of this diet, from Maximilian's perspective is Wiesflecker, *Kaiser Maximilian I*, 201–56.

51. For Maximilian's trials in Italy during the 1490s, see Wiesflecker, *Kaiser Maximilian I*, 9–122.

52. Joseph Grünpeck, "Prognostikon für 1496–1499," 4r–4v.

53. Sections in his *judicium* that deal with this: "Principes alemanie insurgent contra regem romanorum, Ca. x," "De tribulacione quam habebunt electores imperij. episcopus coloniensis et treverensis, Ca. xv," "De prelatis sew Episcopis qui erunt rebelles sumo pontifici, Ca. xvi." Joseph Grünpeck, *Pronosticon sive (ut alij volunt) Judicium Ex coniuntione Saturni et Jovis Decennalique revolutione Saturni Ortu et fine antichristi ac alijs quibusdam interpositis prout exsequentibus claret praeambulis hic inseritur* (Vienna: Johannes Winterburg, 1496), Vienna, Österreichische Nationalbibliothek Ink 17.H.13., n.p.

54. The literature on prophecy in the Middle Ages is immense. The standard work remains Marjorie Reeves, *The Influence of Prophecy in the Later Middle Ages: A Study in Joachimism* (South Bend, IN: University of Notre Dame, 1993). See also Kurze, "Johannes Lichtenberger (†1503)" and Michael Shank, "Academic Consulting in 15th-Century Vienna: The Case of Astrology," in *Texts and Contexts in Ancient and Medieval Science. Studies on the Occasion of John E. Murdoch's Seventieth Birthday*, ed. Edith Sylla and Michael McVaugh (Leiden: Brill, 1997), 245–70, esp. 258–59 and 266–67; Marie Tanner, *The Last Descendant of Aeneas. The Habsburgs and the Mythic Image of the Emperor* (New Haven, CT: Yale University Press, 1993), ch. 6.

55. Joseph Grünpeck, "Prognostikon für 1496–1499," 10r.

56. Joseph Grünpeck, *Pronosticon*, n.p.

57. "und yetzo bey unsern zeiten, als offenbar ist, dergleich vil und menigerley plagen und strafen gevolgt haben und sunderlich in disen tagen swer krankheiten und plagen der menschen, genant die pösen plattern, die vormals bey menschengedechtnüs nye gewesen noch gehört sein." Heinz Angermeier, ed., *Deutsche Reichstagsakten unter Maximilian I.*, vol. 5: *Reichstag von Worms 1495* (Göttingen: Vandenhoek & Ruprecht, 1981), 576. The Latin text is available in C. H. Fuchs, ed., *Die ältesten Schriftsteller über die Lustseuche in Deutchland, von 1495 bis 1510, nebst mehreren Anecdotis späterer Zeit, gesammelt und mit literarhistorischen Notizen und einer kurzen Darstellung der epidemischen Syphilis in Deutschland* (Göttingen: Dieterischschen Buchhandlung, 1843), 305–6.

58. There is a brief analysis of the Latin version in Jon Arrizabalaga, John Henderson, and Roger French, *The Great Pox: The French Disease in Renaissance Europe* (New Haven, CT: Yale University Press, 1997), 107–12. Paul Albert Russell has examined the German version in "Syphilis, God's Scourge or Nature's Vengeance? The German Printed Response to a Public Problem in the Early Sixteenth Century," *Archiv für Reformationsgeschichte* 80 (1989): 286–306; Russel, "Astrology as Popular Propaganda."

59. The most famous of these *concilia* was written by the renowned Italian physician Gentile da Foglio. Nancy Siraisi, *Medieval and Early Renaissance Medicine: An Intro-*

duction to Knowledge and Practice (Chicago: University of Chicago Press, 1990), 128–29. For an analysis of the contents of some early plague tractates, see Anna Montgomery Campbell, *The Black Death and Men of Learning* (New York: AMS Press, 1966), 34–92. Russell compares Grünpeck's work to Gentile's in "Syphilis, God's Scourge or Nature's Vengeance?" 290–91. On Gentile, see Roger French, *Canonical Medicine. Gentile da Foglio and Scholasticism* (Leiden: Brill, 2001).

60. Joseph Grünpeck, *Tractatus de pestilentiali Scorra sive mala de Franzos. Originem. Remediaque eiusdem continens. compilatus a venerabili viro Magistro Joseph Grunpeck de Burckhausenn. super Carmina quedam Sebastiani Brant vtriusque iuris professoris* (Augsburg: Hans Schauer, 1496), A6r–A6v.

61. Grünpeck, *Tractatus de pestilentiali Scorra*, A6v.

62. For a lucid discussion of how the plague could be seen both as a punishment from God and a product of natural causes, see Thilo Esser, "Die Pest—Strafe Gottes oder Naturphänomen? Eine frömmigkeitsgeschichtliche Untersuchung zu Pesttraktaten des 15. Jahrhunderts," *Zeitschrift für Kirchengeschichte* 108 (1997): 32–57. Esser mischaracterizes Grünpeck's tracts as about the plague: "Der Traktat handelt nicht, wie der Titel vermuten ließe, nur von der Franzosenkrannkheit (Sphilis), sondern auch von der Pest." Esser, "Die Pest," 39n.

63. Grünpeck, *Tractatus de pestilentiali Scorra*, A8r.

64. Julius Firmicus Maternus, *Matheseos libri VIII*, bk. 3, chap. 1.

65. For a fuller account of d'Ailly's *thema mundi*, see Smoller, *History, Prophecy, and the Stars. The Christian Astrology of Pierre d'Ailly, 1350–1420,* 65–67.

66. Joseph Grünpeck, *Ein hübscher Tractat von dem ursprung des bösen Franzos. das man nennet die Wylden wärzen. Auch ein Regiment wie man sich regiren soll in diser zeyt* (Augsburg: Hans Schauer, 1496), B3r; Joseph Grünpeck, *Tractatus de pestilentiali Scorra*, B1r–B2r.

67. Grünpeck, *Ein hübscher Tractat*, B7r.

68. Grünpeck, *Ein hübscher Tractat*, B7r–B7v.

69. Grünpeck, *Ein hübscher Tractat*, B8r; Joseph Grünpeck, *Tractatus de pestilentiali Scorra*, C2r–C2v.

70. Grünpeck, *Ein hübscher Tractat*, B8v.

71. "die grossen übel. amm ersten des groß hunger. der wol siben jar geweret hat und noch kein ennde hat. Darnach die grawßam pestilenz. die auch noch regiret. Und der groß kryeg mit dem künige auß Franckreych. über die übel alle kommet nun die erschrockenlich grawßam krankeyt das vorgemelt böß Franzos." Grünpeck, *Ein hübscher Tractat*, C2v.

72. Grünpeck, *Ein hübscher Tractat*, B8v; and Grünpeck, *Tractatus de pestilentiali Scora*, C2v.

73. Grünpeck, *Ein hübscher Tractat*, C1r–C1v; and Grünpeck, *Tractatus de pestilentiali Scora*, C3r–C3v.

74. On humors in treating diseases, see Siraisi, *Medieval and Early Renaissance Medicine*, 115–52.

75. Grünpeck, *Ein hübscher Tractat*, B8v–C1v; Grünpeck, *Tractatus de pestilentiali Scorra*, C2v–C3r.

76. On Celtis's *Sodalitas litteraria Danubiana*, see Lewis W. Spitz, *Conrad Celtis. The German Arch-Humanist* (Cambridge, MA: Harvard University Press, 1957), 55–93. For the role of *sodalitas eruditorum*, see Jan-Dirk Müller, *Gedechtnus. Literatur und Hofgesellschaft um Maximilian I.* (Munich: Wilhelm Fink Verlag, 1982).

77. "cum dignissimi viri, omni virtute, sapientia atque doctrina praediti." Grünpeck, *Tractatus de pestilentiali Scorra*, C2r.

78. Grünpeck, *Tractatus de pestilentiali Scorra*, C4v.

79. For the introduction of humanist studies to Ingolstadt, see the now dated but still useful Gustav Bauch, *Die Anfänge des Humanismus in Ingolstadt: Eine litterarische Studie zur deutschen Universitätsgeschichte* (Munich: R. Oldenbourg, 1901). On Celtis in general, see Lewis W. Spitz, *Conrad Celtis. The German Arch-Humanist* (Cambridge, MA: Harvard University Press, 1957).

80. Jon Arrizabalaga and Roger French, "Coping with the French Disease: University Practitioners' Strategies and Tactics in the Transition from the Fifteenth to the Sixteenth Century," in *Medicine from the Black Death to the French Disease*, ed. Roger French, Jon Arrizabalaga, Andrew Cunningham, and Luis García-Ballester (Brookfield, VT: Ashgate, 1998), 251.

81. This was not the last time a courtier would cast Maximilian in the role of Hercules deciding between virtue and vice. In 1515 during the festivities surrounding the Habsburg double marriage in Vienna, the humanist Benedictus Chelidonius staged a similar play calling on both Maximilian and his grandson Charles to participate. On Chelidonius, see Franz Posset, *Renaissance Monks. Monastic Humanism in Six Biographical Sketches* (Leiden: Brill, 2005).

82. William C. McDonald, "Maximilian I of Habsburg and the Veneration of Hercules: On the Revival of Myth and the German Renaissance," *Journal of Medieval and Renaissance Studies* 6 (1976): 141–42; Dieter Wuttke, *Die Histori Herculis des Nürnberger Humanisten und Freundes der Gebrüder Vischer, Pangratz Bernhaubt gen. Schwenter* (Cologne: Böhlau Verlag, 1964), 207–8.

83. See Hermann Wiesflecker, "Joseph Grünpecks Redaktion der lateinischen Autobiographie Maximilians I," *Mitteilungen des Instituts für österreichische Geschichtsforschung* 78 (1970): 416–31; and his *Joseph Grünpecks Commentaria und Gesta Maximiliani Romanorum Regis. Die Entdeckung eines verlornen Geschichtswerkes* (Graz: Jos. A. Kieneich, 1965).

84. Maximilian, Undated Letter from Maximilian, in "Urkunden und Regesten," ed. Heinrich Zimerman and Franz Kreyczi in *Jahrbuch der Kunsthistorischen Sammlungen des Allerhochsten Kaiserhauses* 3(1885), XVI #2419.

85. See Wiesflecker, *Kaiser Maximilian I*, 261–65. On Hölzl's literary activities, see Jan-Dirk Müller, *Gedechtnus. Literatur und Hofgesellschaft um Maximilian I.* (Munich: Wilhelm Fink Verlag, 1982).

86. Johann Ramminger, "Humanist Poetry and Its Classical Models: A Collection from the Court of Emperor Maximilian I," in *Acta Conventus Neo-Latini Torontonensis. Proceedings of the Seventh International Congress of Neo-Latin Studies, Toronto 8 August to 13 August 1988*, ed. Alexander Dalzell, Charles Fantazzi, and Richard J. Schoeck, Medieval & Renaissance Texts & Studies, (Binghamton, NY: Arizona Center for Medieval and Renaissance Studies, 1991), 581–93.

87. See Dieter Wuttke, "Wunderdeutung und Politik. Zu den Auslegungen der sogenannten Wormser Zwillinge des Jarhes 1495," in *Landesgeschichte und Geistesgeschichte,* ed. Kaspar Elm, Eberhard Gönner, and Eugen Hillenbrand (Stuttgart: W. Kohlhammer Verlag, 1977), 217–44; Niccoli, *Prophecy and People,* esp. ch. 2. See also Lorraine Daston and Katherine Park, *Wonders and the Order of Nature, 1150–1750* (New York: Zone Books, 1998).

88. Joseph Grünpeck "Prodigiorum portentorum, ostentorum et monstrorum, quae in saeculum Maximilianeum inciderunt," Innsbruck, Universitätsbibliothek codex 314, 6v–8r.

89. Grünpeck, "Prodigiorum portentorum," 8r–15r.

90. Grünpeck, "Prodigiorum portentorum," 16r–24r.

91. Wuttke, "Wunderdeutung und Politik," 226–30.

92. In his "Prodigiorum potentorum" Grünpeck repeatedly pointed to these conjoined twins, though he relocated their birth from Worms to a field outside of Laudenburg. For example, Grünpeck, "Prodigiorum portentorum," 15v.

93. Grünpeck, "Prodigiorum portentorum," 15v–16r.

94. In March 1506 Maximilian arranged to pay Konrad Peutinger and Grünpeck 100 gulden for their work on his tomb. Undated letter from Maximilian, in "Urkunden und Regesten," ed. Heinrich Zimerman and Franz Kreyczi, *Jahrbuch der Kunsthistorischen Sammlungen des Allerhochsten Kaiserhauses* 3(1885), XXXI, #2592.

95. For Maximilian's agenda, see Hermann Wiesflecker, *Kaiser Maximilian I. Das Reich, Österreich und Europa an der Wende zur Neuzeit,* vol. 3: *Auf der Höhe des Lebens. 1500–1508. Der große Systemwechsel. Politischer Wiederaufstieg* (Munich: R. Oldenbourg Verlag, 1977), 350–79.

96. Peter Diederichs, *Kaiser Maximilian I als politischer Publizist* (PhD diss., University of Heidelberg, 1931), esp. 48–51. See also Stephan Füssel, *Gutenberg und seine Wirkung* (Frankfurt: Insel Verlag, 1999), 154–57.

97. Joseph Grünpeck, *Ein newe Außlegung der seltzamen Wunderzaichen und Wunderpürden* (Augsburg: Erhard Oeglin, 1507), a1v.

98. Grünpeck, *Ein newe Außlegung,* a2v–a4r.

99. For the failure of Maximilian's efforts, see Wiesflecker, *Kaiser Maximilian I,* 372–79.

100. On Maximilian's crowning in Trent, see Hermann Wiesflecker, *Kaiser Maximilian I. Das Reich, Österreich und Europa an der Wende zur Neuzeit,* vol. 4: *Gründung des habsburgischen Weltreiches, Lebensabend und Tod, 1508–1519* (Munich: R. Oldenbourg Verlag, 1981), 6–15.

101. Joseph Grünpeck, *Speculum naturalis cœlestis & propheticæ visionis: omnium calamitatum tribulationum & anxietatum: quæ super omnes status: stirpes & nationes christianæ reipublice: presertim quæ cancro & septimo climati subiecte sunt: proximis temporibus venture sunt* (Nuremberg: Georg Stuchs, 1508), a2v.

102. For example, Grünpeck, *Speculum naturalis cœlestis,* a2v–a3r, b2v–b4r. About this time, Maximilian seemed less convinced by singular portents; he crossed out the illustrations in the draft of Grünpeck's *Historia* that foregrounded prodigious singular events such as rains of blood and crosses.

103. Grünpeck, *Speculum naturalis cœlestis,* a3r–a5v.

104. For an excellent study of Julius's efforts to reestablish Roman authority, see Ingrid D. Roland, *The Culture of the High Renaissance. Ancients and Moderns in Sixteenth-Century Rome* (Cambridge: Cambridge University Press, 1998), esp. ch. 6.

105. Grünpeck, *Speculum naturalis cœlestis*, a6v–b1r.

106. Willibald Pirckheimer, Willibald Pirckheimer to Maximilian, 1508, in *Willibald Pirckheimers Briefwechsel*, ed. Emil Reicke (Munich: C. H. Beck'sche Verlagsbuchhandlung, 1956), 190–91.

107. Grünpeck's *Spiegel* appeared in 1508 (in Nuremberg), 1510 (in Augsburg), 1513 (in Augsburg), 1515 (in Augsburg), and 1522 (in Leipzig and Augsburg). His two later works based on the *Spiegel* were *Ain nuzliche Betrachtund der natürlichen, hymlischen und prophetischen, ansehungen aller trübsalen, angst, und not, die über all stände geschlechte und gemainden der Christenhait in kurzen tagen geen werden* (Augsburg: Hans Schönsperger, 1522) and *Practica der gegenwertigen grossen Trübsalen, und vilfaltiger Wunder* (Strasbourg: M. Jakob Cammerlander, 1533).

108. Grünpeck's horoscope for Regensburg is preserved in Joseph Grünpeck, "Horoscope for Regensburg," Bayerische Staatsbibliothek Cgm 1502.

109. The copy of this horoscope preserved in ÖNB cod. lat. 8489 was to be a richly illustrated presentation copy, but was not completed. Only the first illustration was added to the text. Joseph Grünpeck, "Maximiliani I Imperatoris Genethliacon germanicum cum praefatione et cum tabula picta familiam caesarem repraesentante," Vienna, Österreichische Nationalbibliothek cod. lat. 8489.

110. Joseph Grünpeck, *Pronostication Doctor Joseph Grünpecks Vom zway und dreyssigsten Jar an biß auff das Vierzigst Jar des aller durchleüchtigsten groß mächtisten Kaiser Carols des fünfften rc. und Begreifft in jr vil zükünfftiger Historien* (n.p., 1532), a4r. This work appeared simultaneously in Latin and in German.

111. "Nec te rursus (ut arbitror) latet, me abhinc decennium circiter haec cum saepe tum vel maximie in Fragmentis nostris Narragonicis de perverso rerum ordine liquido praecinuisse, quin figuram constellationis huius pestiferae coniunctionis errantium siderum superiorum praescripsi. Qui sive significationis aliquid habeant astrorum influxus, ut multi, sive nihil eorum coniuncti vel oppositi affectus portendant, ut Pico nostro placuit, certe utcumque in hoc heu vates nimium verus fui." Sebastian Brant to Konrad Peutinger, July 1504 in Erich König, ed., *Konrad Peutingers Briefwechsel* (Munich: C. H. Beck, 1923), 32–33.

Chapter Three. Teaching Astrology

1. See, for example, letters dating from 1495–97 by Johannes Burge, Johannes Stabius, Joannes Krachenberger, Conrad Amicus, Hieronymous Balbus in Hans Rupprich, ed., *Der Briefwechsel des Konrad Celtis* (Munich: C. H. Beck, 1934), #112, 113, 134, 135, 150, 154. See also, Lewis W. Spitz, *Conrad Celtis. The German Arch-Humanist* (Cambridge, MA: Harvard University Press, 1957), 63–72.

2. Maximilian to Celtis, 7 March 1497, in Rupprich, *Der Briefwechsel des Konrad Celtis*, 261.

3. See Gustav Bauch, *Die Anfänge des Humanismus in Ingolstadt: Eine litterarische Studie zur deutschen Universitätsgeschichte* (Munich: R. Oldenbourg, 1901).

4. In the scholarship on the history of the University of Vienna, Celtis's *Collegium* looms large, usually as an institutional expression of Maximilian's efforts to introduce humanism into the curriculum. The advent of humanism remains the dominant framework for histories of the university. See, for example, Rudolf Kink, *Geschichte der kaiserlichen Universität zu Wien*, vol. 1: *Geschichliche Darstellung* (Vienna: Carl Gerold & Sohn, 1854), 184–230; Joseph Ritter von Aschbach, *Die Wiener Universität und ihre Humanisten im Zeitalter Kaiser Maximilians I.* (Vienna: Wilhelm Braumüller, 1877), esp. 41–122; Gustav Bauch, *Die Rezeption des Humanismus in Wien: Eine litterarische Studie zur deutschen Universitätsgeschichte* (Breslau: M. & H. Marcus, 1903), esp. 55–170; Karl Großmann, "Die Frühzeit des Humanismus in Wien bis zu Celtis Berufung 1497," *Jahrbuch für Landeskunde von Niederösterreich* 22 (1929): 152–325, esp. 220–34, 309–23; Helmuth Grössing, *Humanistische Naturwissenschaft: Zur Geschichte der Wiener mathematischen Schulen des 15. und 16. Jahrhunderts* (Baden-Baden: Valentin Koerner, 1983); Franz Graf-Stuhlhofer, *Humanismus zwischen Hof und Universität. Georg Tannstetter (Collimitius) und sein wissenschaftliches Umfeld im Wien des frühen 16. Jahrhunderts* (Vienna: WUV-Universitätsverlag, 1996); Helmuth Grössing, "Die Wiener Universität im Zeitalter des Humanismus von der Mitte des 15. bis zur Mitte des 16. Jahrhunderts," in *Das alte Universitätsviertel in Wien, 1385–1985*, ed. Günther Hamann, Kurt Mühlberger, and Franz Skacel, Schriftenreihe de Universitätsarchivs (Vienna: Universitätsverlag für Wissenschaft und Forschung, 1985), 37–45. Kurt Mühlberger, "Die Gemeinde der Lehrer und Schüler—Alma Mater Rudolphina," in *Wien. Geschichte einer Stadt. Band I: Von den Anfängen bis zur ersten Wiener Türkenbelagerung (1529)*, ed. Peter Csendes and Ferdinand Opll (Vienna: Böhlau, 2000), 395–98.

5. Traditionally, scholars have referred to this period as the "second Viennese mathematical school." See Grössing, *Humanistische Naturwissenschaft*, 145–47.

6. On this woodcut, see Peter Luh, *Der Allegroische Reichsadler von Conrad Celtis und Hans Burgkmair* (Frankfurt am Main: Peter Lang, 2002).

7. Michael Shank has pointed to how the fifteenth-century Habsburg court increasingly considered the university as a pool of talent to be consulted as needed. Michael Shank, "Academic Consulting in 15th-Century Vienna: The Case of Astrology," in *Texts and Contexts in Ancient and Medieval Science. Studies on the Occasion of John E. Murdoch's Seventieth Birthday*, ed. Edith Sylla and Michael McVaugh (Leiden: Brill, 1997), 245–70.

8. See Aschbach, *Die Wiener Universität und ihre Humanisten im Zeitalter Kaiser Maximilians I*, 374–76; Grössing, *Humanistische Naturwissenschaft*, 175.

9. All of Stiborius's examples use data for Vienna. Some of them are quite specific, for example: "Describatur itaque primo Meridianus secundum gradus declinationis cancri scilicet 27 computatis a meridie ad occasum qualiter est superficies meridiana turris collegij Vienne." Andreas Stiborius, "Liber umbrarum," Munich, Bayerische Staatsbibliothek, Clm 24103, 5r.

10. Stiborius, "Liber umbrarum," 9v. Now thought to be spurious, Messahalla's text on the astrolabe was one of the most common in the medieval and Renaissance periods. See the edition and translation in R. T. Gunther, *Chaucer and Messahalla On the Astrolabe*, vol. 5: *Early Science at Oxford* (Oxford: Oxford University Press, 1929). For the Greek text and a discussion of the spurious ascription, see Anne Tihon, Régine Leurquin,

and Claudy Scheuren, *Une version Byzantine du traité sur l'astrolabe du Pseudo-Messahalla* (Louvain-la-Neuve: Bruylant-Academia, 2001).

11. Stiborius, "Liber umbrarum," 1v–7r.

12. Empty spaces for the illustrations are found throughout Ziegler's copy of the "Liber umbrarum" and the text frequently refers to the illustrations, e.g., Stiborius, "Liber umbrarum," 12v.

13. For example, Stiborius, "Liber umbrarum," 5v, 7r, 13r.

14. Had many students owned brass astrolabes, some of those instruments would have to survive. No metal astrolabe survives that can be identified with this early sixteenth century Viennese *milieu*. One possible candidate is the astrolabe dated ca. 1521 that resembles a Georg Hartmann instrument in the collection at the Museum of the History of Science, Oxford. See MHS Inventory no. 47657.

15. In his "Canones astrolabij" Stiborius explained how to color the front of an astrolabe to correspond to the motion of the four humors. See Andreas Stiborius, "Canones astrolabij Magistri Andree Stiborij boij partim ex veteribus ordinate partim nov invencione additi," Munich, Bayerische Staatsbibliothek, Clm 19689, 75r–75v.

On functional paper instruments, see Suzanne Karr Schmidt, "Printed Instruments," in *Prints and the Pursuit of Knowledge in Early Modern Europe*, ed. Susan Dackerman (New Haven, CT: Yale University Press, 2011), 267–315; Suzanne Karr Schmidt, *Altered and Adorned. Using Renaissance Prints in Daily Life* (New Haven, CT: Yale University Press, 2011), 73–92; Suzanne Karr Schmidt, "Art. A User's Guide: Interactive and Sculptural Printmaking in the Renaissance" (PhD diss., Yale University, 2006), esp. 180–274.

16. There are a few surviving paper-wood astrolabes. See, for example, the Georg Hartmann astrolabe from 1542 in the collection at the Museum of the History of Science, Oxford: MHS Inventory no. 49296.

17. Stiborius, "Liber umbrarum," 13r.

18. Claudia Kren, "Astronomical Teaching at the Late Medieval University of Vienna," *History of Universities* 3 (1983): 21–22.

19. On the importance of extraordinary and private lectures at the University of Vienna and the income masters realized from these lectures, see the early statute that outlines some fees. A later statute from 1509 points to other supplementary lectures and the fees to hear them. See Rudolf Kink, *Geschichte der kaiserlichen Universität zu Wien*, vol. 2: *Statutenbuch der Universität* (Vienna: Carl Gerold & Sohn, 1854), 215, 317. On the difference in fees at Paris and Vienna, see Hastings Rashdall, *The Universities of Europe in the Middle Ages*, vol. 2: *Italy, Spain, France, Germany, Scotland* (Oxford: Clarendon Press, 1936), 243n1.

20. Grössing suggests that Stiborius along with Georg Tannstetter and Johannes Stabius were the most important members of the "second Viennese mathematical school." See Grössing, *Humanistische Naturwissenschaft*, 174.

21. Both extant copies of this text include the comment: "et advertens poeta laureatus et Imperatoris Romanorum Maximiliani historiographus atque Cosmographus Joannes Stabius, vir numeri seculi doctissimus." Stabius was crowned poet laureate sometime in 1503/1504. Stiborius, "Canones astrolabij," 77v.

22. Stiborius, "Canones astrolabij," 67r–73v.

23. Stiborius, "Canones astrolabij," 76r–79r.

24. The student noted in the margin: "Secuntur utilitates addite A. Magistro Andrea Stiborio que in antequis exemplarijs non habentur et iudiciarie deservient." Stiborius, "Canones astrolabij," 74v.

25. Although instruments embodied particular theoretical commitments, they were practical devices aimed at providing a solution to a problem or response to a particular question. See Jim Bennett, "Presidential Address. Knowing and Doing in the Sixteenth Century: What Were Instruments For?," *British Journal for the History of Science* 36 (2003): 129–50, esp. 135–36. For a discussion of the changes in expectations about instruments, see also Deborah Jean Warner, "What Is a Scientific Instrument, When Did It Become One, and Why?," *British Journal for the History of Science* 23 (1990): 83–93.

26. Andreas Stiborius, "Canones astrolabij," 79r–86r.

27. Stiborius, "Canones saphee," Munich, Bayerische Staatsbibliothek Clm. 19689, 293v–294r.

28. Andreas Stiborius, "Canones saphee," 292v.

29. Stiborius, "Canones saphee," 284r–312r.

30. Andreas Stiborius, "Canones super instrumento universali quod organum ptholomei vocant," Clm 19689, 317r.

31. "Sinus altitudinis solis" and "Distancia miliariorum duorom locorum" are Stiborius's two examples of how to use the *organum ptholomei*. Stiborius, "Canones super instrumento," 321v.

32. Vadian's notes that include the "Canones astrolabij" also include a fragment of Stiborius's "Liber horologium."

33. Stiborius, "Canones saphee," e.g., 311v, 312r.

34. Surviving copies of Stiborius's lectures are bound in the order in which the students would have attended them. For example, the copies in Bayerische Staatsbibliothek Clm 24103: Stiborius, "Liber umbrarum;" Stiborius, "Canones saphee;" Stiborius, "Canones super instrumento."

35. Although the Acts are incomplete for the years after 1500, Stiborius fails to appear in any of the official lists. See "Acta facultatis artium Universitatis Vindobonensis," vol. IV, 1497–1555, Universitätsarchiv UAW Cod. Ph 9.

36. Ziegler's copy of the "Liber umbrarum" is today in Andreas Stiborius, "Liber umbrarum," Munich, Bayerische Staatsbibliothek, Clm. 24103, 1r-21r.

37. Vadian's copy of the "Canones astrolabij" is found in Andreas Stiborius, "Canones astrolabij," St. Gall, Vadianische Sammlung Ms. 66, 43r–83r.

38. See Rheticus's dedicatory letter to Emperor Ferdinand I in his edition of Johannes Werner, *De triangulis sphericis libri quatuor. De meteoroscopiis libri sex* (Krakow: Lazarus Andreae, 1557), A3v.

39. This argument is found in Stiborius's two prefatory letters in Tannstetter's edition of Georg Peuerbach, *Tabulae eclypsium Magistri Georgii Peurbachii* (Vienna: Johannes Winterberger, 1514), aa8r–aa8v and AB3r–AB3v.

40. "Pendent item ex hac primi mobilis scientia instrumenta pene infinita. Astrolabium: saphea: organum Prolaemei: metheoroscopion: armilae: torquetum: rectanguls: aequatoria:

compassi: quadrantes: & alia id genus multa. O quam ampla: quam nobilis: quam necessaria omnibus astronomiae studiosis: & tanquam alphabetaria & præliminaris scientia. sine
qua nihil perfectum: nihil consummatum in hac astronomica præcellenti discplina. Age
igitur quisquis es cœli verus amator has tabulas ductas diligentia & sollicitatione confratris
mei charissimi doctoris Georgii Tannstetter impressioni datas." Andreas Stiborius, Prefatory
Letter to Regiomontanus, *Tabula primi mobili* in Peuerbach, *Tabulae eclypsium*, AB3v.

41. Schöner, *Mathematik und Astronomie an der Universität Ingolstadt im 15. und 16.
Jahrhundert*, 255–61.

42. "Acta facultatis artium," 32v. Tannstetter had previously been admitted to exam-
ine students. See "Acta facultatis artium," 30v.

43. The book lists for 1505 and 1511 include Tannstetter as lecturing on the *Theorica
planetarum*, "Acta facultatis artium," 41v, 75v.

44. On his *practica* and wall calendars, see chapter 5 in this volume.

45. For an analysis of Tannstetter's works see Graf-Stuhlhofer, *Humanismus zwischen
Hof und Universität*, 84–94.

46. Tannstetter, "Viri mathematici quos inclytum Viennense gymnasium ordine celebres habuit," in Peuerbach, *Tabulae eclypsium*, aa3v–aa6v.

47. On Tannstetter in general, see Franz Graf-Stuhlhofer's biography, *Humanismus
zwischen Hof und Universität. Georg Tannstetter (Collimitius) und sein wissenschaftliches
Umfeld im Wien des frühen 16. Jahrhunderts* (Vienna: WUV-Universitätsverlag, 1996).

48. "Cuius perraris inventis Invictiss. & illustriss. Caesar Maximilianus quotidie
oblecatur. Et eius Stiboriique (de quo paulo infra dicam) ingeniam miratus: lectiones pub-
licas in astronomia & mathematica Viennae nouo stipendio instituit." Tannstetter, "Viri
mathematici quos inclytum Viennense gymnasium ordine celebres habuit" in Georg Peuerbach, *Tabulae eclypsium Magistri Georgii Peurbachii* (Vienna: Lucas Atlantse, 1514), aa5r.

49. "Debent inde tibi gratias æternales authoris manes sepulti. Debet & tota illustrium litteratorum turba gymnastica: quibus & famam decusque cum tua ingenti
gloria restituere paras. Equidem sepe numero mecum ipse voluebam: & te & Stabium
Stiboriumque: Rosinum: Angelum: Ericium Mathematicos nobiles & multo litterarum
splendore nitentes in nostris oris germanicis a deo optimo maximo conservatos, ut essent
per quos incluta & preclarissima Mathematices studia aliquandiu & turpiter & barbare
perhabita subsisterent et respirarent." Thomas Resch, Dedicatory Letter in Peuerbach,
Tabulae eclypsium, aa6v. On Thomas Resch, see Aschbach, *Die Wiener Universität und ihre
Humanisten im Zeitalter Kaiser Maximilians I.*, esp. 105–12; Alois Schmid, "'Poeta et orator
a Caesare laureatus.' Die Dichterkrönungen Kaiser Maximilians I," *Historisches Jahrbuch*
109 (1989): 100–101.

50. In addition to his edition of Peuerbach's *Tabulae eclypsium* and Regiomontanus's
Tabula primi mobiles, Tannstetter produced editions of other texts by Peuerbach and
Gmunden as well as Johannes de Muris, Bradwardine, and Sacrobosco. Many of these
texts were intended to be university texts; for example, Johannes de Muris, *Arithmetica
communis. Proportiones breves. De latitudinibus formarum. Algorithmus M. Georgii Peur-
bachii in integris. Algorithmus Magistri Joannis de Gmunden de minuciis phisicis* Johannes
Singrenius, 1515), Vienna, Österreichische Nationalbibliothek 72.G.28 Alt Prunk;
Johannes de Sacrobosco, *Opusculum de sphaera clarissimi philosophi Ioannis de Sacro Busto.*

Theoricae planetarum Georgii Purbachii (Vienna: Johannes Singrenius, 1518), Vienna, Österreichische Nationalbibliothek 72.G.10(2) Alt. Prunk. On the attribution of these to Tannstetter, see Graf-Stuhlhofer, *Humanismus zwischen Hof und Universität*, 87–94.

51. On Maximilian's efforts, see Larry Silver, *Marketing Maximilian. The Visual Ideology of a Holy Roman Emperor* (Princeton, NJ: Princeton University Press, 2008); and Christopher S. Wood, "Maximilian I as Archeologist," *Renaissance Quarterly* 58 (2005): 1128–74. Borchardt's study is still useful: Frank L. Borchardt, *German Antiquity in Renaissance Myth* (Baltimore, MD: Johns Hopkins University Press, 1971).

52. Georg Tannstetter, Dedicatory Letter, Proclus Diadochus, *Sphaera. Astronomiam discere incipientibus utilissima. Thoma Linacro Britanno interprete* (Vienna: Hieronimus Vietor, 1511), A1v–A2r.

53. Various students worked carefully through Tannstetter's edition. See the annotated copies: Proclus Diadochus, *Sphaera. Astronomiam discere incipientibus utilissima. Thoma Linacro Britanno interprete.* (Vienna: Johannes Singrenius, 1511), London, British Library 8561.b.6; and Proclus Diadochus, *Sphaera. Astronomiam discere incipientibus utilissima. Thoma Linacro Britanno interprete* (Vienna: Hieronimus VietoremIohannes Singrenium, 1511), Oxford, Bodleian Library Bod Vet D1 e.16. Proclus Diadochus, *Sphaera. Astronomiam discere incipientibus utilissima. Thoma Linacro Britanno interprete* (Vienna: Johannes Singrenius, 1511), Bodleian Library Bod Vet D1 e.16.

54. Tannstetter's summary of Campanus's *Theorica* is preserved in Georg Tannstetter, "Theoricarum planetarum Compositiones ex Campano denuo abbreviatae per clarissimum virum artium, & medicinae Doctyorem Georgium Collimitium Tanstetter," Munich, Universitätsbibliothek 4 Cod. ms. 743, 155r–160r. This text was apparently written sometime after 1513, when Tannstetter assumed teaching responsibilities in the medical faculty. The brevity of the summary suggests that it was not used as a core text in Tannstetter's lectures. Campanus's text is available in a modern edition: Francis S. Benjamin Jr. and G. J. Toomer, eds., *Campanus of Novara and Medieval Planetary Theory. Theorica planetarum* (Madison: University of Wisconsin Press, 1971).

55. Albert Brudzewo, *Commentum in theoricas planetarum Georgii Purbachii* (Milan: Ulrich Scinzenzeler, 1494), Columbia University Library Goff B460. Tannstetter's copy of Brudzewo's text was originally part of a *Sammelband,* which the rare book dealer H. P. Kraus separated to sell the individual tracts. Columbia University purchased Tannstetter's copy of Brudzewo. See H. P. Kraus, *Choice Books and Manuscripts from Distinguished Library, Catalog 126* (New York, 1971), 112–15. See also H. P. Kraus's sale catalog, *H. P. Kraus, Important Works in the Field of Science, Catalog 137* (New York, 1973), #1, 2, 4, 5,10, 12, 14, 17, 65, 71, 73. I thank Jennifer B. Lee of the Rare Books Collection at the Columbia University for a description of Tannstetter's notes.

56. In his lectures on ephemerides, Tannstetter cites Peuerbach no less than six times, often quoting his *Theoricae*. See, for example, Georg Tannstetter, "Georgij Collimitij Dictata in Ephemerides," Vadianische Sammlung Ms. 66, 28v

57. Arthur Goldman, *Die Wiener Universität, 1519–1740* (Vienna: Adolf Holzhausen, 1916), 158.

58. On Tannstetter's absence from the acts, Franz Graf-Stuhlhofer, "Das Weiterbestehen des Wiener Poetenkollegs nach dem Tod Konrad Cetis' (1508)," *Zeitschrift für*

historische Forschung 26, no. 2 (1999): 393–407; Graf-Stuhlhofer, *Humanismus zwischen Hof und Universität*, 63–66; Franz Stuhlhofer, "Georg Tannstetter (Collimitius). Astronom, Astrologe und Leibarzt bei Maximilian I. und Ferdinand I," *Jahrbuch des Vereins der Stadt Wien* 37 (1981): 7–49, esp. 28.

59. Graf-Stuhlhofer, *Humanismus zwischen Hof und Universität*, 50–53.

60. Siegmund Günther, *Geschichte des mathematischen Unterrichts*, vol. 3: *Monumenta Germaniae Paedagogica* (Berlin: A. Hofmann & Comp., 1887), 253; Aschbach, *Die Wiener Universität und ihre Humanisten im Zeitalter Kaiser Maximilians I*, 85–87.

61. Graf-Stuhlhofer, *Humanismus zwischen Hof und Universität*, 73–75.

62. There was remarkably little change in the official lectures between the early fifteenth century and the early sixteenth century. Compare the lists found in Kren, "Astronomical Teaching at the Late Medieval University of Vienna" with the list of books for 1517 in "Acta facultatis artium," 101v.

63. References to an early printed edition of these lectures seem to be spurious. See Anton Mayer, *Wiens Buchdruckergeschichte, 1482–1882*, vol. 1 (Vienna: F. Jasper, 1882–87), 34; M. Denis, *Wiens Buchdruckergeschichte bis zum Jahre 1560* (Vienna, 1782), 199.

64. Vadian's notes are found in Georg Tannstetter, "Georgij Collimitij Dictata in Ephemerides." If Perlach's comment about the scope of Tannstetter's lectures is correct, Vadian's copy is incomplete or represents an early stage in the development of Tannstetter's lectures. It is possible that Perlach attended these lectures, for he arrived at the University of Vienna in 1511 and by 1513 was working closely with Tannstetter.

65. Tannstetter, "Georgij Collimitij Dictata in Ephemerides," 1v–3v.

66. Tannstetter, "Georgij Collimitij Dictata in Ephemerides," 3v–16v.

67. Tannstetter, "Georgij Collimitij Dictata in Ephemerides," 17v–18v.

68. Ptolemy discussed the effects of eclipses at length in book two of his *Quadripartitum*. For these two points, see II. 6.

69. Tannstetter, "Georgij Collimitij Dictata in Ephemerides," 16v–23v.

70. Tannstetter, "Georgij Collimitij Dictata in Ephemerides," 26r–26v.

71. Tannstetter, "Georgij Collimitij Dictata in Ephemerides," 27v.

72. Tannstetter, "Georgij Collimitij Dictata in Ephemerides," 28v–30v.

73. "De eclipsibus solis et lunae a. 1523–1544 notatu digna," Vienna, Österreichische Nationalbibliothek, cod. lat. 10455, 165r–7r, esp. 165r.

74. Tannstetter, "Georgij Collimitij Dictata in Ephemerides," 13r.

75. One contemporary attempted to solve the problems in both Peuerbach's and Regiomontanus's texts but usually gave up before he succeeded. See Georg Peuerbach, *Tabulae eclypsium Magistri Georgii Peurbachii* (Vienna: Johannes Winterberger, 1514), London, Welcome Institute Inc. 5.b.2.

76. Pliny's own sentiments toward astrology were rather ambiguous. See Tamsyn Barton, *Ancient Astrology* (London: Routledge, 1994), 55–56; Valerie Flint, *The Rise of Magic in Early Medieval Europe* (Princeton, NJ: Princeton University Press, 1991), 109–33. For a different understanding of Pliny, see Stephen C. McCluskey, *Astronomies and Cultures in Early Medieval Europe* (Cambridge: Cambridge University Press, 1998), esp. 16–18 and 133–40; Olaf Pedersen, "Some Astronomical Topics in Pliny," in *Science in the Early Roman Empire: Pliny the Elder, His Sources and Influence*, ed. Roger French and Frank Greenaway

(Totowa, NJ: Barnes & Noble, 1986), 162–96; B. S. Eastwood, "Plinian Astronomy in the Middle Ages and Renaissance," in ibid., 197–251.

77. Charles G. Nauert, "Caius Plinius Secundus," *Catalogus Translationum et Commentariorum* 4 (1980): 297–422. On the printing history of Pliny's *Historia naturalis*, see Martin Davies, "Making Sense of Pliny in the Quatrocento," *Renaissance Studies* 9, no. 2 (1995): 240–57.

78. This quarrel has long attracted attention of historians of science and medicine, as well as intellectual historians. See Lynn Thorndike, *A History of Magic and Experimental Science*, vol. IV: *Fourteenth and Fifteenth Centuries* (New York: Columbia University Press, 1958), 594; Arturo Castigloni, "The School of Ferrara and the Controversy on Pliny," in *Science, Medicine and History: Essays on the Evolution of Scientific Thought and Medical Practice Written in Honour of Charles Singer*, ed. E. Ashworth Underwood (London, 1953), 269–79; Roger French, "Pliny and Renaissance Medicine," in *Science in the Early Roman Empire*, French and Greenaway, 252–81.

79. To make this argument Leoniceno drew on his unrivalled collection of Greek manuscripts. See Daniela Mugnai Carrar, *La biblioteca di Nicolò Leoniceno. Tra Aristotle e Galeno: cultura e libri di un medico umanista* (Florence, 1991).

80. Vivian Nutton has argued this point forcefully for the case of medicine. Vivian Nutton, "Hellenism Postponed: Some Aspects of Renaissance Medicine, 1490–1530," *Sudhoffs Archiv* 81, no. 2 (1997): 158–70; French, "Pliny and Renaissance Medicine," esp. 255–63.

81. Graf-Stuhlhofer suggests that Tannstetter's interests were humanist and philological, while Eastwood characterizes his interests as scientific and pedagogical: Graf-Stuhlhofer, *Humanismus zwischen Hof und Universität*, 109–12; Eastwood, "Plinian Astronomy in the Middle Ages and Renaissance," 217.

82. Georg Tannstetter, "Item, Georgii Collimitii, et Joachimi Vadiani, in eundem secundum Plinia scholia quaedam. Ad haec Index rerum quae hic disputantur praecipuarum, utilis," in *In C. Plinii de Naturali Historia librum secundum commentarius, quo difficultates Plinianae, praesertim astronomicae, omnes tolluntur* (Basel: Henricus Petrus, 1531), 400.

83. Tannstetter, "Scholia quaedam," 401.

84. See Pedersen for a useful summary of Pliny's planetary theory, "Some Astronomical Topics in Pliny," 182–87.

85. Tannstetter, "Scholia quaedam," 420–21, 430–36.

86. Tannstetter, "Scholia quaedam," 436 [misnumbered 434].

87. Tannstetter, "Scholia quaedam," 444–45 [numbered 435].

88. "Pro Plinio li. 7, ca. xlix, a Georgio Tanstetter Collimitio, rogante Gundelio, explanationes." Anonymous student in Pliny, *C. Plinii secundi liber septimus naturalis historiae* (Vienna: Johannes Singrenius, 1519), Houghton Library GC5.G9554.519p, n.p.

89. Contrast this with Anthony Grafton's characterization of Cardano's approach to reinterpreting the history of astrology. Anthony Grafton, *Cardano's Cosmos. The Worlds and Works of a Renaissance Astrologer* (Cambridge, MA: Harvard University Press, 1999), ch. 8.

90. Tannstetter, "Scholia quaedam," 405–6.

91. Milich's commentary was first published in 1538 as Jakob Milich, *Liber secundus C. Plinii de mundi historia cum commentariis Iacobi Milichii diligenter conscriptis & recognitis* (Schwäbisch Hall: Peter Brubach, 1538). Typically, Milich's commentary is thought to have been largely the work of Philip Melanchthon.

92. Sachiko Kusukawa, *The Transformation of Natural Philosophy: The Case of Philip Melanchthon* (New York: Cambridge University Press, 1995), 137. On Milich and his activities at Wittenberg, see Claudia Brosseder, *Im Bann der Sterne. Caspar Peucer, Philipp Melanchthon und andere Wittenberger Astrologen* (Berlin: Akademie Verlag, 2004). See also Eastwood, "Plinian Astronomy in the Middle Ages and Renaissance"; French, "Pliny and Renaissance Medicine"; and Lynn Thorndike, *A History of Magic and Experimental Science*, vol. V: *Sixteenth Century* (New York: Columbia University Press, 1958), 385–89.

93. During the 1549–50 academic year one student extensively annotated his copy of Milich's text, noting at the beginning that "M. Sebastianus began his exposition of Pliny's book 2 on 24 October 1549 at Wittemberg and finished on 10 June 1550." See Jakob Milich, *Liber secundus C. Plinii de mundi historia cum commentariis Iacobi Milichii diligenter conscriptis & recognitis* (Frankfurt: Peter Brubach, 1543), Bayerische Staatsbibliothek Sig. 4 A.lat.b 400.

94. Maximilian's Habsburg predecessors had relied on physician-astrologers, though his father seems to have increasingly relied on astrologers who had appointments in the arts faculty. His contemporary princes, such as the Sforza dukes in Milan, preferentially employed physician-astrologers. An interesting contrast to this pattern was the Hungarian king, Matthias Corvinus who seems to have relied on astrologers who had little training in medicine. Later in the century, at least in the Germanies, it became increasingly common for academic astrologers to hold appointments in the medical faculty.

On the Habsburgs and the Sforza, see Shank, "Academic Consulting in 15th-Century Vienna"; Monica Azzolini, "The Politics of Prognostication: Astrology, Political Conspiracy and Murder in Fifteenth-Century Milan," *History of Universities* 22 (2008): 6–34; Monica Azzolini, *The Duke and the Stars. Astrology and Politics in Renaissance Milan* (Cambridge, MA: Harvard University Press, 2012); Monica Azzolini, "The Political Uses of Astrology: Predicting the Illness and Death of Princes, Kings and Popes in the Italian Renaissance," *Studies in History and Philosophy of Biological and Biomedical Sciences* 41, no. 2 (2010): 135–45; Monica Azzolini, "Reading Health in the Stars: Politics and Medical Astrology in Renaissance Milan," in *Horoscopes and Public Spheres. Essays on the History of Astrology*, ed. Günther Oestmann, H. Darrel Rutkin, and Kocku von Stuckrad, vol. 42, Religion and Society Series (Berlin: Walter de Gruyter, 2005), 183–205. For the contrasting case of Hungary, see Darin Hayton, "Expertise *Ex Stellis*: Comets, Horoscopes, and Politics in Renaissance Hungary," *Osiris* 25 (2010): 27–46. For the later sixteenth century, see Robert S. Westman, "The Astronomer's Role in the Sixteenth Century: A Preliminary Study," *History of Science* 18 (1980): 118–19.

95. Tannstetter matriculated on May 10, 1508. *Acta facultatis medicinae*, 68. His request for exemption was in 1513: *Acta facultatis medicinae*, 84. See also Graf-Stuhlhofer, *Humanismus zwischen Hof und Universität*, 73–75.

96. Graf-Stuhlhofer has suggested that these lectures were delivered in 1526/27. Graf-Stuhlhofer, *Humanismus zwischen Hof und Universität*, 147.

97. On Stainpeis, see Christian Pawlik, *Martin Stainpeis: Liber de modo studendi seu legendi in medicina. Bearbeitung und Erläuterung einer Studienanleitung für Mediziner im ausgehenden Mittelalter* (PhD diss., Technische Universität Munich, 1980); Richard J. Durling, "An Early Manual for the Medical Student and the Newly Fledged Practitioner: Martin Steinpeis, 'Liber de modo studendi seu legendi in medicina' (Vienna, 1520)," *Clio Medica* 5 (1970): 7–33; Harry Kühnel, *Mittelalterliche Heilkunde in Wien*, vol. 5, Studien zur Geschichte der Universität Wien (Graz: Verlag Hermann Böhlaus Nachf., 1965), 84–86; Malcolm Lee Crystal, "Medicine in Vienna in the Sixteenth and Seventeenth Centuries" (PhD diss., University of Virginia, 1994), 218–19.

98. Martin Stainpeis, *Liber de modo studendi seu legendi in medicina* (Vienna, 1520), 20r-v, 21v.

99. Stainpeis, *Liber de modo studendi*, 7v, 16v. On the *Articella* in the medical curriculum, see Luke Demaitre, "The Art and Science of Prognostication in Early University Medicine," *Bulletin of the History of Medicine* 77 (2003): 765–88. On Galen's *De criticis diebus*, see Cornelius O'Boyle, *Medieval Prognosis and Astrology. A Working Edition of the Aggregationes de crisi et creticis diebus, with Introduction and English Summary* (Cambridge: Wellcome Unit for the History of Medicine, 1991).

100. In 1521, Tannstetter had recommended Stainpeis's recipe as a means of warding off the plague. Georg Tannstetter, *Regiment für den Lauff der Pestilentz durch Georgen Tannstetter von Rain der siben freyen künst unnd Ertzney doctor: kurtzlich beschriben. Anno. 1521* (n.p., 1521), Vienna, Österreichische Nationalbibliothek Sig. 31.V.67, b2r.

101. Georg Tannstetter, *Artificium de applicatione astrologiæ ad medicinam* (Strasbourg: Georgius Ulricherus, 1531), 3r–8r.

102. Tannstetter, *Artificium de applicatione*, 7v.

103. Roger French argues that one of the primary reasons learned physicians incorporated predictive astrology was to compete in a medical market place by attracting patients with their technical apparatus. Roger French, "Foretelling the Future: Arabic Astrology and English Medicine in the Late Twelfth Century," *Isis* 87, no. 3 (1996): 453–80, esp. 475. The importance of the medical market place in shaping learned medical practice is an important part of the general argument in Roger French, *Medicine before Science. The Business of Medicine from the Middle Ages to the Enlightenment* (Cambridge: Cambridge University Press, 2003).

104. Stainpeis, *Liber de modo studendi*, 102v–111v.

105. Georg Tannstetter, *Artificium de applicatione*, 29v–31v. See Nancy Siraisi, *Medieval and Early Renaissance Medicine: An Introduction to Knowledge and Practice* (Chicago: University of Chicago Press, 1990), 135–36; Roger French, "Astrology in Medical Practice," in *Practical Medicine from Salerno to the Black Death*, ed. Luis García-Ballester, Roger French, and Jon Arrizabalaga (Cambridge: Cambridge University Press, 1994), 50–53.

106. Tannstetter, *Artificium de applicatione*, 31r–1v.

107. Tannstetter, *Artificium de applicatione*, 32v–3r.

108. Tannstetter, *Artificium de applicatione*, 33r–4v. On Tannstetter's wall calendars and yearly *practica*, see chapter 5 in this volume.

109. Tannstetter, *Artificium de applicatione*, 36r–7r.

110. Tannstetter, *Artificium de applicatione*, 37v–8r.

111. Tannstetter, *Artificium de applicatione*, 39v–46r. On the importance of predicting the death of a patient, see Siraisi, *Medieval and Early Renaissance Medicine*, 133–36. On the problems physicians faced when their medicine seemed to fail them, see Jon Arriza-balaga, John Henderson, and Roger French, *The Great Pox: The French Disease in Renaissance Europe* (New Haven, CT: Yale University Press, 1997); Jon Arrizabalaga and Roger French, "Coping with the French Disease: University Practitioners' Strategies and Tactics in the Transition from the Fifteenth to the Sixteenth Century," in *Medicine from the Black Death to the French Disease*, ed. Roger French, Jon Arrizabalaga, Andrew Cunningham, and Luis García-Ballester (Brookfield, VT: Ashgate, 1998), 248–87.

112. For example, Stiborius pointed out that the astrolabe should be colored to correspond to the motions of the four humors in his canon "Mociones 4or humorum." Andreas Stiborius, "Canones astrolabij," 75r–75v.

113. Tannstetter describes an astrolabe that has been painted so that the unequal hours are colored to correspond to the four humours. Tannstetter, *Artificium de applicatione*, 10v.

114. Physicians at German universities increasingly enjoyed such rewards. Vivian Nutton, "Medicine at the German Universities, 1348–1500: A Preliminary Sketch," in French et al., *Medicine from the Black Death to the French Disease*, 96.

115. Graf-Stuhlhofer, *Humanismus zwischen Hof und Universität*, 80–81.

116. For a brief treatment of Perlach's relationship to Tannstetter, see Franz Graf-Stuhlhofer, "Andrej Perlach kot ucenec jurija Tannstetterja na dunaju," *Casopis za Zgodovino in Narodopisje St.* 2 (1991): 280–83. The disintegration of their relationship can be followed in the Acts of the medical faculty, Karl Schrauf, ed., *Acta facultatis medicae universitatis Vindobonensis*, vol. 3, *1490–1558* (Vienna: Verlag des Medicinischen Doktorenkollegiums, 1904), 179–218.

117. Graf-Stuhlhofer has suggested that Perlach wrote out the manuscript copy of Stiborius and Tannstetter's text: Georg Tannstetter and Andreas Stiborius, "Super requisitione Leonis Papae X et Maximiliani I Imperatoris de Romani calendarii correctione consilium in studio Viennensi conscriptum et editum," Vienna, Österreichische Nationalbibliothek cod. lat. 10358, 113r–20r.

118. The 1550 matriculation records indicate that Perlach had been teaching for thirty-four years. This suggests that his lectures in 1517 were his earliest. Willy Szaivert and Franz Gall, eds., *Die Matrikel der Universität Wien*, vol. 3 (Graz: Hermann Böhlaus, 1971), 91; "Acta facultatis artium," 101v, 136r.

119. "Almanach arabum, Diale/Diurnale latinum, Ephemeridis grecum est liber in quo astra de die in diem describuntur." Andreas Perlach, "A magistro Andrea Perlachio Stiro super almanach collectanea," Vienna, Österreichische Nationalbibliothek Ser. n. 4265 , 307v.

120. Unfortunately, these lecture notes are fragmentary. They end abruptly after an example of converting astronomical time to "usual time." The notes for this lecture span only four folios. Perlach, "A magistro Andrea Perlachio Stiro super almanach," 307v–309r.

121. The classic account of the medieval university remains Rashdall. On the division of lectures, see Hastings Rashdall, *The Universities of Europe in the Middle Ages*, vol. 1: *Salerno, Bologna, Paris* (Oxford: Clarendon Press, 1936), 205–7, 433–35, 490–96. For Vienna, see Kink, *Geschichte der kaiserlichen Universität zu Wien*, 216; Joseph Ritter von

Aschbach, *Geschichte der Wiener Universität im ersten Jahrhunderte ihres Bestehens*, vol. 1 (Vienna, 1865), 62, 85. More recently and with a focus on Italian universities, see Paul Grendler, *The Universities of the Italian Renaissance* (Baltimore, MD: Johns Hopkins University Press, 2002), 144–46.

122. On Perlach's *Usus almanach*, see chapter 6 in this volume.

123. Andreas Perlach, *Ephemerides Andreae Perlachii Stiri ex Witschein, Artium & Philosophiae magistri, Mathematicarum disciplinarum studij Viennesis professoris ordinarij, pro Anno domini & Saluatoris nostri Iesu Christi M.D.XXXI. cum configurationibus & habitudinibus planetarum inter se, & cum stellis fixis, cum ex secuno, tum ex primo mobili contingentibus. Insuper adiunximus his nostris in hunc annum Ephemeridibus, Prognosticon, superioris anni eclipsium, quarum effectus hoc Anno apparebunt* (Vienna, 1530), A3v.

124. Andreas Perlach, *Commentaria ephemeridium clarissimi viri D. Andreæ Perlachii Stiri, Medicae artis doctoris, ac in academia Viennensi Ordinarij quondam Mathematici, ad usum stuiosorum ita fideliter conscripta, ut quisque absque Præceptore, ex sola lectione integram indo artem consequi possit* (Vienna: Egidius Aquila, 1551), 1r–8r.

125. Perlach, *Commentaria ephemeridium*, 8v–62r.

126. Perlach, *Commentaria ephemeridium*, 62v–78r.

127. Perlach, *Commentaria ephemeridium*, 78r–83r.

128. Perlach, *Commentaria ephemeridium*, 83r–86r.

129. Hayton, "Expertise *Ex Stellis*."

130. Anthony Grafton, "Geniture Collections, Origins and Uses of a Genre," in *Books and the Sciences in History*, ed. Marina Frasca-Spada and Nick Jardine (Cambridge: Cambridge University Press, 2000), 49–68.

131. Perlach, *Commentaria ephemeridium*, 86v–93r.

132. On Perlach's use of ephemerides as a vehicle of pro-Habsburg propaganda, see chapter 6 in this volume.

133. On Peter Apian, see Karl Röttel, ed., *Peter Apian. Astronomie, Kosmographie und Mathematik am Beginn der Neuzeit* (Eichstätt: Polygon-Verlag, 1995). On Georg Hartmann, see Hans Gunther Klemm, *Georg Hartmann aus Eggolsheim (1489–1564). Leben und Werk eines Fränkischen Mathematikers und Ingenieurs* (Forchheim: Gürtler-Druck, 1990). See also Hartmann's notes from his time at the University of Vienna: Georg Hartmann, "Collectanea mathematica praeprimis gnomonicam spectantia," Vienna, Österreichische Nationalbibliothek cod. lat. 12768, 1r–151v. John Lamprey has produced an edition and translation of this manuscript: Georg Hartmann, *Hartmann's Practika: A Manual for Making Sundials and Astrolabes with the Compass and Rule,* ed. John Lamprey (Bellvue, CO: John Lamprey, 2002).

134. Notker Hammerstein, "Relations with Authority," in *Universities in Early Modern Europe (1500–1800)*, ed. Hilde de Ridder-Symoens, A History of the University in Europe, vol. 2 (Cambridge: Cambridge University Press, 1996), 114–53.

Chapter Four. Instruments and Authority

1. On Maximilian II's interest in medicine and natural history, see Paula Sutter Fichtner, *Emperor Maximilian II* (New Haven, CT: Yale University Press, 2001), 99–104.

For Maximilian's use of various sciences to advance his irenic court policies, see Howard Louthan, *The Quest for Compromise. Peace Makers in Counter Reformation Vienna* (Cambridge: Cambridge University Press, 1997), esp. 53–105.

2. Schöner, *Gnomonice*, A5v. Historians use the term prince practitioner to describe princes who combined an interest in the technical and occult arts with an ability and expertise. On prince practitioners, see Bruce T. Moran, "German Prince-Practitioners: Aspects in the Development of Courtly Science, Technology, and Procedures in the Renaissance," *Technology and Culture* 22 (1981): 253–74; Bruce T. Moran, "Patronage and Institutions: Courts, Universities, and Academies in Germany. An Overview 1550–1750," in *Patronage and Institutions. Science, Technology, and Medicine at the European Court, 1500–1750*, ed. Bruce T. Moran (Woodbridge: Boydell Press, 1991), 169–83. See also Tara Nummedal, *Alchemy and Authority in the Holy Roman Empire* (Chicago: University of Chicago Press, 2007).

3. "Igitur multi nunc sunt Principes Illustrissimi, qui omnes Mathematicarum disciplinarum scientia etiam excellunt, & a negociis ac occupationibus publicis subinde ad has tanquam quedam laxamenta animi sese recipiunt, his sese oblectant, instrumenta pingunt, qualia multa vidi a Principibus picta, motus syderum observant, & similia Mathematicorum officia faciunt." Andreas Schöner, *Gnomonice Andreae Schoneri Noribergensis, Hoc est: De descriptionibus horologiorum sciotericorum Omnis Generis Proiectionibus circulorum Sphaericorum ad superficies* (Noribergae: Montanus & Neuberus, 1562), A5v–6r.

4. Schöner, *Gnomonice*, A3v. Schöner also mentioned Andreas Kunhofer, who had studied with Stiborius and Stabius in Ingolstadt and, apparently, wrote a text on constructing sundials. Little is known about Kunhofer. See Ernst Zinner, *Astronomische Instrumente des 11. bis 18. Jahrhunderts* (Munich: C. H. Beck, 1956), 66, 421.

5. Tannstetter, "Viri mathematici quos inclytum Viennense gymnasium ordine celebres habuit," in Georg Peuerbach, *Tabulae eclypsium Magistri Georgii Peurbachii* (Vienna: Lucas Atlantse, 1514), aa5r.

6. On Stiborius's lectures at the university, see chapter 3 in this volume.

7. For an account of the final two months of the conflict and the negotiations, see Hermann Wiesflecker, *Kaiser Maximilian I. Das Reich, Österreich und Europa an der Wende zur Neuzeit*, vol. 5: *Der Kaiser und seine Umwelt. Hof, Staat, Wirtschaft, Gesellschaft und Kultur* (Munich: R. Oldenbourg Verlag, 1986), 328–37.

8. Stiborius's *Clipeus Austrie* is an early example of the type of project Moran points to when he says: "Most projects were tied to the political or economic ambitions of the patron and depended upon personal relationships that were, by no means, limited to the court itself." Moran, "Patronage and Institutions," 169.

9. Andreas Stiborius, "In Clipeum Austrie," Munich, Bayerische Staatsbibliothek Clm 19689, 286r.

10. "Cum autem dive Rex Maximiliane, Antique celi venate revoluciones et gesta tua illustrissima bella fortissima Triumphi Clarissimi per universum orbem conclamati: quem et oriente et occidente septentrione atque meridie a monstris tirrannorum inde sinentur purgare nec cesses: humano generi alterum te herculem tanquam Jove ipso celicis demissum auxiliatorem mortalibus lune clarius confirment." Stiborius, "In Clipeum Austrie," 286r.

11. Stiborius, "In Clipeum Austrie," 286v.

12. For a good treatment of this topic, see Frank L. Borchardt, *German Antiquity in Renaissance Myth* (Baltimore, MD: Johns Hopkins University Press, 1971), esp. ch. 3; Christopher S. Wood, "Maximilian I as Archeologist," *Renaissance Quarterly* 58 (2005): 1128–74.

13. Larry Silver has detailed how Maximilian used his visual projects to establish his inheritance. Larry Silver, *Marketing Maximilian. The Visual Ideology of a Holy Roman Emperor* (Princeton, NJ: Princeton University Press, 2008), esp. 41–108; Larry Silver, "Power of the Press: Maximilian's Arch of Honor," in *Albrecht Dürer in the Collection of the National Gallery of Victoria*, ed. Irena Zdanowicz (Melbourne: National Gallery of Victoria, 1994), 45–64.

14. Andreas Stiborius, "In Clipeum Austrie," 286r–286v.

15. "nova quadam inventione qua celestis globi circuli tam varie super planum qualecumque In quintumque libris meis umbrarum (prope diem pariter tue maiestati dictandis) depicti proiecti atque conculati eciam in rectas lineas extenduntur." Stiborius, "In Clipeum Austrie," 286v. Stiborius later compared his canons on the *Clipeus* to his lectures on the saphea, a universal astrolabe. See Stiborius, "In Clipeum Austrie," 288v.

16. On Stiborius's *Liber umbrarum*, see chapter 3 in this volume.

17. While the first two canons describe the lines on the surface of the *Clipeus Austrie* and the areas of its surface, there is no overall description of the instrument. Andreas Stiborius, "In Clipeum Austrie," 287v–288v.

18. This could help make sense of a statement Stiborius made in his preface, claiming that he decided to "draw" a certain mathematical and physical device. I have been unable to find any drawing of the *Clipeus Austrie* or any instrument that corresponds to Stiborius's brief description. Stiborius, "In Clipeum Austrie," 286r.

Paper instruments have begun attracting more attention. See Owen Gingerich's article for an introduction to paper instruments in late sixteenth century. Owen Gingerich, "Astronomical Paper Instruments with Moving Parts," in *Making Instruments Count. Essays on Historical Scientific Instruments Presented to Gerard L'Estrange*, ed. R. J. W. Anderson, J. A. Bennett, and W. F. Ryan (Aldershot: Variorum, 1993), 63–74. More recently, Suzanne Karr Schmidt has worked on paper astronomical instruments. See Suzanne Karr Schmidt, "Printed Instruments," in *Prints and the Pursuit of Knowledge in Early Modern Europe*, ed. Susan Dackerman (New Haven, CT: Yale University, 2011), 267–315, and her dissertation, "Art. A User's Guide: Interactive and Sculptural Printmaking in the Renaissance" (PhD diss., Yale University, 2006).

19. Stiborius, "In Clipeum Austrie," 288v.

20. Stiborius, "In Clipeum Austrie," 289r.

21. Stiborius, "In Clipeum Austrie," 289v–291r.

22. Stiborius, "In Clipeum Austrie," 291v–292r. This manuscript seems to be unfinished. In addition to ending abruptly, it does not include canons Stiborius listed on his list of uses, namely, how to determine critical days.

23. Cuspinian recorded the time of signing in his diary, Johannes Cuspianus, *Tagebuch, 1502–1527*, vol. 1: *Fontes Rerum Austriacarum. Österreichische Geschichts-Quellen. Scriptores* (Vienna: Kaiserl. Königl. Hof- und Staatsdruckerei, 1855), 400.

24. On astrology's role in revealing the course of politics, see Alfred Schmid, *Augustus*

und die Macht der Sterne. Antike Astrologie und die Etablierung der Monarchie in Rom (Cologne: Böhlau, 2005). As applied to religion, see Laura Ackerman Smoller, *History, Prophecy, and the Stars. The Christian Astrology of Pierre d'Ailly, 1350–1420* (Princeton, NJ: Princeton University Press, 1994), ch. 4; J. D. North, "Astrology and the Fortunes of Churches," in *Stars, Minds and Fate: Essays in Ancient and Medieval Cosmology*, ed. J. D. North (London: Hambledon, 1989), 59–89. In general, see Dietrich Kurze, "Prophecy and History: Lichtenberger's Forecasts of Events to Come (From the Fifteenth to the Twentieth Century); Their Reception and Diffusion," *Journal of the Warburg and Courtauld Institutes* 21 (1958): 63–85; Eugenio Garin, *Astrology in the Renaissance. The Zodiac of Life*, trans. Carolyn Jackson (London: Routledge & Kegan Paul, 1976), ch. 1.

25. On Stiborius's and Tannstetter's opinion about the proposed calendar reform, see chapter 5 in this volume.

26. Moran emphasizes the importance of sundials and related horological instruments in the rise of prince practitioners in the later sixteenth century. See Moran, "German Prince-Practitioners: Aspects in the Development of Courtly Science, Technology, and Procedures in the Renaissance," 258–59.

27. In the eighteenth century French sovereigns tried to extend their control by asserting their authority to standardize measurements. See Ken Alder, "A Revolution to Measure: The Political Economy of the Metric System in France," in *The Values of Precision*, ed. M. Norton Wise (Princeton, NJ: Princeton, 1995), 39–71. In general, standardization has been linked to governmental efforts to expand dominion and to regulate society. See, for example, the essays in Marie-Noëlle Bourguet, Christian Licoppe, and H. Otto Sibum, eds., *Instruments, Travel and Science. Itineraries of Precision from the Seventeenth to the Twentieth Century* (London: Routledge, 2002); M. Norton Wise, ed., *The Values of Precision* (Princeton, NJ: Princeton University Press, 1995).

28. Alois Schmid, "'*Poeta et orator a Caesare laureatus.*' Die Dichterkrönungen Kaiser Maximilians I," *Historisches Jahrbuch* 109 (1989): 56–108, esp. 88–89.

29. The best biography of Stabius is Helmuth Grössing, "Johannes Stabius. Ein Oberösterreicher im Kreis der Humanisten um Kaiser Maximilian I," *Mitteilungen des Oberösterreichischen Landesarchiv* 9 (1968): 239–64. This work should be supplemented with his dissertation on Stabius: Helmuth Grössing, "Johannes Stabius. Ein Beitrag zur Kulturgeschicht der Zeit Kaiser Maximilians I" (PhD diss., University of Vienna, 1964).

30. On Maximilian's *Ehrenpforte* and *Triumphzug* see Silver, *Marketing Maximilian*, esp. chs. 1 and 2.

31. Maximilian's interest in the printing press as a political tool has received considerable attention. In particular, see Peter Diederichs, *Kaiser Maximilian I als politischer Publizist* (Heidelberg, 1931); Peter Diederichs, "Kaiser Maximilian I. (1459–1519)," in *Deutsche Publizisten das 15. bis 20. Jahrhunderts*, ed. Heinz-Dietrich Fischer (Munich: Verlag Dokumentation, 1971), 35–42; Günther Lammer, "Literaten und Beamte im publizistischen Dienst Kaiser Maximilians I. 1477–1519" (PhD diss., Karl-Franzens Universität, 1983), esp. 1–42, 493–577; Jan-Dirk Müller, *Gedechtnus. Literatur und Hofgesellschaft um Maximilian I.* (Munich: Wilhelm Fink Verlag, 1982), esp. 251–80. See also Larry Silver, "Prints for a Prince: Maximilian, Nuremberg, and the Woodcut," in *New Perspectives on the Art of Renaissance Nuremberg*, ed. Jeffrey Chips Smith (Austin: University of Texas at

Austin, 1985), 7–21; Larry Silver, "Paper Pageants: The Triumphs of Emperor Maximilian I," in *"All the World's a Stage": Art and Pageantry in the Renaissance and Baroque*, ed. Barbara Wollesen-Wisch and Susan Munshower, vol. 1 (University Park: Pennsylvania State University Press, 1990), 292–331; Silver, *Marketing Maximilian*.

32. These celestial maps are unusual in that the ecliptic rather than the celestial equator forms the perimeter of the map. This mode of projection has forerunners in a couple of manuscripts—most notably ÖNB, cod. lat. 5415—but is otherwise unusual. On these maps, see the dated but extremely useful W. Voss, "Eine Himmelskarte vom Jahre 1503 mit den Wahrzeichen des Wiener Poettenkollegiums als Vorlage Albrecht Dürers," *Jahrbuch der Preußischen Kunstsammlungen* 64 (1943): 89–150. See also Edmund Weiss, "Albrecht Dürers Geographische, Astronomische und Astrologische Tafeln," *Jahrbuch der kunsthistorischen Sammlungen des allerhöchsten Kaiserhauses* 7 (1888): 207–20.

33. Weiss, "Albrecht Dürers Geographische, Astronomische und Astrologische Tafeln," 209.

34. This picture certainly changed later in the century, when instruments increasingly played important political roles, usually in patronage relationships. See Helen Watanabe-O'Kelly, *Court Culture in Dresden from Renaissance to Baroque* (Houndmills: Palgrave Macmillan, 2002); Mario Biagioli, *Galileo Courtier. The Practice of Science in an Age of Absolutism* (Chicago: University of Chicago Press, 1993); Richard Westfall, "Science and Patronage. Galileo and the Telescope," *Isis* 76 (1985): 11–30; Bruce T. Moran, "Princes, Machines and the Valuation of Precision in the 16th Century," *Sudhoffs Archiv* 61 (1977): 209–28; Moran, "German Prince-Practitioners."

35. Only two copies of the dedication and the canons survive: Johannes Stabius, "Horoscopion universale pro multiplici diversarum gentium ritu: diei: noctisque horas & momenta distinguens," Vienna, Österreichische Nationalbibliothek cod. lat. 5280, 26r–27r; Johannes Stabius, "Horoscopion universale," Würzburg, Universitätsbibliothek I.t.q. 236, 44r–47r. The copy from the ÖNB has been reprinted in *Jahrbuch der kunsthistorischen Sammlungen des allerhöchsten Kaiserhauses* 7 (1888): 214–16.

36. Agostino Chigi is thought to have taken particular pleasure in pointing out to his visitors that the ceiling in his Sala was, in fact, his horoscope. See his fundamental essay on this: Fritz Saxl, "The Villa Farnesina," in *Lectures I* (London: Warburg Institute, 1957), 189–99. For more recent scholarship, see Mary Quinlan-McGrath, "The Villa Farnesina, Time-Telling Conventions and Renaissance Astrological Practice," *Journal of the Warburg and Courtauld Institutes* 58 (1995): 52–71; Kristen Lippencott, "Two Astrological Ceilings Rconsidered: The Sala di Galatea in the Villa Farnesina and the Sala del Mappamundo at Caprarola," *Journal of the Warburg and Courtauld Institutes* 53 (1994): 185–207; Mary Quinlan-McGrath, "The Astrological Vault of the Villa Farnesina Agostino Chigi's Rising Sign," *Journal of the Warburg and Courtauld Institutes* 47 (1984): 91–105; Peter Schiller, "Die himmelskundliche Ikonographie der Decke der Sala di Galatea in der Villa Farnesina in Rom," in *Die okkulten Wissenschaften in der Renaissance*, ed. August Buck (Wolfenbüttel: Herzog August Bibliothek, 1992), 255–88.

37. Stiborius, "In Clipeum Austrie," 286v.

38. "mundi orbem imitans: forme circularis existit. Illudque sub divi Maximiliani Cæsaris maximo in terris numine in lucem edidi. Quoniam sicut species orbicularis.

Mathematicarum figurarum omnium existit capacissima & perfectissima. Ita amplitudo Maiestatis Cæsaree cunctas in terrarum orbe celsitudines: & quodcunque in eo præclarum: seu opinione mortalium excellens est: aut complectitur: aut exuperat." Johannes Stabius, "Dedicatory Letter to Matthäus Lang" Johannes Stabius, *Horoscopion omni generaliter* (1512).

39. Marie Tanner, *The Last Descendant of Aeneas. The Habsburgs and the Mythic Image of the Emperor* (New Haven, CT: Yale University Press, 1993), ch. 12.

40. Johannes Stabius, "Horoscopion universale pro multiplici diversarum gentium ritu: diei: noctisque horas & momenta distinguens," Vienna, Österreichische National-bibliothek cod. lat. 5280, 26r.

41. The original copy of this instrument in the Albertina, in Vienna, is possibly Maximilian's own copy. It has been partially colored according to the canons.

42. Franz Graf-Stuhlhofer has shown that a number of events in Maximilian's life were arranged to correspond to astrologically propitious times. See Helmuth Grössing and Franz Stuhlhofer, "Versuch einer Deutung der Rolle der Astrologie in den persönlichen Entscheidungen einiger Habsburger des Spätmittelalters," *Anzeiger der phil.-hist. Klasse der Österreichischen Akademie der Wissenschaften* 117 (1980): 267–83.

43. See J. D. North, "Astrolabes and the Hour-Line Ritual," *Journal for the History of Arabic Sciences* 5 (1981): 113–14. See also Francis Maddison and Anthony Turner, "The Names and Faces of the Hours," in *Between Demonstration and Imagination. Essay in the History of Science and Philosophy Presented to John D. North*, ed. Lodi Nauta and Arjo Vanderjagt (Leiden: Brill, 1999), 124–56.

44. For example, Martin Bylica's suggestion that the astrologer use these planetary hour lines to construct horoscope. See Darin Hayton, "Expertise *Ex Stellis*: Comets, Horoscopes, and Politics in Renaissance Hungary," *Osiris* 25 (2010): 27–46. The best explanation for calculating houses remains J. D. North, *Horoscopes and History* (London: Warburg Institute, 1986).

45. On Lang's role as political advisor to Maximilian, see Hermann Wiesflecker, *Kaiser Maximilian I. Das Reich, Österreich und Europa an der Wende zur Neuzeit*, vol. 4: *Gründung des habsburgischen Weltreiches, Lebensabend und Tod, 1508–1519* (Munich: R. Oldenbourg Verlag, 1981), 90–113.

46. Stabius failed to point out that to use this instrument Lang would have had to affix two threads, one to the apex of each triangle. These threads served as rules or measures.

47. This was not the only one of Stabius's works to undergo a shift in dedication. The celestial maps that he, Dürer, and Heinfolgel produced were originally intended for Bannisius, but when printed were dedicated to Lang. There is what appears to be an early copy of these celestial map in the Bayerische Staatsbibliothek. It has been trimmed so that the dedication, if there was one, can no longer be read. All other examples that I have found are from the later, 1515 printing. On these maps, and their dedications, see Weiss, "Albrecht Dürers Geographische," 209. Campbell Dodgson, "Drei Studien," *Jahrbuch der kunsthistorischen Sammlungen des Allerhochsten Kaiserhauses* 29 (1910/11): 7–9.

48. In 1513 Peutinger wrote to Bannisius asking him to support a Christoph Wesler's application for imperial support. See Konrad Peutinger to Jakob Bannisius, February

13, 1513 in Erich König, ed., *Konrad Peutingers Briefwechsel* (Munich: C. H. Beck, 1923), 185–86.

49. Benedictus Chelidonius, a monk from St. Aegidius's monastery in Nuremberg, dedicated a play to Bannisius and wrote a book recounting the celebrations around the double marriage. See Franz Posset, *Renaissance Monks. Monastic Humanism in Six Biographical Sketches* (Leiden: Brill, 2005), 84. On the importance of the Congress of Vienna for Maximilian, see Wiesflecker, *Kaiser Maximilian I*, vol. 4, 181–204.

50. "Edicto Caesareo vetitum est. sub mulcta nummorum aureorum quinquaginta: ne quis alius: organum hoc: eiusque canones: vel cudat: aut venditet." Johannes Stabius, *Horoscopion omni generaliter congruens climati (dedicated to Jakob Bannisius).*

51. Johannes Stabius, *Astrolabium Imperatorium* (1515), Munich, Bayerische Staatsbibliothek Einbl. VIIII, 12.

52. Stabius, *Astrolabium Imperatorium.*

53. "idem denique imperatorium Astrolabium artificio impressorm in plurima derivavi exemplaria, ut alii etiam quicumque coelestium contemplatione circuitionum delectantur." Stabius, *Astrolabium Imperatorium.*

54. The number of printed instruments must have been lower than the estimated 700 copies of the *Ehrenforte* that were printed before Maximilian's death in 1519.

55. In many ways, Maximilian was at the forefront of developments in northern Europe that produced prince-practitioners. Bruce Moran has traced these developments in the later sixteenth-century German territorial courts, especially in Hessen-Kassel. See Bruce T. Moran, *The Alchemical World of the German Court. Occult Philosophy and Chemical Medicine in the Circle of Moritz of Hessen (1572–1632)* (Stuttgart: Franz Steiner Verlag, 1991); Moran, "Patronage and Institutions"; Moran, "Princes, Machines and the Valuation of Precision"; Robert S. Westman, "The Astronomer's Role in the Sixteenth Century: A Preliminary Study," *History of Science* 18 (1980): 105–47; Watanabe-O'Kelly, *Court Culture*; Tara Nummedal, *Alchemy and Authority in the Holy Roman Empire* (Chicago: University of Chicago Press, 2007).

56. Maximilian, Letter from Maximilian, September 1514, in "Urkunden und Regesten," ed. Heinrich Zimerman, in *Jahrbuch der kunsthistorischen Sammlungen des allerhöchsten Kaiserhauses* 1 (1883), LVII #332 Compare this sum to the 500 gulden Maximilian gave Grünpeck and Peutinger in 1506 for their work on his tomb or the 1,000 gulden he gave the University of Vienna in 1500.

57. Maximilian, Letter from Maximilian to Jakob Villiner, in "Urkunden und Regesten," ed. Heinrich Zimerman, in *Jahrbuch der kunsthistorischen Sammlungen des allerhöchsten Kaiserhauses* 1 (1883), LX #352.

58. Silver, "Prints for a Prince," 11. For Stabius's income, see Maximilian, Letter from Maximilian, 1514, in "Urkunden und Regesten," ed. Heinrich Zimerman, in *Jahrbuch der kunsthistorischen Sammlungen des allerhöchsten Kaiserhauses* 1 (1883), LXXII #446. Nuremberg's city council, however, stopped paying both men after Maximilian's death, claiming that the agreement had expired with the emperor. For Dürer, see Silver, "Prints for a Prince," 11; for Stabius, see Maximilian, Letter from Maximilian, in "Urkunden und Regesten," ed. Hans Petz in *Jahrbuch der kunsthistorischen Sammlungen des allerhöchsten Kaiserhauses* 10 (1889) XLIV #5827.

59. With his crowning in 1503, Stabius had been awarded 30 gulden. Maximilian, Letter from Maximilian, in "Urkunden und Regesten," edited by Heinrich Zimerman and Franz Kreyczi in *Jahrbuch der kunsthistorischen Sammlungen des allerhöchsten Kaiserhauses* 3 (1885) XXVII #2551.

60. Grössing, "Johannes Stabius," 248–49.

61. Tannstetter, "Viri mathematici quos inclytum Viennense gymnasium ordine celebres habuit," in Peuerbach, *Tabulae eclypsium*, aa5r.

62. On their activities at the University of Vienna, see chapter 3 in this volume.

63. The emperor met with considerable opposition to many of his reforms and ultimately had to concede authority to different councils; for example, the *Reichsregiment* and the *Reichskammergericht* that the German princes and electors established to weaken central authority.

64. On the Habsburg practice of viewing the university as a corporate body of experts, see Michael Shank, "Academic Consulting in 15th-Century Vienna: The Case of Astrology," in *Texts and Contexts in Ancient and Medieval Science. Studies on the Occasion of John E. Murdoch's Seventieth Birthday*, ed. Edith Sylla and Michael McVaugh (Leiden: Brill, 1997), 245–70.

65. Stiborius, "In Clipeum Austrie," 291r.

66. On the revival of astronomy and astrology at the University of Vienna under Maximilian, see Helmuth Grössing, *Humanistische Naturwissenschaft: Zur Geschichte der Wiener mathematischen Schulen des 15. und 16. Jahrhunderts* (Baden-Baden: Valentin Koerner, 1983), 145–201.

67. For Frederick III's understanding of the university as a corporate body of experts, see Shank, "Academic Consulting."

Chapter Five. Wall Calendars and *Practica*

1. "Erat mihi quondam disputatio in aula tempore coenae cum quodam doctore, qui coepit vituperare studium mathematices. Ego, quum mihi proxime assidebat, quaesivi, an distributio anni non sit necessaria. Respondit non esse adeo necessariam. Nam rusticos suos etiam scire, quando sit dies, quando nox, quando hyems, quando aestas, quando meridies sine cognitione tali. Ego vicissim dixi: Illa profecto non est doctoralis responsio. Ey das ist ein schöner doctor, ein grober narr, man solt im ein dreck ins paret scheissen und uffsetzen. Quanta est haec insania! Est magnum beneficium dei, quod unusquilibet potest literas Calendarum in pariete." Philipp Melanchthon, *Melanchthoniana paedagogica. Eine Ergänzung zu den Werken Melanchthons im Corpus Reformatorum*, ed. Karl Hartfelder (Leipzig: B. G. Teubner, 1892), 182. I thank J. B. Trapp for pointing out this reference.

2. See Leonhard Hoffmann, "Almanache des 15. und 16. Jahrhunderts und ihre Käufer," in *Beiträge zur Inkunabelkunde*, ed. Hans Lülfing, Ursula Altmann, and Heinrich Roloff (Berlin: Akademie Verlag, 1983), 130–43. On calendars in Vienna, see Josef Seethaler, "Das Wiener Kalenderwesen von seinen Anfängen bis zum Ende des 17. Jahrhunderts. Ein Beitrag zur Geschichte des Buchdrucks" (PhD diss., University of Vienna, 1982); Helmut Lang, "Wiener Wandkalender des 15. und 16. Jahrhunderts," *Biblos* 17 (1968): 40–50; Joh.

Wussin, "Alte Wiener Drucke," *Berichte und Mittheilungen des Alterthums-Vereines zu Wien* 26 (1890): 75–82; Josef Wünsch, "Wiener Kalender-Einblattdrucke des XV., XVI. und XVII. Jahrhunderts," *Berichte und Mittheilungen des Alterthums-Vereines zu Wien* 44 (1911): 67–81. On calendar production in Nürnberg, see K. Mathäus, "Zur Geschichte des Nürnberger Kalenderwesens. Die Entwicklung der in Nürnberg gedruckten Jahreskalender in Buchform," *Archiv für Geschichte des Buchwesens* 9 (1969): 965–1396. See also Francis B. Brévart, "The German *Volkskalender* of the Fifteenth Century," *Speculum* 63, no. 2 (1988): 312–42; Francis B. Brévart, "Johann Blaubirers Kalender von 1481 und 1483: Traditionsgebundenheit und experimentelle Innovation," *Gutenberg Jahrbuch* 63 (1988): 74–83. The early work by Gustav Hellmann remains important: Gustav Hellmann, *Versuch einer Geschichte der Wettervorhersage im xvi. Jahrhundert* (Berlin: Akademie der Wissenschaften, 1924); and his *Die Meteorologie in den deutschen Flugschriften und Flugblättern des XVI. Jahrhunderts. Ein Beitrag zur Geschichte der Meteorologie* (Berlin: Akademie der Wissenschaften, 1921).

3. Masters at universities throughout Europe produced annual *practica*, but those at German universities produced the greatest number. See Robert Westman's discussion in *The Copernican Question. Progonostication, Skepticism, and Celestial Order* (Berkeley: University of California Press, 2011), 62–75.

4. *Practica* were also used by Lutheran authors to express despair at the irreligion and degradation of society. See Robin Bruce Barnes, "Hope and Despair in Sixteenth-Century Almanacs," in *Die Reformation in Deutschland und Europa: Interpretationen und Debatten*, ed. Hans Guggisberg, Gottfried G. Krodel, and Hans Füglister (Gütersloh: Gütersloher Verlagshaus, 1993), 440–61 and C. Scott Dixon, "Popular Astrology and Lutheran Propaganda in Reformation Germany," *History* 84 (1999): 403–18.

5. No single prince had a monopoly on these interpretations. Astrologers at rival courts often struggled to assert their interpretation of a particular phenomenon. For one example, see the contest between Perlach and Johannes Carion in chapters 6 and 7 in this volume.

6. I have adopted the term *practica* to refer to both the Latin and vernacular annual astrological pamphlets.

7. Jonathan Green, *Printing and Prophecy. Prognostication and Media Change 1450–1550* (Ann Arbor: University of Michigan Press, 2012), esp. ch. 5.

8. As early as 1456 single-sheet calendars were printed. See Falk Eisermann and Volker Honemann, "Die ersten typographischen Einblattdrucke," *Gutenberg Jahrbuch* 75 (2000), 90. See also Eckhard Simon, *The Türkenkalender (1454) Attributed to Gutenberg and the Strasbourg Lunation Tracts* (Cambridge, MA: Medieval Academy of America, 1988), and Zinner's bibliography for a partial listing of early printed calendars and wall calendars: Ernst Zinner, *Geschichte und Bibliographie der Astronomischen Literatur in Deutschland zur Zeit der Renaissance* (Stuttgart: Anton Hiersemann, 1964).

9. For an overview of medieval calendars, see Teresa Pérez-Higuera, *Medieval Calendars* (London: Weidenfeld and Nicolson, 1998). On astrological medicine and almanacs, see Hilary M. Carey, "What Is the Folded Almanac? The Form and Function of a Key Manuscript Source for Astro-Medical Practice in Later Medieval England," *Social History of Medicine* 16 (2003): 481–509; Hilary M. Carey, "Astrological Medicine and the

Medieval English Folded Almanac," *Social History of Medicine* 17 (2004): 345–63; Cornelius O'Boyle, "Astrology and Medicine in Later Medieval England. The Calendars of John Somer and Nicholas of Lynn," *Sudhoffs Archiv* 89 (2005): 1–22.

10. There are a few exceptions, including Regiomontanus's 1474 *Kalendar*, which was produced as a pamphlet. See also the 1490 *Kalendar* and a 1495 *Kalendar*, both in German and printed in Augsburg. All three are available in facsimile reprints. See Regiomontanus, *Der deutsche Kalender des Johannes Regiomontan,* ed. Ernst Zinner (Leipzig: Otto Harrassowitz, 1937); "Teutscher Kalender," (1922); *Teutsche Kalender* (n.p., n.d.). See also Brévart, "Johann Blaubirers Kalender von 1481 und 1483: Traditionsgebundenheit und experimentelle Innovation."

11. The best source for comparing a wide variety of fifteenth-century calendars is collection of facsimile calendars printed in Paul Heitz and Konrad Haebler, *Hundert Kalender Inkunabeln* (Strasbourg: Heitz und Miendel, 1905). My account is drawn also from examining calendars in the following libraries: Österreichische Nationalbibliothek, Vienna; Herzog August Bibliothek, Wolfenbüttel; Bayerische Staatsbibliothek, Munich; British Library, London; as well as the Photo Collection at the Warburg Institute, London.

12. The *Monatsregeln* were codified in both Latin and German manuscripts as early as the fourteenth century and continued to play an important role in daily life well into the sixteenth century. On the *Monatsregeln*, see Otrun Riha, *Meister Alexanders Monatsregeln. Untersuchung zu einem spätmittelalterlichen Regimen duodecim mensium mit kritischer Textausgabe* (Würzburg: Böhler Verlag, 1985).

13. Some of the more famous fifteenth-century astrologers who composed calendars include Wenceslaus de Budweis, Johannes Angelus, Johannes Glogovia, Johannes Neuman, and Johannes Virdung. Although certainly not complete, Zinner's bibliography is helpful here in estimating the numbers of calendars produced by different people. Examples of calendars by Budewies, Angelus, Sulzpach, and Neuman can be found in Heitz and Haeber, *Hundert Kalender*: Budweis, #57 and 60; Angelus, #62 (missing from Zinner's bibliography); Sulzpach, #58 and 59; Neuman #90. There is a copy of Budweis's 1496 calendar and a copies of Virdung's 1492 and 1495 calendars in the Warburg Photo Collection, Drawer "Prognostica," Folders "Prognostica, Dated—German," each filed under the appropriate year. See also the descriptions of fifteenth-century calendars in Wünsch, "Wiener Kalender-Einblattdrucke des XV., XVI. und XVII. Jahrhunderts."

14. In his *Libellus consolatorius* Tannstetter claims to have been producing *practica* for nineteen years. It is possible that he composed a calendar as early as 1504, but the earliest copy I can identify is from 1505. See Georg Tannstetter, *Libellus consolatorius, quo opinionem iam dudum annis hominum ex quorundam Astrologastrorum divinatione insidentem, de futuro diluuio & multis alijs horrendis periculis XXIIII. anni a fundamentis extirpare conatur* (Vienna: Johannes Singrenius, 1523), A2r.

It was not uncommon for the local prince to require his astrologers at the university to produce yearly calendars. Paul Fabricius was required to produce calendars as part of his duties at the university. See Seethaler, "Wiener Kalenderwesens," 245–45 and 446n17. See also Barbara Bauer, "Die Rolle des Hofastrologen und Hofmathematicus als fürsterlicher Berater," in *Höfischer Humanismus*, ed. August Buck (Weinheim: VCH Verlagsgesellschaft, 1989), 97; Mathäus, "Zur Geschichte des Nürnberger Kalenderwesens,"

1087–92; Hellmann, *Versuch einer Geschichte der Wettervorhersage im XVI. Jahrhundert,*
4.

15. On Muntz and Rosinus, see Seethaler, "Das Wiener Kalenderwesen von seinen
Anfängen bis zum Ende des 17," 64.

16. Seethaler, "Das Wiener Kalenderwesen von seinen Anfängen bis zum Ende des 17,"
72. See also Zinner, *Geschichte und Bibliographie der Astronomischen Literatur in Deutschland zur Zeit der Renaissance,* as well as the older bibliographic and publishing histories,
including Emil Weller, *Repertorium typographicum. Die deutsche Literatur im ersten Viertel
des sechzehnten Jahrhunderts,* 2 vols. (Nordlingen: Beck'schen Buchhandlung, 1864–1885);
Michael Denis, *Wiens Buchdruckergeschichte bis zum Jahre 1560* (Vienna, 1782); Anton
Mayer, *Wiens Buchdruckergeschichte, 1482–1882,* vol. 1 (Vienna: F. Jasper, 1882–87).

17. For the best introduction to using this data for calculating the different variables,
see Hermann Grotefend, *Taschenbuch der Zeitrechnung des deutschen Mittelalters und der
Neuzeit,* 2 vols. (Hannover: Hahnische Buchhandlung, 1891).

18. By the end of the century wall calendars provided a significant supplement to
the author's professorial salary. Josef Seethaler, "Das Wiener Kalenderwesen von seinen
Anfängen bis zum Ende des 17," 260–61.

19. The best introduction to this history of calendar reform is J. D. North, "The Western Calendar-'Intolerabilis, horribilis, et derisiblis': Four Centuries of Discontent," in *Gregorian Reform of the Calendar. Proceedings of the Vatican Conference to Commemorate Its
400th Anniversary, 1582–1982,* ed. G. V. Coyne, M. A. Hoskin, and O. Pedersen (Vatican:
Pontifical Academy of Sciences, 1983), 75–113. See also Christine Gack-Scheiding, *Johannes
de Muris Epistola super reformatione antiqui kalendarii. Ein Beitrag zur Kalendarreform im
14. Jahrhundert* (Hannover: Hahnische Buchhandlung, 1995).

20. On Regiomontanus, see E. Glowatzki and Helmut Göttsche, *Die Tafeln des
Regiomontanus: Ein Jahrhundertwerk* (Munich: Institut für Geschichte der Naturwissen-
schaften, 1990); Ernst Zinner, *Leben und Werken des Joh. Müller von Königsberg gennant
Regiomontanus* (Osnabrück: Otto Zeller, 1968); and James Steven Byrne, "The Stars, the
Moon, and the Shadowed Earth: Viennese Astronomy in the Fifteenth Century" (PhD
diss., Princeton University, 2007), 36–41.

21. On Middelburg's suggestions, see North, "The Western Calendar," 96.

22. "Studiosius committimus, ut his receptis, exacta diligentia huic perquisitioni,
& eo vigiliantius incumbatis, & iuditio & censura vestra decus & famam Germaniae
nostrae nationis adaugeatis & amplificetis. Quam quidem discussionem & cousilium
verstrum nobis in scriptis ad manus Consiliarii & Secretarii nostri Iacobi de Bannissi tras-
mittetis. Si ad dictam sessionem deccimam ad Calen. Decemb. iuxta Pontificis requisitio-
nem proficisci non poteritis." Andreas Stiborius and Georg Tannstetter, *Super requisitione
Sanctissimi Leonis Papae.X. et Divi Maximiliani Imp. P. F. Aug. De Romani Calendarii
correctione Consilium in Florentissimo studio Viennensi Austriae conscriptum & aeditum.*
(Vienna: Johannes Singrenius, 1514), A1v

23. "in eo solvendo nodo, & errore eliminando, qui involvuit omnem celevriorum
religionis nostrae festivitate. Datis in oppido nostro Inspruck quarta Octob. M.D.XIII.
Regni nostri Romani XXVIIII. per Regem per se." Stiborius and Tannstetter, *De Romani
Calendarii,* A1v.

24. The closing letter in their work is dated: "Datis in oppido nostro Inspruckh. Die sextadecima Mensis Decembris Anno dñi. M.D.XIIII. Regni nostro Ro. Vicesimonono." Stiborius and Tannstetter, *De Romani Calendarii*, B4r. A manuscript copy of the text survives: Vienna, Österreichische Nationalbibliothek, cod. lat. 10358, fols. 113–120. This copy was perhaps written down by Andreas Perlach. See Franz Graf-Stuhlhofer, "Andrej Perlach kot ucenec jurija Tannstetterja na dunaju," *Casopis za Zgodovino in Narodopisje St.* 2 (1991), 281.

25. Stiborius and Tannstetter, *De Romani Calendarii*, A4r.

26. Stiborius and Tannstetter, *De Romani Calendarii*, B1r.

27. Stiborius and Tannstetter, *De Romani Calendarii*, A5v.

28. Stiborius and Tannstetter, *De Romani Calendarii*, B3r.

29. Stiborius and Tannstetter, *De Romani Calendarii*, A4r.

30. "in hac saeculorum foelicitate qua se vertitas luce clarius aperit adminiculo artis quo tot iam codices mira celeritate impraesi circuferuntur, adeo ut mediocriter docti ac rudes etiam, ac pene minimi, videre & legere possint quae ante aliquot saecula vix doctissimi potuerunt." Stiborius and Tannstetter, *De Romani Calendarii*, B1v.

31. Stiborius and Tannstetter, *De Romani Calendarii*, B1v.

32. Seethaler, "Das Wiener Kalenderwesen von seinen Anfängen bis zum Ende des 17," 76.

33. The fragment of Tannstetter's 1519 wall calendar suggests that he took data from Perlach's 1519 *Almanach novum*, which Perlach had calculated for Vienna.

34. In 1522 and 1523 Tannstetter produced calendars for Vienna and Salzburg; in 1524 and 1525 he produced them for Vienna and Passau; for 1527 he produced them for Vienna and Olmutz. For a description of this calendar, see Lang, "Wiener Wandkalender," 46.

35. On the *Holzkalender*, see the now dated but still useful Alois Riegle, "Die Holzkalender des Mittelalters und der Renaissance," *Mittheilungen des Instituts für österreichische Geschichtsforschung* 9 (1888): 82–103.

36. The literature on *Bauernkalender* is substantial, though now a bit old. Representative of this literature is Hellmut Rosenfeld, "Die Titelholzschnitte der Bauernpraktik von 1508–1600 als soziologische Selbsinterpretation," in *Festschrift für Joseph Benzing zum sechzigsten Geburtstag. 4. Februar 1964*, ed. Elizabeth Geck and Guido Pressler (Wiesbaden: Guido Pressler Verlag, 1964), 373–89; Hellmut Rosenfeld, "Bauernkalender und Mandlkalender als literarisches Phänomen des 16. Jahrhunderts und ihr Verhältnis zur Bauernpraktik," *Gutenberg Jahrbuch* 38 (1963): 88–96; Hellmut Rosenfeld, "Kalender, Einblattkalender, Bauernkalender und Bauernpraktik," *Bayerisches Jahrbuch für Volkskunde* (1962): 7–24. Seethaler makes the point that the *Bauernkalender* were intended for illiterate people; Seethaler, "Das Wiener Kalenderwesen von seinen Anfängen bis zum Ende des 17," 298–301. See also Otta Wenskus, "Columellas Bauernkalender zwischen Mündlichkeit und Schriftlichkeit," in *Gattungen wissenschaftlicher Literatur in der Antike*, ed. Wolfgang Kullmann, Jochen Althoff, and Markus Asper (Tübingen: Gunter Narr Verlag, 1996), 253–62.

37. Rosenfeld, "Die Titelholzschnitte der Bauernpraktik von 1508–1600 als soziologische Selbsinterpretation."

38. This value is an educated guess, based on the 1,000 copies of Regiomontanus's

calendars printed in 1472 and estimates that in the late sixteenth century as many as 8,000–10,000 copies were printed each year. Conservative estimates by Seethaler suggest that print runs were approximately 400 copies per year, Josef Seethaler, "Das Wiener Kalenderwesen von seinen Anfängen bis zum Ende des 17," 270–72. For Regiomontanus's calendar, see Regiomontanus, *Der deutsche Kalender des Johannes Regiomontan*. The later estimate is drawn from Mathäus, "Zur Geschichte des Nürnberger Kalenderwesens. Die Entwicklung der in Nürnberg gedruckten Jahreskalender in Buchform." Unfortunately, our knowledge of the history of printing in Vienna remains incomplete. Little information survives about the size of print runs, the costs of printing, or sales prices. For the history of printing in Vienna, see Denis, *Wiens Buchdruckergeschichte bis zum Jahre 1560*; Mayer, *Wiens Buchdruckergeschichte, 1482–1882*; and Eduard Langer, *Bibliographie der österreichischen Drucke des 15. und 16. Jahrhunderts* (Vienna, 1913).

39. On the population of Vienna, see Richard Perger, "Der organisatorische und wirtschaftliche Rahmen," in *Wien. Geschichte einer Stadt*vol, vol. I: *Von Anfängen bis zur ersten Wiener Türkenbelagerung (1529)* (Vienna: Böhlau, 2001), 206.

40. Literacy rates in southern German cities have been estimated at 3–5 percent. If this percentage holds for Vienna, there were perhaps 650–1,100 literate residents in the city. Lang first claimed that there were sufficient numbers of calendars printed for each household in Vienna. See Lang, "Wiener Wandkalender," 49. Recent work on the size of Vienna at the end of the fifteenth century suggests there were approximately 1,300 inhabited buildings in Vienna and another 900 in the suburban villages. See Perger, "Der organisatorische und wirtschaftliche Rahmen," 206.

41. For the prices of single-sheet calendars in late fifteenth century, see Brévart, "The German *Volkskalender*," 335.

42. In 1530 the price of a wall calendar was still about one day's wages for a farm laborer. Seethaler, "Das Wiener Kalenderwesen von seinen Anfängen bis zum Ende des 17," 301–3.

43. On the importance of bathing in the late medieval period, see Georg Zappert, "Über das Badewesen mittelaltlicher und späterer Zeit," *Archiv für Kunde österreichischer Geschichtsforschung* 21 (1859): 3–166, esp. 33–42.

44. Andreas Perlach, *Usus almanach seu Ephemeridum: ex commentarijs Georgii Tannsteteri Colimitii, Preceptoris sui decerpti, & in quinquaginta propositiones, per Magistrum Andream perlachium stirum redacti* (Vienna: Hieronimus Vietor, 1519), C1r–C3v.

45. Representative of the vernacular handbooks is the *Temporal Des weitberümpten M. Johann Künigspergers / Natürlicher kunst der Astronomei / kurzer begriff / Von natürlichem einfluß der Gestirn / Planeten / und Zeichen / etc. Von den Complexionen / natur und Eigenschafft der menschen / Regiment durchs Jar über / mit Essen / Schlaffe / Baden / Purgieren / Aderlassen / etc. Auffs ordenlichst zugericht* (Erfurt: Wolffgang Sthürmer, n.d.).

46. The back of one copy of Johann Virdung's 1495 calendar has been extensively annotated by a contemporary reader. Among the topics treated were both weather and harvest predictions. See the copy Johannes Virdung, *[Wall calendar]* (1494), Warburg Institute Photo Collection Drawer: Prognostica; Folder: Dated Prognostica—1495.

47. In the Latin version of his calendar: "Portenta et significationes illarum in prognostico annali exponentur." Georg Tannstetter, *[Wall Calendar for 1523]* (1522), London, War-

burg Institute Photo Collection Drawer: Almanacs, Folder: Almanacs and Calendars—
European.

48. Seethaler, "Das Wiener Kalenderwesen von seinen Anfängen bis zum Ende des 17," 86.

49. François Rabelais, *Pantagrueline prognostication pour l'an 1533. Avec Les almanachs pour les ans 1533, 1535 et 1541. La grande et vraye Pronostications nouvelle de 1544*, ed. M.-A. Screech (Geneva: Librairie Droz, 1974).

50. The earliest surviving English translation of the *Pantagruéline prognostication* is from the mid-seventeenth century; see F. A. Wilson, "Introduction," in *Pantagruel's Prognostication* (Oxford: Luttrell Society, 1947), iv. For a useful introduction to English comic prognostica, see F. A. Wilson, "Some English Mock-Prognostications," *Library* 19 (1938): 6–43.

51. As early as 1527 Aretino had produced such prognostica: *Judicio over pronostico de mastro Pasquino quinto evangelista de anno 1527*. For a brief account of this text, see Lynn Thorndike, *A History of Magic and Experimental Science*, vol. V: *Sixteenth Century* (New York: Columbia University Press, 1958), 203, 312. For a later prognosticon in which Aretino calls Guarico a "buffalo along with the other false ox-astronomers," see also Pietro Aretino, *Un Pronostico satirico di Pietro Aretino (MDXXIIII)*, ed. Alessandro Luzio (Bergamo: Instituto Italiano d'Arti Graciche, 1900), 15.

52. Seethaler, "Das Wiener Kalenderwesen von seinen Anfängen bis zum Ende des 17," 85. Zinner's earliest entry is a multiyear prognostication printed in 1470. Zinner, *Geschichte und Bibliographie der Astronomischen Literatur in Deutschland zur Zeit der Renaissance*, #22.

53. Only the title page from Stabius's 1499 *practica* survives in the Ratsschulbibliothek in Zwickau.

54. On Stabius's activities at the University of Ingolstadt, see Christoph Schöner, *Mathematik und Astronomie an der Universität Ingolstadt im 15. und 16. Jahrhundert*, vol. 13, Ludovico Maximlianea (Berlin: Duncker & Humblot, 1994), esp. 272–84.

55. Johannes Stabius, *Practica Teutsch von Ingoldstat Joannis Stabij Philosophi und Mathematici auff Tausent funffhundert und ain Jar* (n.p., 1500), London, British Library, A1v–A2r.

56. Stabius, *Practica Teutsch*, A4v–A5r, A6r.

57. Shortly after coming to Vienna Stabius produced a learned and elite prognostication based on the Saturn-Jupiter conjunction of 1503/4. On this text, see chapter 7 in this volume. On Stabius's other erudite astrological projects, see chapter 2 in this volume.

58. "Quare ego magister Georgius Lanstetter Collimitius ad honorem omnipotentis dei & ornamentum florentissime Vienensis achademie communemque utilitatem hominum annale prognosticon super anno .M.D.V. stilo facili & plano scribere institui." Georg Tannstetter, *Iudicium Viennese Magistri Georgii Lanstetter Collimitii Super anno M.ccccv.* (n.p.: 1504), A1r.

59. Tannstetter, *Iudicium Viennese . . . M.ccccv*, A1r–A1v.

60. Tannstetter, *Iudicium Viennese . . . M.ccccv*, A1v.

61. Tannstetter, *Iudicium Viennese . . . M.ccccv*, A1v.

62. For a desription of this as it pertains to Maximilian, see Hermann Wiesflecker, *Kaiser Maximilian I. Das Reich, Österreich und Europa an der Wende zur Neuzeit*, vol.

3: *Auf der Höhe des Lebens. 1500–1508. Der große Systemwechsel. Politischer Wiederaufsteig* (Munich: R. Oldenbourg Verlag, 1977), 164–98.

63. Tannstetter, *Iudicium Viennese . . . M.cccc.v*, A2r.

64. Tannstetter, *Iudicium Viennese . . . M.cccc.v*, A2r.

65. Tannstetter, *Iudicium Viennese . . . M.cccc.v*, A2r–A2v.

66. Tannstetter, *Iudicium Viennese . . . M.cccc.v*, A2v.

67. "Invictissimus & serenissimus Romanorum Rex semper Augustus hoc anno bonam fortunam habebit. victoriam in bellis obtinebit populus de aliis principibus ad sacram maiestatem suam fugiet & honorabit. multos in carceres mittet. illis tamen gratiam magnam exhibebit." Tannstetter, *Iudicium Viennese . . . M.cccc.v*, A3r.

68. Tannstetter, *Iudicium Viennese . . . M.cccc.v*, A3r.

69. Tannstetter, *Iudicium Viennese . . . M.cccc.v*, A3r–A4v.

70. Georg Tannstetter, *Practica Teutsch auff das M.cccc. und eylff Jar. durch Mayster Georgen Tannstetter* ([Vienna], 1510), A1r–A1v.

71. Tannstetter, *Practica Teutsch auff das M.cccc. und eylff Jar*, A3r–A3v.

72. Tannstetter, *Practica Teutsch auff das M.cccc. und eylff Jar*, A3r–A3v.

73. Tannstetter, *Practica Teutsch auff das M.cccc. und eylff Jar*, B1r.

74. Johann Virdung's *practica* were extremely popular in the early sixteenth century, and in each he devoted more than half the space to weather predictions. For example, see Johannes Virdung, *Practica deutsch uff das 1523. jare* (Nuremberg, 1522); Johannes Virdung, *Practica deutsch Meister Hansen Virdung von Haßfurt / uff das M.cccc. und xxiij. jare* and Johann Virdung, *Practica deütsch / des würdigen hochgelerten und weytbrümpted Doctor Johansen Virdungs von Haßfurt / auff das M.CCCC. und. XXXI. jare gemact* (n.p., 1530).

75. Efforts to centralize and control the production of precise, technical knowledge is an exercise in political authority. See the essays in M. Norton Wise, ed., *The Values of Precision* (Princeton, NJ: Princeton University Press, 1995).

76. Ulirch Hutten in Georg Tannstetter, *Judcium Viennense Anni Millesimiquingentesimi duodecimi. Ad illustrissimum principem Ludovicum bavarie Ducem Comitemque palatinum rheni* (Nuremberg: Wolfgang Huber, 1511), Munich, Bayerische Staatsbibliothek Res/4° Astr.p. 510, 22, A2v–A3r. All quotations are from this edition.

77. For a description of the woodcut used on Tannstetter's calendar, see Wünsch, "Wiener Kalender-Einblattdrucke des XV., XVI. und XVII. Jahrhunderts," 78.

78. There seems to have been little recycling of woodcuts used in *practica* and calendars, perhaps because the woodcuts wore out after printing *practica*, *judicia*, and wall calendar for a given year. In cases when a woodcut has been recycled, it was often originally used to illustrate a text that was not printed in such large numbers. For example, the woodcut from the title page of Stabius's 1503/4 prognosticon is reused on a Chistopher Hochstetter, *Practica Theutsch Christoffero Hochstetter Doctor der Ertzeney Beschriben auff das Tausend Fünffhundert und achtundzwantzigst jar* (Augsburg: Ulhart, 1527), Most woodcuts seem to have been made or at least selected for the specific text they accompanied.

79. On pirated texts, see Adrian Johns, *The Nature of the Book. Print and Knowledge in the Making* (Chicago: University of Chicago Press, 1998), esp. 324–79 and 444–542.

80. One owner bound the two editions together in the same *Sammelband*: Georg Tannstetter, *Judcium Viennense Anni Millesimiquingentesimi duodecimi. Ad illustrissimum*

principem Ludovicum bavarie Ducem Comitemque palatinum rheni; Georg Tannstetter, *Judcium Viennense Anni Millesimiquingentesimi duodecimi. Ad illustrissimum principem Ludovicum bavarie Ducem Comitemque palatinum rheni* (Cologne: Henry de Nussia, 1511), Munich, Bayerische Staatsbibliothek Res/4° Astr.p. 510, 21.

81. Nussia's edition was made quickly and cheaply. The text is more compressed and fills more of the page. Further, text formatting, particularly paragraph divisions, are nearly absent from the Nussia edition.

82. On Ptolemy's division of the zodiac, see Ptolemy, *Quadripartitum*, II, 3.

83. Tannstetter, *Judcium Viennense Anni Millesimiquingentesimi duodecimi. Ad illustrissimum principem Ludovicum bavarie Ducem Comitemque palatinum rheni.*, A4v.

84. By 1517 Tannstetter was employing this approach in his *practica*. See Georg Tannstetter, *Judicium Astronomicum Viennense. anni M.CCCC.xvij. Ad nobiles et providentissimos dños Magistrum civium et universum inclite urbis Vienne Senatum per Georgium Tannstetter Collimitium artium et medicine doctorem diligenter elaboratum* (Vienna: Johannes Singrenius, 1516), A4r.

85. Tannstetter owned a copy of Jacob Schonheintz *Apologie astrologie* (Nuremberg: Georg Schenk, 1502), which he surely read. This tract was part of a *Sammelband* that contained at least eleven different astrological tracts. The rare book dealer H. P. Kraus purchased it in 1966 and subsequently separated the tracts and sold them. Schonheintz *Apologie* was sold into a private collection and is no longer accessible. For a description of the original *Sammelband*, see H. P. Kraus's *Catalogue 126*. I thank J. Lipton, a former employee of H. P. Kraus, for the information about this *Sammelband*.

On the influence of Pico's *Disputationes adversus astrologiam divinatricem* on astrology in the sixteenth century, see Steven Vanden Broecke, *The Limits of Influence: Pico, Louvain, and the Crisis of Renaissance Astrology* (Leiden: Brill, 2003).

86. Tannstetter, *Judicium Astronomicum Viennense. anni M.CCCC.xvij*, A2v.

87. See Tannstetter's long discussion in the opening pages of his *Judicium*. Tannstetter, *Judicium Astronomicum Viennense. anni M.CCCC.xvij*, A2v.

88. Georg Tannstetter, *Judicium astronomicum Viennense anni M.D.xix. ad Reverendissimi in choro patrem et illusttrissimi principem et d. d. Matteum S. Angeli sa. sa. Ro. ecclesie Cardinalem Burcennse coadiutorem Salisburgennse et in germania legatum. etc. do. suum gratiosissimi Geo. Tannstetter* (Vienna: Johannes Singrenius, 1518), A3r.

89. For example, in his *practica* for 1522 and 1524 Tannstetter found corulers for all the seasons. See Georg Tannstetter, *Ad laudem et gloriam am Serenissimi et Ptoentissimi principis Ferdinandi. Principis Hispaniarum. Archiducis Austrie. Ducis Burgundie etc. et ad usum commoditatemque provinciarum Austrie. Georgius Tannstetter Collimitius Lycoripensis Artium et Medicine Doctor. Vienne hoc prognosticum pro anno Christi 1522 edidit* (Landshut: Johann Weßemburger, 1521); Georg Tannstetter, *Practica gemacht zu Wienn in Osterreych aus der Kunst Astronomia / durch Georgen annstetter von Rain am Lech / der freyen künsten und erzney Doctor / und ist wie ain außzugund ain sundere erklerung des grosseren Judicij so vorhin auff das M.D.XXIIII. Jar von im ausgangen ist* (Vienna: Johannes Singrenius, 1523).

90. For a discussion of this, see chapter 6 in this volume.

91. On the war in Italy as it pertains to Maximilian, see Hermann Wiesflecker, *Kaiser Maximilian I. Das Reich, Österreich und Europa an der Wende zur Neuzeit*, vol. 4:

Gründung des habsburgischen Weltreiches, Lebensabend und Tod, 1508–1519 (Munich: R. Oldenbourg Verlag, 1981), 1–153.

92. Georg Tannstetter, *Judcium Viennense Anni Millesimiquingentesimi duodecimi. Ad illustrissimum principem Ludovicum bavarie Ducem Comitemque palatinum rheni.*, A3v–A4r.

93. Georg Tannstetter, *Judicium Astronomicum pro anno Christi. M.cccc.xiij. per Magister Georgium Tannstetter collimitium astronomie professorem Viene Pannonie editum.* (Nuremberg: Wolfgang Huber, 1512), A3r.

94. On the relationship between Maximilian's politics and the French Disease, see chapter 2 in this volume. See also Darin Hayton, "Joseph Grünpeck's Astrological Explanation of the French Disease," in *Responding to Sexual Disease in Early-Modern Europe*, ed. Kevin Siena (Toronto: Centre for Renaissance and Reformation Studies, 2005), 241–74.

95. He warned of outbreaks in 1512, 1513, 1517, 1519, 1522, and 1524.

96. The literature on the 1524 conjunction is immense. See "Aus der Blütezeit der Astrometeorologie," in Gustav Hellmann, *Beiträge zur Geschichte der Meteorologie* (Berlin: Behrend, 1917); Thorndike, *A History of Magic*, 173–233; Paolo Zambelli, "Fine del mondon o inizio della propaganda? Astrologia, filosofia della storia e propaganda politico-religiosa ne dibattito sulla congiunzione del 1524," in *Scienze, credenze occulte, livelli di cultura* (Florence: 1982), 291–368; Paolo Zambelli, "Many Ends for the World. Luca Gaurico Instigator of the Debate in Italy and in Germany," in *"Astrologi hallucinati." Stars and the End of the World in Luther's Time*, ed. Paolo Zambelli (Berlin: Walter de Gruyter, 1986), 239–63; Heike Talkenberger, *Sintflut. Prophetie und Zeitgeschehen in Texten und Holzschnitten astrologichser Flugschriften, 1488–1528* (Tübingen: Max Niemeyer Verlag, 1990).

97. For an analysis of these works, see chapter 7 in this volume.

98. Tannstetter, *Practica gemacht zu Wienn*, A2r.

99. Tannstetter, *Practica gemacht zu Wienn*, B1r.

100. "Am Frreitag in der opposicion ein wunderlich regnung selzam unser wetter gewesen." Marginal note in Tannstetter, *Practica gemacht zu Wienn*, B1r.

101. "Ist diß ganz wircl schon gewesen." Marginal note in Tannstetter, *Practica gemacht zu Wienn*, B1r.

102. For an extended treatment of this work, see chapter 7 in this volume.

103. " . . . / ich werde alles das ich vor mals in dem grössern Iudicio auff obgemelts jar gemacht hab / yez in diser *practica* widerrüffenn /. . . " Tannstetter, *Practica gemacht zu Wienn*, A1v.

104. Tannstetter, *Practica gemacht zu Wienn*, A4r.

105. On Stöffler's attack, see chapter 7 in this volume.

106. "Hey inn diser *practica* hast du ambermals mein Judicium unnd maynung von den künftigen züfellen des M.D.XXiiij. jars. Darauß du leiderlich versteen magst / mit was unbillichkait und lugen mein namen schmehen thün." Tannstetter, *Practica gemacht zu Wienn*, A1v.

107. Tannstetter, *Practica gemacht zu Wienn*, B3v–B4r.

108. Compare the copies: Tannstetter, *Practica gemacht zu Wienn*, Regensburg, Staatliche Bibliothek, Phil. 2280, angeb. 1 and Georg Tannstetter, *Practica gemacht zu Wienn*

in Osterreych aus der Kunst Astronomia / durch Georgen annstetter von Rain am Lech / der freyen künsten und erzney Doctor / und ist wie ain außzugund ain sundere erklerung des grosseren Judicij so vorhin auff das M.D.XXIIII. Jar von im ausgangen ist (Augsburg: 1523), Augsburg, Staats- und Stadtsbibliothek, 4 Kult 186–131.

109. On the relation of Vögelin's activities to Tannstetter's, see Franz Graf-Stuhlhofer, "Das Weiterbestehen des Wiener Poetenkollegs nach dem Tod Konrad Cetis' (1508)," *Zeitschrift für historische Forschung* 26, no. 2 (1999): 399; Josef Seethaler, "Das Wiener Kalenderwesen von seinen Anfängen bis zum Ende des 17." On Perlach's adoption of the *practica* format, see chapter 6 in this volume.

110. Georg Tannstetter, *Juditium Astronomicum Viennense Georgij Tannstetter Collimitij Lycoripensis artium et medicine doctoris in annum christi. 1525* (Vienna: Johannes Singrenius, 1524), Augsburg, Staats- und Stadtsbibliothek 4 Math. 575.

111. See the discussion of Schreibkalender in Seethaler, "Das Wiener Kalenderwesen," 293–98.

112. On Perlach's ephemerides, see chapter 6 in this volume.

113. After this, another of Tannstetter's students, Johannes Vögelin, largely assumed responsibility for producing wall calendars. Seethaler, "Das Wiener Kalenderwesen," 77.

Chapter Six. Ephemerides and Their Uses

1. "A magistro Andrea perlachio Stiro Super Almanach collectanea 1519 die jovis ante Esto mihi. Almanach arabum, Diale/Diurnale latinum, Ephemerides grecum est liber in quo astra de die in diem destribunture. Nota Quilibet planeta eciam quodlibet zodiaci signum quandam habent commemenciam [*sic*] seu similitudinem ad suum characterem qui ipsum denotat non ergo fortuitu et a casu illi characteres imposti sunt. Est igitur arietis character talis ♈ qui videtur duo cornua qualia arietis sunt representare." Andreas Perlach, "A magistro Andrea Perlachio Stiro super almanach collectanea," Vienna, Österreichische Nationalbibliothek Ser. n. 4265, 307v.

2. On wall calendars and *practica*, see chapter 5 in this volume.

3. Perlach's subtitle includes the following line: "Ex tabulis doctiss. viri magistri Johannis de Gmunden studii quondam Viennensis alumni."

4. The subtitle ends with "in officina Collimitiana."

5. On Tannstetter's learned texts, see chapter 4 in this volume. For more on wall calendars, *practica*, and prognostications, see chapters 5 and 7 in this volume.

6. For a brief discussion of Johann Cuspinian's use of ephemerides, see Hans Ankwicz Kleehoven, *Der Wiener Humanist Johannes Cuspinian. Gelehrter und Diplomat zur Zeit Kaiser Maximilian I.* (Graz: Hermann Böhlaus, 1959), 24–25.

7. Readers often annotated their ephemerides. See, for example, some of the annotated copies of Perlach's almanacs: e.g., Andreas Perlach, *Almanach novum super anno Christi saluatoris nostri: M.D.XVIII. Ex tabulis doctiss. viri magistri Joannis de Gmunden studij quondam Wiennensis* (Vienna: Hieronimus Vietor, 1517), Munich, Bayerische Staatsbibliothek Res/4° Eph. Astr.38; Andreas Perlach, *Almanach novum super anno Christi saluatrois nostri. M.D.xviiij. Ex tabulis doctiss. viri magistri Joannis de Gmunden studij quondam Viennense alumni in officina Collimitiana* (Vienna: Johannes Singrenius, 1518), Kraków, Biblioteka Jagiellonska Cim 4735.

8. "Cum Ephemerides anni imminentis computatione sedula per me compositas edidissem, dignum existimavi studiosis lectoribus laborem, Si Canones nonnullos generales exeruditissimis & copiosissimis comentariis, non minus ducentis propositionibus digestis praeceptoris mei & colendissimi & fidelissimi Georgii Tannstetter Collimitii Licoripensis, decerptos, publice commoditatis spe provocatus / subiungerem." Andreas Perlach, *Usus almanach seu Ephemeridum: ex commentarijs Georgii Tannsteter Colimitii, Preceptoris sui decerpti, & in quinquaginta propositiones, per Magistrum Andream perlachium stirum redacti* (Vienna: Hieronimus Vietor, 1519), A1v.

9. Andreas Perlach, *Almanach novum super anno . . . M.D.XVIII*, A1v. On Bishop Georg Slatkonia, see Theophil Antonicek, Elisabeth T. Hilscher, and Harmut Krones, eds., *Die Wiener Hofmusikkapelle. Georg von Slatkonia und die Wiener Hofmusikkapelle* (Vienna: Böhlau, 1999); Gernot Gruber, "Beginn der Neuzeit," in *Musikgeschichte Österreichs. Von den Anfängen zum Barock*, ed. Rudolf Flotzinger (Vienna: Böhlau Verlag, 1995), 169–213, esp. 186–94.

10. On the interaction of professors and the wider public and the practical use of elite knowledge, see Jacques Verger, *Men of Learning in Europe at the End of the Middle Ages* (South Bend, IN: University of Notre Dame, 2000), esp. ch. 2.

11. On letters of dedication, see Kevin Sharpe, *Reading Revolutions. The Politics of Reading in Early Modern England* (New Haven, CT: Yale University Press, 2000), 52–55, and Brian Richardson, *Printing, Writers and Readers in Renaissance Italy* (Cambridge: Cambridge University Press, 1999), 50–57.

12. Johannes Stöffler and Jakob Pflaum, *Almanach nova plurimis annis venturis inservientia* (Venice: Peter Lichtenstein, 1507), A3r–A8r.

13. Stöffler and Pflaum, *Almanach nova*, A8r–B8r.

14. In 1548 Peter Pitatus published an edition of Stöffler's ephemerides that covered the years 1544 to 1556. To this set of ephemerides Pitatus attached a lengthy guide to their use. He also rearranged the introduction and significantly expanded its content. This offered one of the best guides to using the ephemerides available and is comparable with Perlach's *Commentaria*. For a discussion of Perlach's text, see chapter 3 in this volume.

15. One reader corrected some of the errors in the tables of horoscopic houses. See Johannes Stöffler and Jakob Pflaum, *Almanach nova plurimis annis venturis inservientia* (Venice: Peter Lichtenstein, 1506), Munich, Bayerische Staatsbibliothek Res/4° Eph. Astr.126. For a particularly interesting copy of Stöffler and Pitatus's *Ephemeridum reliquiae* in which the reader left scarcely a page unmarked, see Johannes Stöffler and Peter Pitatus, *Ephemeridum reliquiae Ioannis stoeffleri germani, superadditis novis usque ad annum Christi 1556. durantibus, Petri Pitatis Veronensis Mathematicis, una cum additionibus longe utilissimis eiusdem* (Tubingen: Ulrich Morhard, 1548), Munich, Bayerische Staatbibliothek 4°Eph. astr. 132.

16. On Tannstetter's lectures on ephemerides, see chapter 4 in this volume. The claim that Perlach's text is the same as Tannstetter's rests on a spurious citation in Meyer: "Denis, W. B., S. 139, Nr. 199, kannte nur die Ausgabe vom Jahre 1519. 'Der Commentarius ist gleich dem *Usus* Almanach Adreae Perlachii." Anton Mayer, *Wiens Buchdruckergeschichte, 1482–1882*, vol. 1 (Vienna: F. Jasper, 1882–87), 34. Meyer's reference is to Michael Denis, *Wiens Buchdruckergeschichte bis zum Jahre 1560* (Vienna: 1782).

17. Perlach, *Usus Almanach*, A2r.

18. Perlach, *Usus Almanach*, A2v–B1v.

19. Perlach, *Usus Almanach*, B1v.

20. Perlach, *Usus Almanach*, B1v–B2v, B4v.

21. Perlach, *Usus Almanach*, A3v.

22. Perlach, *Usus Almanach*, A3v.

23. "Figuram coeli ad quodvis momentum artificialiter constituere." Perlach, *Usus Almanach*, c2r.

24. See *Canon Joannis de monte regio in Ephemerides* (Vienna, 1512). Vienna, Universitätsbibliothek Sig. I.233.605 Adl.4.

25. For a study on the "bookish" nature of natural philosophy in the sixteenth century, particularly the act of excerpting quotations to use later as authorities, see Ann Blair, *The Theater of Nature* (Princeton, NJ: Princeton University Press, 1997), esp. ch. 2.

26. Perlach, *Usus Almanach*, E4r.

27. "Optimum autem, ut Juppiter in ascendente & in medio coeli sol existat." Perlach, *Usus Almanach*, E4r.

28. If the pattern of the late fifteenth century continued, it is likely that Ptolemy's *Centiloquium* made it on the official course list every few years, and might have been a more common text in other lectures. On astronomical teaching at the University of Vienna in the late fifteenth century, see Claudia Kren, "Astronomical Teaching at the Late Medieval University of Vienna," *History of Universities* 3 (1983): 15–30. For teaching in general at this time, see Claudia Kren, "Patterns in Arts Teaching at the Medieval University of Vienna," *Viator* 18 (1987): 321–27.

29. Perlach, *Usus Almanach*, E1v–E4r.

30. Ptolemy makes this distinction in book 2, chapter 1 of his *Tetrabiblos*. Ptolemy, *Quadripartiti*, 25v.

31. See Eugenio Garin, *Astrology in the Renaissance. The Zodiac of Life*, trans. Carolyn Jackson (London: Routledge & Kegan Paul, 1976).

32. For a detailed discussion of the use of intermediaries in patronage relationships, see Mario Biagioli, *Galileo Courtier. The Practice of Science in an Age of Absolutism* (Chicago: University of Chicago Press, 1993). For a critique of Biagioli's work, see Michael Shank, "How Shall We Practice History? The Case of Mario Biagioli's *Galileo Courtier*," *Early Science and Medicine* 1, no. 1 (1996): 106–50.

33. See also Franz Graf-Stuhlhofer, "Andrej Perlach kot ucenec jurija Tannstetterja na dunaju," *Casopis za Zgodovino in Narodopisje St.* 2 (1991): 280–83.

34. Tannstetter remarked in the preface to his *Ad laudem et gloriam Serenissimi et Potentissimi principis Ferdinandi* (1522) that he had observed the motions of Mars "not infrequently with my Perlach." Two years later in *In gratiam serenissimi ac potentissimi Principis & domini Ferdinandi* and its German translation, Tannstetter's tracts on the great conjunction of 1524, he again noted that Perlach had assisted him "in no little degree."

35. For an account and discussion of the types of marginalia, see William H. Sherman, *John Dee. The Politics of Reading and Writing in the English Renaissance* (Amherst: University of Massachusetts Press, 1995), 53–100; Blair, *The Theater of Nature*.

36. In 1520 the plague was again a problem in Vienna, which prompted Tannstetter to write his *Regiment für den Lauff der Pestilentz*.

37. See the interlineations in Andreas Perlach, *Usus almanach seu Ephemeridum: ex commentarijs Georgii Tannsteter Colimitii, Preceptoris sui decerpti, & in quinquaginta propositiones, per Magistrum Andream perlachium stirum redacti* (Vienna: Hieronimus Vietor, 1518), Österreichische Nationalbibliothek Sig. 72.V.77, E1v and E2r.

38. See Andreas Perlach, *Usus almanach seu Ephemeridum: ex commentarijs Georgii Tannsteter Colimitii, Preceptoris sui decerpti, & in quinquaginta propositiones, per Magistrum Andream perlachium stirum redacti* (Vienna: Hieronimus Vietor, 1518), Munich, Bayerische Staatsbibliothek Sig. 4 Eph. Astr. 88d.

39. See Perlach, *Usus Almanach*, BSB Sig. 4 Eph. Astr. 88d, b2v. This copy is in sad shape: it is missing a number of pages, and others have been bound together out of order.

40. Next to the line, "ab occasu solis" this reader wrote "ut bohemis"; next to "ab ortu" he wrote "ut nirumbergensis"; next to "dimidium" he wrote "ut Wiennesis." See *Utilitasdecima* in Perlach, *Usus Almanach*, BSB Sig. 4 Eph. Astr. 88d, b1v.

41. "Quando dua corpora veniunt in eundem locum." See Perlach, *Usus Almanach*, BSB Sig. 4 Eph. Astr. 88d, a2r.

42. Perlach, *Usus Almanach*, BSB Sig. 4 Eph. Astr. 88d, b4r.

43. Perlach, *Usus Almanach*, BSB Sig. 4 Eph. Astr. 88d, b4r.

44. Ptolemy's *Quadripartitum* was published first in 1484. This was the Aegidius of Tebaldis's translation and included his translation of Haly's commentary as well as Ptolemy's *Centiloquium*. This edition was reprinted in 1493 and 1519. See Lynn Thorndike, *A History of Magic and Experimental Science*, vol. I: *During the First Thirteen Centuries of Our Era* (New York: Columbia University Press, 1923), 110.

45. Lisa Jardine and Anthony Grafton, "'Studied for Action': How Gabriel Harvey Read His Livy," *Past and Present* 129 (1990): 30–78.

46. "pilosa signa." Andreas Perlach, *Usus Almanach*, Vienna Österreichische Nationalbibliothek 72.v.77, E2r.

47. "Signum plantandum / seminandum." Perlach, *Usus Almanach*, Österreichische Nationalbibliothek Sig. 72.v.77, E2r.

48. "Mania est insania vel morbus furoris. Nostro unsinnig." Perlach, *Usus Almanach*, Österreichische Nationalbibliothek Sig. 72.v.77, D1r.

49. "Item Cuspis—dicitur initium alicuius domus Et planeta dicitur esse in cuspide quando est infra 5 gradus a principio alicuius domus." Perlach, *Usus Almanach*, Österreichische Nationalbibliothek Sig. 72.v.77, D4r. On the *medium coeli* and planetary aspects, see Perlach, *Usus Almanach*, Österreichische Nationalbibliothek Sig. 72.v.77, D1r and E1r.

50. "24 dies februaris est locus Bisexti diei aiunt Bisextus quia bis in eo mense diurnus 6to kalendris martij fieri." Andreas Perlach, *Usus almanach seu Ephemeridum: ex commentarijs Georgii Tannsteter Colimitii, Preceptoris sui decerpti, & in quinquaginta propositiones, per Magistrum Andream perlachium stirum redacti* (Vienna: Hieronimus Vietor, 1518), Österreichische Nationalbibliothek Sig. *69.H.66(3), t.p.

51. Perlach, *Usus Almanach*, Österreichische Nationalbibliothek Sig. *69.H.66(3), t.p.

52. See, for example, his note at Perlach, *Usus Almanach*, Österreichische Nationalbibliothek Sig. *69.H.66(3), A3r.

53. Perlach, *Usus Almanach*, Vienna, Österreichische Nationalbibliothek Sig. *69.H.66 (3),B2v.

54. On practical and goal-oriented reading, see Jardine and Grafton, "'Studied for Action'"; Anthony Grafton, *Bring Out Your Dead. The Past as Revelation* (Cambridge, MA: Havard University Press, 2001), 1–15.

55. On this conjunction and Tannstetter's prognostication stemming from it, see chapter 7 in this volume.

56. On wall calendars, see chapter 5 in this volume; on prognostications see chapter 7 in this volume.

57. "ut Ephemerides in annum 29 locupletiores / maioreque ingenio et industria compositas / cum ad gloriam T. M. tum ad studij Viennensis / quod T. M. iam et stipendijs liberalioribus tum etiam lectoribus singularum facultatum clarissimis restaurare cepit . . . ad imprimendum darem." Andreas Perlach, *Ephemerides Andree Perlachij Stiri ex Witscheyn, Artium & Philosophiae magistri magnificae Urbis Viennensis collegae, pro Anno domini & Saluatoris nostri Iesu christi M.D.XXIX. cum configurationibus & habitudinibus planetarum inter se & cum stellis fixis utique insignioribus cum ex secundo tum ex primo mobili contingentibus, sub talis forma hactenus nemini visae* (Vienna: Hieronimus Vietor, 1528), Österreichische Nationalbibliothek 72.T.95, A1v.

58. On Ferdinand's reforms at the University of Vienna, see Kurt Mühlberger, "Die Gemeinde der Lehrer und Schüler—Alma Mater Rudolphina," in *Wien. Geschichte einer Stadt. Band I: Von den Anfängen bis zur ersten Wiener Türkenbelagerung (1529)*, ed. Peter Csendes and Ferdinand Opll (Vienna: Böhlau, 2000), 395–96; Rudolf Kink, *Geschichte der kaiserlichen Universität zu Wien*, vol. 1. Geschichtliche Darstellung (Vienna: Carl Gerold & Sohn, 1854), 252–57.

59. "His etiam annectere placuit confutationem vaticinij Joannis Charionis Brandenburgensis astrologi / qui edito nuper iudicio / multas Austrie minas et lachrymas presagit et vaticinatur / . . . / quem quidvis aliud quam verum vatem esse Austriacis malim / haud bene imminentibus huic provintie periculis prospexisse declarerem / et metum animis quorundam ob illius vaticinium citra rationem incussum eximere." Perlach, *Ephemerides . . . M.D.XXIX*, A1v.

60. Perlach, *Ephemerides . . . M.D.XXIX*, A3v.

61. "Ex his quisque mediocriter ingenij lector intelligere poterit" Perlach, *Ephemerides . . . M.D.XXIX*, A4r. On the use of astrology to predict the weather, see Charles Burnett and Gerrit Boss, *Scientific Weather Forecasting in the Middle Ages: The Writings of al-Kindi* (London and New York: Kegan Paul International, 2000); Stuart Jenks, "Astrometeorology in the Middle Ages," *Isis* 74, no. 2 (1983): 185–210; Joëlle Ducos, *La méteorologie en Français au Moyen Âge (XIII–XIVe siècles)* (Paris: Champion, 1998). On weather prediction in ephemerides and other ephemeral literature, Gustav Hellmann's early work remains important, e.g., Gustav Hellmann, *Meteorologische Volksbücher : Ein Beitrag zur Geschichte der Meteorologie und zur Kulturgeschichte* (Berlin: H. Paetel, 1895); Gustav Hellmann, *Beiträge zur Geschichte der Meteorologie* (Berlin: Behrend & Co., 1917); Gustav Hellmann, *Die Meteorologie in den Deutschen Flugschriften und Flugblattern des XVI. Jahrhunderts; Ein Beitrag zur Geschichte der Meteorologie* (Berlin: Akademie der Wissenschaften, 1921).

62. Perlach, *Ephemerides . . . M.D.XXIX*, A4v. This belief was found in both Ptolemy and his Arab commentators. The Arab writers developed this into the theory of great conjunctions that enjoyed a wide currency in the Latin west from the twelfth to the seventeenth centuries. On the theory of great conjunctions, see J. D. North, "Astrology and the Fortunes of Churches," in *Stars, Minds and Fate: Essays in Ancient and Medieval*

Cosmology, ed. J. D. North (London: Hambledon, 1989), 59–89. See also the recent edition of Albumasar's *On Great Conjunctions*, Abū Maʿšar, *On Historical Astrology. The Book of Religions and Dynasties (On the Great Conjunctions)* (Leiden: Brill, 2000).

63. Perlach, *Ephemerides . . . M.D.XXIX*, M2r.

64. Perlach, *Ephemerides . . . M.D.XXIX*, M2v.

65. Perlach, *Ephemerides . . . M.D.XXIX*, M4r. Perlach seemed almost prescient in this prediction. On September 26, a Turkish army led by Suleiman the Great surrounded Vienna and besieged the city for nearly a month. A brief account of the siege of Vienna is found in Peter Csendes and Ferdinand Opll, eds., *Wien. Geschichte einer Stadt*, vol. 1: *Von den Anfängen bis zur Ersten Türkenbelagerung* (Vienna: Böhlau, 2001), 187–90. See also Paula Sutter Fichtner, *Ferdinand I of Austria: The Politics of Dynasticism in the Age of the Reformation*, East European Monographs (New York: Columbia University Press, 1982), 84; Geoffrey Parker, "The Political World of Charles V," in *Charles V and His Time. 1500–1558*, ed. Hugo Soly (Antwerp: Mercatorfonds, 1999), 155–56.

66. On Carion, see Reiner Reisinger, *Historische Horoskopie. Das iudicium magnum des Johannes Carion für Albrecht Dürers Patenkind* (Wiesbaden: Harrassowitz, 1997); Dietmar Fürst and Jürgen Hamel, *Johann Carion (1499–1537): Der erste Berliner Astronom* (Berlin: Archenhold-Sternwarte, 1988).

67. Gustav Hellmann, *Versuch einer Geschichte der Wettervorhersage im xvi. Jahrhundert* (Berlin: Akademie der Wissenschaften, 1924), 26.

68. See the bibliography in Fürst and Hamel, *Johann Carion (1499–1537): Der erste Berliner Astronom*.

69. Johannes Carion, *Beteutnuss und Offenbarung* (Berlin: 1526), B2v–B3r.

70. He was certainly not wrong in this assumption. At least one reader was particularly concerned with Perlach's refutation of Carion's work, underlining the relevant lines in the preface where Perlach initially attacked Carion. See Andreas Perlach, *Ephemerides Andree Perlachij Stiri ex Witscheyn, Artium & Philosophiae magistri magnificae Urbis Viennensis collegae, pro Anno domini & Saluatoris nostri Iesu christi M.D.XXIX. cum configurationibus & habitudinibus planetarum inter se & cum stellis fixis utique insignioribus cum ex secundo tum ex primo mobili contingentibus, sub talis forma hactenus nemini visae* (Vienna: Hieronimus Vietor, 1528), Boston Public Library *E.189.51, A1v.

71. This exchange took place in the preface to Perlach's 1531 tract on the significance of a recent comet. See chapter 7 in this volume.

72. Perlach, *Ephemerides . . . M.D.XXIX*, M4v.

73. On Ptolemy's development of this, see Ptolemy, *Tetrabiblos*, I, 19. See his discussion in the contemporary edition: Ptolemy, *Quadripartiti*, 19r–19v.

74. These quadrangles were based on Ptolemy's division of the signs into solstitial and equinoctial, solid, and bicorporeal signs. See Ptolemy, *Tetrabiblos*, I, 12. See Ptolemy, *Quadripartiti*, 15v–16v.

75. Perlach, *Ephemerides . . . M.D.XXIX*, M4v.

76. Perlach, *Ephemerides . . . M.D.XXIX*, M5r.

77. Perlach, *Ephemerides . . . M.D.XXIX*, M5v.

78. Perlach, *Ephemerides . . . M.D.XXIX*, M5v.

79. Perlach, *Ephemerides . . . M.D.XXIX*, M6r.

80. Fichtner, *Ferdinand I*, 79–91.

81. Parker, "The Political World of Charles V," esp. 213–14.

82. "qui singulari quodam amore complectitur omnes bonarum literarum studiosos / utique eos qui in rem astronomicam incunbunt / et precipue me /. . . ." Perlach, *Ephemerides Andreae Perlachii Stiri ex Witschein, Artium & Philosophiae magistri, Mathematicarum disciplinarum studij Viennesis professoris ordinarij, pro Anno domini & Saluatoris nostri Iesu Christi M.D.XXXI. cum configurationibus & habitudinibus planetarum inter se, & cum stellis fixis, cum ex secuno, tum ex primo mobili contingentibus. Insuper adiunximus his nostris in hunc annum Ephemeridibus, Prognosticon, superioris anni eclipsium, quarum effectus hoc Anno apparebunt* (Vienna, 1530), Österreichische Nationalbibliothek 72.T.95*, A1r.

83. Perlach, *Ephemerides . . . M.D.XXXI*, A4r.

84. Perlach, *Ephemerides . . . M.D.XXXI*, A2v.

85. Perlach, *Ephemerides . . . M.D.XXXI*, A2v–A4r.

86. Ottavia Niccoli, *Prophecy and People in Renaissance Italy* (Princeton, NJ: Princeton University Press, 1990), 159–60.

87. "Prodijt etiam ante paucos annos / ex stirpe Maximiliani Cesaris / perpetua memoria digni quidam / qui dictam stellam in Horoscopo / et quod equidem miraculi loco duxerim alteram infortunam in loco stelle / sub qua cepit secta Mahumetica habet / Is si deus vitam prolongaverit / proprijs / ac suorum amicorum utetur viribus ad extirpandam / profligandam / et pellandam hanc sectam ex ijs locis / que antecessoribus suis antea per vim et iniuriam ademit / et erit vindex atque ultor omnium malorum / que anteactis temporibus patrata sunt." Perlach, *Ephemerides . . . M.D.XXXI*, A1r.

88. Fichtner, *Ferdinand I*, 88.

Chapter Seven. Prognostications

1. On Girolamo Cardano's attempt to rectify Luther's horoscope, see Anthony Grafton, *Cardano's Cosmos. The Worlds and Works of a Renaissance Astrologer* (Cambridge, MA: Harvard University Press, 1999), 76–77.

2. "Genesin istam meam iam ante videram ex Italia huc missam. Sed cum sic sint hoc anno hallucinati astrologi, nihil mirum, si sit qui & hoc nugari ausus sit." Quoted in Martin Luther, *D. Martin Luthers Briefwechsel*, vol. 3: *1523–1525* (Weimar: Hermann Böhlaus, 1933), 260.

3. The best discussion of this is Alois Schmid's article, "*'Poeta et orator a Caesare laureatus'*: Die Dichterkrönungen Kaiser Maximilians I," *Historisches Jahrbuch* 109 (1989), esp. 65–83. Hermann Wiesflecker makes much the same point in *Kaiser Maximilian I. Das Reich, Österreich und Europa an der Wende zur Neuzeit*, vol. 5: *Der Kaiser und seine Umwelt. Hof, Staat, Wirtschaft, Gesellschaft und Kultur* (Munich: R. Oldenbourg Verlag, 1986), 452–66. See also Peter Diederichs, *Kaiser Maximilian I als politischer Publizist* (Heidelberg, 1931); Edeltraud Hönig, "Kaiser Maximilian I als politischer Publizist" (PhD diss., University of Graz, 1970).

4. Schmid, "*'Poeta et orator a Caesare laureatus*,'" 99. Stabius's *Martyrium S. Colo-*

manni Martyris Austriaci sapphico carmine conscriptum accompanied a woodcut of the Saint by Hans Springinklee, and was finally printed in 1513.

5. On Maximilian's efforts to establish his ancestors and his own sanctity, see Larry Silver, *Marketing Maximilian. The Visual Ideology of a Holy Roman Emperor* (Princeton, NJ: Princeton University Press, 2008), esp. chs. 2 and 4.

6. Conrad Celtis, *Ludus Diane in modum Comedie coram Maximiliano Rhomanorum Rege Kalendis Martijs et Ludis saturnalibus in arce Linsiana danubij actus: Clementissimo Rege et Regina ducibusque illustribus Mediolani totaque Regia curia spectatoribus: per petrum Bonumum Regi: Cancel. Joseph Grunpeckium Reg. Secre. Conradum Celten: Reg: Poe. Ulsenium Pfrisium: Vincentium Longinum in hoc Ludo Laurea donatum foeliciter et iucundissime representatus* (Nuremberg: Hieronymus Hölecius, 1501). On Celtis's *Ludus Diane*, see Lewis W. Spitz, *Conrad Celtis. The German Arch-Humanist* (Cambridge, MA: Harvard University Press, 1957), 72–76; Virginia Gingerick, "The *Ludus Diane* of Conrad Celtis," *Germanic Review* 15 (1940): 159–80.

7. Johannes Stabius, "Carmen Sapphicum: ad Max. Ro. re. Trenice" in Celtis, *Ludus Diane*, A5v–A6v.

8. There are two copies of the broadsheet in the Bayerische Staatsbibliothek, Johannes Stabius, *Pronosticon Ioann. Stanbii. Au. vatis. & Poc. Lau. ad annos domini. M.D.III.&IIII.* (Nuremberg: Johannes Weissenburger, 1502), Munich, Bayerische Staatsbibliothek 2 Einblat. IV, 7; Johannes Stabius, *Pronosticon Ioann. Stanbii. Au. vatis. & Poc. Lau. ad annos domini. M.D.III.&IIII.* (Nuremberg: Johannes Weissenburger, 1502), Munich, Bayerische Staatsbibliothek 2 Einblat. IV, 7a. A third copy that has been trimmed considerably is found in the Staats- und Stadtbibliothek, Augsburg, Johannes Stabius, *Pronosticon Ioann. Stanbii. Au. vatis. & Poc. Lau. ad annos domini. M.D.III.&IIII.* (Nuremberg: Johannes Weissenburger, 1502), Augsburg, Stadts- und Staatbibliothek, Sig. 2 Ink 651. Two copies of the quarto-format pamphlet survive: Johannes Stabius, *Pronosticon Ioann. Stanbii. Au. vatis. & Poc. Lau. ad annos domini. M.D.III.&IIII.* (Nuremberg: Johannes Weissenburger, 1502), Halle, Marienbibliothek Q3.105(19); Johannes Stabius, *Pronosticon Ioann. Stanbii. Au. vatis. & Poc. Lau. ad annos domini. M.D.III.&IIII.* (Nuremberg: Johannes Weissenburger, 1502), Bamberg, Staatsbibliothek Bamberg 22/Inc.typ.Ic.I.4#4.

9. In his poem, Stabius borrowed liberally from the epic poets such as Vergil, Lucan, Ovid, as well as from Pliny and Macrobius.

10. On the relationship between astrology and prophecy, see Ottavia Niccoli, *Prophecy and People in Renaissance Italy* (Princeton, NJ: Princeton University Press, 1990); Marjorie Reeves, *The Influence of Prophecy in the Later Middle Ages: A Study in Joachimism* (South Bend, IN: University of Notre Dame, 1993); Dietrich Kurze, "Johannes Lichtenberger (†1503): Eine Studie zur Geschichte der Prophetie und Astrologie," *Historische Studien* 379 (1960); Dietrich Kurze, "Prophecy and History: Lichtenberger's Forecasts of Events to Come (From the Fifteenth to the Twentieth Century); Their Reception and Diffusion," *Journal of the Warburg and Courtauld Institutes* 21 (1958): 63–85; Dietrich Kurze, "Popular Astrology and Prophecy in the Fifteenth and Sixteenth Centuries: Johannes Lichtenberger," in *"Astrologi hallucinati." Stars and the End of the World in Luther's Time*, ed. Paolo Zambelli (Berlin: Walter de Gruyter, 1986), 177–93.

11. "gelidus subsidere cancri Saturnus graditur: modo quem violentus adire Mars parat: ille seni contraria lumina miscet." Johannes Stabius, *Pronosticon Ioann. Stabii.*

12. The best introduction to the tradition of great conjunctions is found in J. D. North, "Astrology and the Fortunes of Churches," in *Stars, Minds and Fate: Essays in Ancient and Medieval Cosmology*, ed. J. D. North (London: Hambledon, 1989), 59–89.

13. Stabius, *Pronosticon Ioann. Stabii.*, ll. 49–62.

14. The best introduction to this tradition remains Reeves, *The Influence of Prophecy*, esp. 295–392. See also Marie Tanner, *The Last Descendant of Aeneas. The Habsburgs and the Mythic Image of the Emperor* (New Haven, CT: Yale University Press, 1993), esp. ch. 6 And for this specific context, see Michael Shank, "Academic Consulting in 15th-Century Vienna: The Case of Astrology," in *Texts and Contexts in Ancient and Medieval Science. Studies on the Occasion of John E. Murdoch's Seventieth Birthday*, ed. Edith Sylla and Michael McVaugh (Leiden: Brill, 1997), 245–70 and Darin Hayton, "Astrology as Political Propaganda: Humanist Responses to the Turkish Threat in Early-Sixteenth-Century Vienna," *Austrian History Yearbook* 38 (2007): 61–91.

15. In the late fifteenth century, the most famous attempt to find a different Last World Emperor was the anonymous *Tractatus de Turcis*, which proposed Mathais Corvinus. The *Tractatus de Turcis* enjoyed a wide readership in both manuscript and printed editions. A typical, late fifteenth-century manuscript, annotated in the 1480s and 1490s, is "Tractatus de Thurcis prout ad presens ecclesia sancta ab eis affligitur collectus," Vienna, Österreichische Nationalibibliothek cod. lat. 4224, fols. 348r–61v. A contemporary printed edition is *Tractatus quidam de Turcis* (Nuremberg: Conrad Zeninger, 1481).

16. Dietrich Kurze's studies remain the fundamental work on Lichtenberger: Kurze, "Johannes Lichtenberger (†1503)"; Kurze, "Popular Astrology and Prophecy."

17. Lichtenberger indicated his astrological commitment in the title to his text: Johannes Lichtenberger, *Prognosticatio latina anno lxxxviij ad magnam coniunctionem Saturni et Jovis qui fuit anno lxxxiiii ac eclipsim solis anni sequentis scilicet lxxxv confecta ac nunc de novo emendata* (Mainz: Jakob Meydenbach, 1499), Bayerische Staatsbibliothek 2 Inc.c.a. 2729. Kurze argues that Lichtenberger had only a simple understanding of astrology and that he invoked it only to lend credibility to his prophetic interpretations. See Kurze, "Popular Astrology and Prophecy," 40–43, 184–85.

18. Reeves characterizes Lichtenberger's position as "Joachimist" because of this focus on the *nova reformatio* rather than on the Antichrist. Reeves, *The Influence of Prophecy*, 349. This needs to be nuanced in light of Joachim of Fiore's emphasis on antichrists and the Antichrist. See Robert E. Lerner, "Antichrists and Antichrist in Joachim of Fiore," *Speculum* 60 (1985): 553–70.

19. Late fifteenth-century Methodian prophecies also looked to Maximilian as the Last World Emperor. See *Opusculum divinarum revelationum et de vita Antichristi* (Basel: Micheal Furter, 1498).

20. There was a long tradition of weaving astrology and prophecy together to support religious and political claims. See Jonathan Green, *Printing and Prophecy. Prognostication and Media Change 1450–1550* (Ann Arbor: University of Michigan Press, 2012); Laura Ackerman Smoller, *"Teste Albumasare cum Sibylla*: Astrology and the Sibyls in Medieval Europe," *Studies in History and Philosophy of Biological and Biomedical Sciences* 41, no. 2

(2010): 76–89; Laura Ackerman Smoller, *History, Prophecy, and the Stars. The Christian Astrology of Pierre d'Ailly, 1350–1420* (Princeton, NJ: Princeton University Press, 1994).

21. Stabius, *Pronosticon Ioann. Stabii.*, ll. 76–85.

22. Stabius, *Pronosticon Ioann. Stabii.*, ll. 88–98.

23. Ernst Sackur, ed., *Sibyllinische Texte und Forschungen: Pseudomethodius, Adso und tiburtinische Sibylle* (Halle: Niemeyer, 1898), 92.

24. Sackur, *Sibyllinische Texts*, 74–75.

25. Sackur, *Sibyllinische Texts*, 91–92. On the tradition of Alexander's Gates, see Andrew Runni Anderson, *Alexander's Gate, Gog and Magog, and the Enclosed Nations* (Cambridge, MA: Medieval Academy of America, 1932), esp. 44–48. See also Andrew Gow, "Gog and Magog on Mappaemundi and Early Printed World Maps: Orientalizing Ethnography in the Apocalyptic Tradition," *Journal of Early Modern History* 2 (1998): 61–88.

26. Stabius, *Pronosticon Ioann. Stabii.*, ll. 99–112.

27. Stabius, *Pronosticon Ioann. Stabii.*, ll. 125–46.

28. Stabius, *Pronosticon Ioann. Stabii.*, ll. 149–60.

29. Diederichs, *Kaiser Maximilian I als politischer Publizist*, 43–44; Hermann Wiesflecker, *Kaiser Maximilian I. Das Reich, Österreich und Europa an der Wende zur Neuzeit*, vol. 3: *Auf der Höhe des Lebens. 1500–1508. Der große Systemwechsel. Politischer Wiederaufsteig* (Munich: R. Oldenbourg Verlag, 1977), 144–63.

30. Diederichs, *Kaiser Maximilian I als politischer Publizist*, 44.

31. See the annotations in Johannes Stabius, *Pronosticon Ioann. Stanbii. Au. vatis. & Poc. Lau. ad annos domini. M.D.III.&IIII.* (Nuremberg: Johannes Weissenburger, 1502), Staatsbibliothek Bamberg 22/Inc.typ.Ic.I.4#4.

32. See Johannes Stabius, *Pronosticon Ioann. Stanbii. Au. vatis. & Poc. Lau. ad annos domini. M.D.III.&IIII.* (Nuremberg: Johannes Weissenburger, 1502), Augsburg, Stadts- und Staatbibliothek Sig. 2 Ink 651.

33. On these projects and their relations to Maximilian's political program, see Silver, *Marketing Maximilian*.

34. On Stabius's paper instruments, see chapter 4 in this volume.

35. On Stabius's title of nobility, see Helmuth Grössing, "Johannes Stabius. Ein Oberösterreicher im Kreis der Humanisten um Kaiser Maximilian I," *Mitteilungen des Oberösterreichischen Landesarchiv* 9 (1968): 248–49. On the sum paid to Stabius, see Maximilian, Letter from Maximilian, September 1514, in *Jahrbuch der kunsthistorischen Sammlungen des allerhöchsten Kaiserhauses*, vol. 1 (1883) #332.

36. See Paolo Zambelli, "Many Ends for the World. Luca Gaurico Instigator of the Debate in Italy and in Germany," in *"Astrologi hallucinati,"* 239; Paolo Zambelli, "Fine del mondo o inizio della propaganda? Astrologia, filosofia della storria e propaganda politico-religiosa ne dibattito sulla congiunzione del 1524," in *Scienze, credenze occulte, livelli di cultura* (Florence: L. S. Olschki, 1982), 291–368.

37. The best bibliography of the works produced remains Gustav Hellmann, "Aus der Blütezeit der Astrometerologie," in his *Beiträge zur Geschichte der Meteorologie* (Berlin: Behrend & Co., 1917). For the tone adopted by various authors, see Zambelli, "Fine del mondon o inizio della propaganda?"

38. Georg Tannstetter, *Libellus consolatorius, quo opinionem iam dudum annis hominum ex quorundam Astrologastrorum divinatione insidentem, de futuro diluuio & multis alijs horrendis periculis XXIIII. anni a fundamentis extirpare conatur* (Vienna: Johannes Singrenius, 1523), A2r. As Jonathan Green points out, sensationalist broadsheets and pamphlets always risked inciting fear and unrest. Consequently, it became increasingly important for local officials to control them or limit their impact. See Jonathan Green, *Printing and Prophecy*, esp. ch. 3.

39. See chapter 5 in this volume.

40. Tannstetter, *Libellus consolatorius*, A2r.

41. Rudolf Kink, *Geschichte der kaiserlichen Universität zu Wien*, vol. 1: *Geschichliche Darstellung* (Vienna: Carl Gerold & Sohn, 1854), 137. Tannstetter had been commended for a short plague tract: Georg Tannstetter, *Regiment für den Lauff der Pestilentz durch Georgen Tannstetter von Rain der siben freyen künst unnd Ertzney doctor: kurtzlich beschriben. Anno. 1521* (n.p., 1521). See Karl Schrauf, ed., *Acta facultatis medicae universitatis Vindobonensis*, vol. 3: *1490–1558* (Vienna: Verlag des Medicinischen Doktorenkollegiums, 1904), 146.

42. One of the few copies of this text to survive is in George Tannstetter, *Regiment für den Lauff der Pestilentz durch Georgen Tannstetter von Rain der siben freyen künst unnd Ertzney doctor: kurtzlich beschriben. Anno. 1521* (n.p., 1521), Vienna, Österreichische Nationalbibliothek Sig. 31.V.67.

43. Tannstetter, *Libellus consolatorius*, A4v.

44. Tannstetter, *Libellus consolatorius*, B1r–B2r.

45. Tannstetter, *Libellus consolatorius*, B2v.

46. Pico's attack on Albumasar's theory of conjunctions is confined to his *Disputationes adversus astrologiam divinatricem*, book V. See the modern edition: G. Pico della Mirandola, *Disputationes adversus astrologiam divinatricem, Libri I–V*, ed. Eugenia Garin (Florence: Vallecchi Editore, 1946).

47. Tannstetter, *Libellus consolatorius*, B3r.

48. See chapter 5 in this volume.

49. Tannstetter, *Libellus consolatorius*, B3r.

50. On this work and how it fits with d'Ailly's larger efforts to understand history and the Western Schism, see Smoller, *History, Prophecy, and the Stars*, 6–85.

51. Tannstetter, *Libellus consolatorius*, B3v.

52. Smoller, *History, Prophecy, and the Stars*, 68.

53. See chapter 6 in this volume. Pico's attack of the "ten revolutions of Saturn" is in book V, chapter III of his *Disputationes*, della Mirandola, *Disputationes*, 538–42.

54. Tannstetter, *Libellus consolatorius*, B4r.

55. Tannstetter, *Libellus consolatorius*, B4r–C1r.

56. Tannstetter, *Libellus consolatorius*, C1v–C2r.

57. The two works are virtually identical. Graf-Stuhlhofer offers a brief comparison of the two. Franz Graf-Stuhlhofer, *Humanismus zwischen Hof und Universität. Georg Tannstetter (Collimitius) und sein wissenschaftliches Umfeld im Wien des frühen 16. Jahrhunderts* (Vienna: WUV-Universitätsverlag, 1996), 138 and notes 546 and 547. Joseph Seethaler claims that Tannstetter's German version was printed first and that the Latin version was printed

NOTES TO PAGES 185–186263

nearly a year later. Franz Graf-Stuhlhofer adopts a more agnostic position. See Josef Seethaler, "Das Wiener Kalenderwesen von seinen Anfängen bis zum Ende des 17. Jahrhunderts. Ein Beitrag zur Geschichte des Buchdrucks" (PhD diss., University of Vienna, 1982),
93 and 357n71 ; Graf-Stuhlhofer, *Humanismus zwischen Hof und Universität*, 138n547.

58. Niccoli, *Prophecy and People*, 159–60.

59. Paola Zambelli remarks that both of Tannstetter's works, the Latin *Libellus conso-
latarius* and the German *Zw eren und gefallen*, are equally reassuring. She uses Tannstetter's works to qualify Ottavia Niccoli's claim about Latin and vernacular tracts adopting
different tones. Zambelli, "Fine del mondo o inizio della propaganda? Astrologia, filosofia
della storia e propaganda politico-religiosa ne dibattito sulla congiunzione del 1524," 334.
See Niccoli's response in *Prophecy and People*, 159n50.

60. On Tannstetter's 1524 *Practica*, see chapter 5 in this volume.

61. For Ferdinand's early efforts to establish his authority in Austria and the empire,
see Paula Sutter Fichtner, *Ferdinand I of Austria: The Politics of Dynasticism in the Age of the
Reformation*, East European Monographs (New York: Columbia University Press, 1982),
esp. 13–39.

62. Niccoli claims that astrologers adopted different messages in their vernacular and
Latin texts because they wanted to communicate the truth to the elites but wanted merely
to calm the broader public. Niccoli, *Prophecy and People*, 159n50.

63. This is most forcefully argued in Sara Schechner Genuth, *Comets, Popular Culture,
and the Birth of Modern Cosmology* (Princeton, NJ: Princeton University Press, 1997), esp.
ch. 2. More broadly, see Katherine Park and Lorraine J. Daston, "Unnatural Conceptions:
The Study of Monsters in Sixteenth- and Seventeenth-Century France and England," *Past
and Present* 92 (1981): 20–54; Lorraine Daston and Katherine Park, *Wonders and the Order
of Nature, 1150–1750* (New York: Zone Books, 1998).

64. See Charles Webster, *From Paracelsus to Newton: Magic and the Making of Modern
Science* (Cambridge: Cambridge University Press, 1982), esp. ch. 1; Tabitta van Nouhuys,
*The Age of Two-Faced Janus. The Comets of 1577 and 1618 and the Decline of the Aristotelian
World View in the Netherlands* (Leiden: Brill, 1998).

65. Peter Creutzer apparently mistook Aurora Borealis for a comet in 1527. See his
Peter Creutzer, *Außlegung Peter Creutzers / etwan des weytberümpten Astrologi M. Jo. Liech-
tenbergers discipels / uber den erschröcklichen Cometen / so im Westrich und umbligenden
grenzen erschinen / am xj. tag Weinmonats / des M.D.xxvij. jars / zü eeren den wolgepornen
Herrn / herr Johan / und Philips Granzen beyde / Will und Reungrauen etc* (Nuremberg:
Georg Wachter, 1528), and Wolfgang Kokott's analysis of this tract: Wolfgang Kokott,
"Peter Creutzers 'Komet' vom 11. Oktober 1527: Zur Langlebigkeit von Fehldatierungen
der Sekundärliteratur," in *Cosmographica et gegraphica: Festschrift für Heribert M. Nobis*,
ed. Bernhard Fritscher and Gerhard Brey (Munich: Institute für Geschichte der Naturwissenschaft, 1994), 249–54.

66. Scholarship celebrates a new emphasis on observation in cometary tracts during
the sixteenth century: Wolfgang Kokott, *Die Kometen der Jahre 1531 bis 1539 und ihre
Bedeutung für die spätere Entwicklung der Kometenforschung* (Stuttgart: Verlag für Geschichte der Naturwissenschaften und der Technik, 1994); Kokott, "The Comet of 1533,"
Journal for the History of Astronomy 12 (1981): 95–111; Kokott, "Peter Creutzers 'Komet'

vom 11. Oktober 1527: Zur Langlebigkeit von Fehldatierungen der Sekundärliteratur";
Jane L. Jervis, *Cometary Theory in Fifteenth-Century Europe* (Boston: Kluwer Academic,
1985); C. Doris Hellman, *The Comet of 1577: Its Place in the History of Astronomy* (New
York: Columbia University Press, 1944); Nouhuys, *The Age of Two-Faced Janus*; J. R.
Christianson, "Tycho Brahe's German Treatise on the Comet of 1577: A Study in Science
and Politics," *Isis* 70 (1979): 110–40.

67. I have found only four works composed specifically about this comet: Jacob
Locher, *De cometa sub septentrionibus viso aquei coloris, carmen* (Augsburg: Hans Froschauer, 1506), British Library C.106.d.13; Gaspar Torrella, *Judicium Universale. De
portentis: præsagiis: & ostentis: rerum admirabilium ac Solis et Lunee defectibus & Cometis*
(Rome: Johannes Besicken, 1507), British Library 1039.k.33; Johannes Virdung, *Explanatio
maximarum et formidabilium rerum futurarum Anno Salutis M. D. VII. que per Cometem
Anno 1506 in Climate nostro conspectum portenduntum* (Strasbourg, 1506); Marco Scribanario, *Iudicio de Marco Scribanario Bolognese supra la expositione de la apparente Cometa
al Illustreier Excel. Meser Iohanne de Bentivolgi da Bologna* 1506), Bayerische Staatsbibliothek Res/4 Astr.p.529,6. The comet did appear in some *practica* for 1507 as one of the signs
to be interpreted for the coming year.

68. Hans Joachim Köhler estimates that 10,000,000 individual pamphlets were
printed between 1501 and 1530. This figure relies primarily on religious pamphlets and
does not take into account the astrological *practica* and prognostications that poured from
the presses during this period. See Hans-Joachim Kohler, "The Flugschriften and Their
Importance in Religious Debate: A Quantitative Approach," in *"Astrologi hallucinati,"*
153–75, esp. 154–15.

Although astrologers across Europe produced cometary tracts, the bulk of them seem
to come from German authors. Kokott provides a representative list of German authors
on cometary tracts. See Kokott, *Die Kometen der Jahre 1531 bis 1539 und ihre Bedeutung für
die spätere Entwicklung der Kometenforschung.*

69. The emphasis on observation was not new in the sixteenth century. Already by the
mid-fifteenth century astrologers opened their texts by giving detailed accounts of when
and where they saw the comet. See Jane L. Jervis, *Cometary Theory in Fifteenth-Century
Europe* (Boston: Kluwer Academic, 1985). See also, for example, Martin Bylica, "Judicium
de cometa que apparuit Anno domini Mcccc 68," Bayerische Staatsbibliothek Clm 18782,
fols. 208r–15r and Martin Bylica, "Judicium de cometa que apparavit Anno Domini
Mcccc 72," Österreichische Nationalbibliothek cod. lat. 4777, fols. 7r–17r.

70. "Am fünffzehenden tage des montas Augusti nach mittag far nahet umb .9. ure
des klaynen zaygers / hab ich gesehen einen Cometen etwa newn grad hoch / uber [*sic*]
dem horizont in dem viertel gegen occident und mitnacht / welchs Cometen corpus und
schwanz gnaw in ainer rechten linia mit den lezten zwayn sternn des wagens / sonst urse
maioris gnant." Johannes Schöner, *Coniectur odder abnemliche Außlegung Joannis Schöners
uber den Cometen so im Augustmonat / des M.D.XXXj. jars erschinen ist / zu ehren einem
erbern Rath / und gmainer burgerschafft der stat Nuremberg außgangen* (Nuremberg: Friderich Peypus, 1531), A2r.

71. Johannes Virdung, *Der gesehen worden ist / im Augstmon im 1531. jare / durch Doctor
Johansen Virdung von Haßfurt zü Eren Dem durchleuchtigisten hochgebornen fürsten unnd*

herren Herrn Ludwigen Palzgrauenn bey Rheyn herzogen in Bayern des heiligen Römischen Reichs Erz durchssessen und Churfürsten. Unnd zü eyner warnung yedermenniglichen / wann es ist nye kein Comet gesehen worden / der nit etwas groß erschröcklichs bracht / und bezeigt wie hernach volgt Capitels weyß. etc. (Speier, 1531), A1v–A2r.

72. Virdung, *Der gesehen worden ist*, A2r.

73. Virdung, *Der gesehen worden ist*, A2v–A4r.

74. Andreas Perlach, *Des Cometen und ander erscheinung in den lüfften / Im XXXI. Jar gesehenn bedütung Durch Andreen Perlach von Witschein / der sibenn freyen / und natürlichen kunst maister / Diser zeyt auff der löblichen hohen schül zü Wien / in der Astronomey / was die himlischen leüff würckung / und / jre einflüß betreffen ist / verordenter Läser* (n.p., 1531), t.p.

75. Perlach, *Des Cometen*, B1r.

76. Perlach, *Des Cometen*, A2r.

77. Perlach, *Des Cometen*, A2r.

78. "FReundlicher lieber Leser / du solst dise mein Prognostication / die ich auß der natürlichen kunst / Astrologia genant / genommen hab / nicht dafür halten / dz es müß also geschehen / Dann als Ptho. der Haydnisch maister / und ein unglaubiger schreibt / Der weyß herschet über dz gestirn / vil mer ein götlicher Christenlicher mensch / mit seinem gepet gegen Got über dz gestirn und seinem einfluß herschen mag / So hab ich alzeyt in meinen Juditijs / disen prauch gehabt / das ich an natrülich [*sic*] ursach / nichts hab wöllen schreiben und an tag geben / da mit ein yetlicher abnem / das ich mich allain des grundts der natürlichen kunst Astrologia genant behilff / und kainer andern / darum kainer mein Juditia / vergleichen sol / gegen des Charion / welcher nit auß dem grundt / der natürlichen kunst Astrologia / sonder auß ainer andern / die er villeicht nit melden darff / genomen hat / . . . / dann het Charion ainen rechten grundt der natürlichen khunst Astrologey / wer nit müglich gewäsen / das er so vil schändlicher irtumb in seinem püchel / wider den grundt der kunst in druck het außgeen lassen / und sollich schändtlich irthum / das auch einem anfahenden schüler züverweysen wär / will schweigen eim solchen welcher der aller gelertist Astrologus geacht wil sein / und von vilen (wie wol von kainem recht gelerten verstendigen man) darfür gahalten [*sic*] wirdt." Perlach, *Des Cometen*, A2v.

79. On Perlach's earlier critique, see chapter 6 in this volume.

80. Perlach, *Des Cometen*, A2v.

81. "das er kain vorstandt hat der finsternus theorick." Perlach, *Des Cometen*, A3r.

82. Johannes Carion, *Beteutnuss und Offenbarung* (Berlin, 1526), B3r.

83. Perlach, *Des Cometen*, A3r.

84. Perlach, *Des Cometen*, A3v.

85. Perlach was not the last to accuse Carion of illicit practices. Toward the end of the century Hermann Wilken accused Carion of necromancy. See Green, *Printing and Prophecy*, 88.

86. According to Trithemius, Pelagius was Libanius Gallus's teacher, who in turn taught Trithemius. Johannes Trithemius, "Annalium Hirsavgiensium, Opus nunquam hactenus editum, & ab Eruditis semper desideratum. Complectens historiam Franciae et Germaniae, gesta imperatorum, regum, principum, episcoporum, abbatum, et illustrium virorum," 2 (1690), 585–86.

87. Trithemius is our only source for this identification. Given his talents for fabricating stories, it is possible that he invented Pelagius to give his own activities a more respectable lineage. For a brief explanation of the relationship between Pelagius, Libanius, and Trithemius, see Noel L. Brann, *Trithemius and Magical Theology. A Chapter in the Controversies over Occult Studies in Early Modern Europe* (Albany: State University of New York Press, 1999), 109–12. For a more extended treatment of Ferdinando de Córdova's life and writings, see John Monfasani, *Fernando of Cordova. A Biographical and Intellectual Profile*, vol. 28, Transactions of the American Philosophical Society (Philadelphia, PA: American Philosophical Society, 1992). A. Bonilla y San Martin, *Fernando de Córdoba (1425–1486?) y los orígenes del Renacimiento filosófico en España* (Madrid, 1911). On Pelagius and his connection to the medieval magical tradition of the *ars notoria*, see Julien Veronese, "L'Ars notoria: une tradition théurgico-magico au Moyen Age (XII–XVIe siècle)" (PhD diss, Paris-X Nanterre, 1999); Jean Dupède, "L'écriture chez l'eremite Pelagius," in Le texte et son inscription," in *Le texte et son inscription*, ed. Roger Lauter (Paris: CNRS, 1989), 113–54; Jean Dupède, "L'Ars notoria et la polémique sur la divination et la magie," in *Divination et controverse religieuse en Frnace au XVIe siècle, Collection de l'École Normale Supérieur de Jeunes Filles* (Paris: École Normale Supérieur de Jeunes Filles, 1987), 123–34.

88. Brann, *Trithemius and Magical Theology*, 110.

89. Brann, *Trithemius and Magical Theology*, 4. The quotation is from Maximilian's *Weisskunig*: "Der jung weiß kunig erfordert ainen sondern gelerten man in der swarzen kunst." Maximilian, *Kaiser Maximilians Weisskunig*, ed. H. Th. Musper, vol. 1: Textband (Stuttgart: W. Kohlhammer Verlag, 1956), 225.

90. "Mich gedunckt auch genzlich er hab sein Juditia genommen / auß den püchern magistri Pelagi heremite in regno maioricarum von der beschwerung der geyst / dann ein solchs zü Perlin abgeschriben ist worden / unnd nit aller zü gehörung gen Osterreich pracht / das ich mit meinen augen gesehen hab / Es sein vorzeyten auch solch leut gewesen / wie Ptholomeus anzaigt / die mit solchen künsten sein gangen / und gesagt / sy nemen yr Juditia auß der Astrologey / so sy doch keinen rechten grundt (wie diser Charion) darinn hetten." Perlach, *Des Cometen*, A3r–A4r.

91. "das diser Charion entweder kain Latein versteet / oder muß mit einem boßhafftigen und tewflischen geyst besessen sein." Perlach, *Des Cometen.*, A4r.

92. Perlach, *Des Cometen*, A4r.

93. Perlach, *Des Cometen*, B2r–B2v.

94. Perlach, *Des Cometen*, C2r–C2v.

95. Perlach, *Des Cometen*, C2v–C3r.

96. "Auß disem allen mag man wol abnemen / das dy ursach / welch die materi dises Cometen übersich zogen hat / den sterben zu Wien und andern orten gemacht hat." Perlach, *Des Cometen*, C3r.

97. Perlach, *Des Cometen*, D2r–D2v.

98. Perlach, *Des Cometen*, D2v.

99. "Dy Machometisch Sect / wirt widerwertigkeit zu fürchten haben / zaigt an der Mon (welcher jr schildt und Helm ist) dann er has Jar schier all monat gar nahent zü jren widerwertigen stern / in des stiers aug sich vergügen wirdt / und etlich mal den selbigen verdecken /

darumb auch der Mon unselig wirdt von dem selbigen stern / Aber doch ist dise Sect nicht zu verachten / dann der Jupiter kumpt in jr zaichen / das ist in Scorpion / zum endt des Jars / kumpt er zu jrn stern / Darumb wirdt dise Zect / vil und grosse sach untersteen / mit kriegen und seltzamen anschlegen wider jre feindt." Perlach, *Des Cometen*, D3v–D4r.

100. Perlach, *Des Cometen*, D4r.

101. They would only begrudgingly provide troops later in the decade. See Paula Sutter Fichtner, *Ferdinand I of Austria*, 102–39.

Conclusion. Astrology and Maximilian's Legacy

1. Johannes Schöner, *De iudiciis nativitatum. Libri tres* (Nuremberg: Ioannem Montanum & Ulricum Neuberum, 1545), A4v

2. Heinrich Rantzau, *Catalogus, Imperatorum, regum, ac virorum illustrium, qui artem astrologicam amarunt, ornarunt & exercuerunt* (Leipzig: Georg Defner, 1584), Bayerische Staatsbibliothek 4 Astr.p.356, 67. On Rantzau, see Günther Oestmann, *Heinrich Rantzau und die Astrologie*, vol. 2: *Disquisitiones Historiae Scientiarum. Braunschweiger Beiträge zur Wissenschaftsgeschichte* (Braunschweig: Braunschweigisches Landesmuseum, 2004).

3. See Jacques Verger, *Men of Learning in Europe at the End of the Middle Ages* (South Bend, IN: University of Notre Dame, 2000).

4. In addition to Moran's work and Shank's, see Notker Hammerstein, "Relations with Authority," in *Universities in Early Modern Europe (1500–1800)*, ed. Hilde de Ridder-Symoens, A History of the University in Europe, vol. 2 (Cambridge: Cambridge University Press, 1996), 114–53.

5. On the political importance of precise, technical knowledge, see the essays in M. Norton Wise, ed., *The Values of Precision* (Princeton, NJ: Princeton University Press, 1995).

6. On the political importance of standardizing and centralizing technical information, see the essays in Wise, *The Values of Precision*. See especially Ken Alder, "A Revolution to Measure: The Political Economy of the Metric System in France," in Wise, *The Values of Precision*, 39–71.

7. The best overview of the Habsburgs' patronage and collecting is found in Hugh Trevor Roper, *Princes and Artists: Patronage and Ideology at Four Habsburg Courts, 1517–1633* (New York: Harper and Row, 1976).

8. On these collections, see Elisabeth Scheicher, "The Collection of Archduke Ferdinand II at Schloss Ambrass: Its Purpose, Composition and Evolution," in *The Origins of Museums. The Cabinet of Curiosities in Sixteenth- and Seventeenth-Century Europe*, ed. Oliver Impey and Arthur MacGregor (Oxford: Clarendon Press, 1985), 29–38.

9. See the suggestive comments in Evans, *Rudolf II and His World*, esp. 14–15, 62, and 182

10. Rudolf and his court have received considerable scholarly attention. The fundamental study remains Evans, *Rudolf II and His World*. For more recent work, see Kaufmann, "Astronomy, Technology, Humanism, and Art at the Entry of Rudolf II into Vienna, 1577"; Thomas Da Costa Kaufmann, "From Mastery of the World to Mastery of Nature: The *Kunstkammer*, Politics, and Science," in *The Mastery of Nature. Aspects of Art, Science, and Humanism in the Renaissance*, ed. Thomas Da Costa Kaufmann (Princeton, NJ: Princeton University Press, 1993), 174–94; Fučíková, "The Collection of Rudolf II

at Prague: Cabinet of Curiosity or Scientific Museum?" in Impey and MacGregor, *The Origins of Museums*; Eliška Fučíková, ed., *Rudolf II and Prague. The Court and the City* (London: Thames and Hudson, 1997).

11. In addition to scholarship on Emperor Rudolf II and Landgrave Moritz of Hessen-Kassel, see Adam Mosley's recent work on Tycho Brahe and the Landgrave Wilhelm IV of Hessen-Kassel: Adam Mosley, *Bearing the Heavens. Tycho Brahe and the Astronomical Community of the Late Sixteenth Century* (Cambridge: Cambridge University Press, 2007), and Tara Nummedal's study of alchemy in later sixteenth-century Germany, *Alchemy and Authority in the Holy Roman Empire* (Chicago: University of Chicago Press, 2007). See also Eric Ash's book on expertise in Elizabethan England: Eric H. Ash, *Power, Knowledge, and Experience in Elizabethan England* (Baltimore, MD: Johns Hopkins University Press, 2004); Pamela Smith's study on Joachim Becher: Pamela H. Smith, *The Business of Alchemy: Science and Culture in the Holy Roman Empire* (Princeton, NJ: Princeton University Press, 1994); and Paula Findlen on *Kunskammer* and collecting in Italy: Paula Findlen, *Possessing Nature. Museums, Collecting, and Scientific Culture in Early Modern Italy* (Berkeley: University of California Press, 1994).

12. On Rudolf II, see R. J. W. Evans, *Rudolf II and His World. A Study in Intellectual History, 1576–1612* (Oxford: Oxford University Press, 1973); Bruce T. Moran, *The Alchemical World of the German Court. Occult Philosophy and Chemical Medicine in the Circle of Moritz of Hessen (1572–1632)* (Stuttgart: Franz Steiner Verlag, 1991). Thomas Da Costa Kaufmann, "Astronomy, Technology, Humanism, and Art at the Entry of Rudolf II into Vienna, 1577," 136–50; Bruce T. Moran, "Princes, Machines and the Valuation of Precision in the 16th Century," *Sudhoffs Archiv* 61 (1977): 209–28; Fučíková, "The Collection of Rudolf II at Prague," 47–53. See also Bruce T. Moran, "Patronage and Institutions: Courts, Universities, and Academies in Germany. An Overview 1550–1750," in *Patronage and Institutions. Science, Technology, and Medicine at the European Court, 1500–1750*, ed. Bruce T. Moran (Woodbridge: Boydell Press, 1991), 169–83.

BIBLIOGRAPHY

Manuscripts, Ephemeral Texts, and Annotated Printed Sources

Augsburg, Germany
Staats- und Stadtbibliothek

Perlach, Andreas. *Des Cometen und ander erscheinung in den lüfften / Im XXXI. Jar gesehenn bedütung Durch Andreen Perlach von Witschein / der sibenn freyen / und natürlichen kunst maister / Diser zeyt auff der löblichen hohen schül zü Wien / in der Astronomey / was die himlischen leüff würckung / und / jre einflüß betreffen ist / verordenter Läser.* N.p., 1531. Sig.: 4 Math 316/Beibd.

Stabius, Johannes. *Pronosticon Ioann. Stanbii. Au. vatis. & Poc. Lau. ad annos domini. M.D.III.&IIII.* Nuremberg: Johannes Weissenburger, 1502. Sig.: 2 Ink 651.

Tannstetter, Georg. *Practica gemacht zu Wienn auff des MCCCC und XIX jar.* Vienna: Johannes Singrenium, 1518. Sig.: 4 Kult 186–110.

Tannstetter, Georg. *Practica gemacht zu Wienn in Osterreych aus der Kunst Astronomia: vnd ist wie ain außzug vnd ain sundere erklerung des grosseren Indici so vorhin auff das MDXXIIII. Jar von im ausgangen ist.* Augsburg, 1523. Sig.: 4 Kult 186–131.

Tannstetter, Georg. *Juditium Astronomicum Viennense Georgij Tannstetter Collimitij Lycoripensis artium et medicine doctoris in annum chriti. 1525.* Vienna: Johannes Singrenium, 1524. Sig.: 4 Math. 575.

Bamberg, Germany
Staatsbibliothek

Stabius, Johannes. *Pronosticon Ioann. Stanbii. Au. vatis. & Poc. Lau. ad annos domini. M.D.III.&IIII.* Nuremberg: Johannes Weissenburger, 1502. Sig.: 22/Inc. typ.Ic.I.4#4.

Cambridge, Massachusetts
Houghton Library, Harvard University

Gundelius, Phillip, ed. *C. Plinii Secundi liber septimus naturalis historiae.* Vienna: Joannes Singrenium, 1519. Call # GC5.G9554.519p.

Bethesda, Maryland
U.S. National Library of Medicine

Tannstetter, Georg. *Compendiariu[m] die[rum] obseruando[rum], pro venis incidendis, balneis, seminando plantandoq[ue] ad annum salutis nostre M.d. xvij.* 1516.

Boston, Massachusetts
Boston Public Library

Perlach, Andreas. *Ephemerides Andree Perlachii pro anno M.D.XXIX. cum configuratiuonibus et habitudinibus planetarum inter se et cum stellis fixis utique insignioribus cum ex secundo tum ex prino mobili contingentibus, sub tali forma hactenus nemin visae.* Vienna: Hieronimum Vietorem, 1528. Sig.: *E.189.51.

Coburg, Germany
Landesbibliothek

Tannstetter, Georg. *Practica teutsch auff das M.ccccc. und eylff Jar durch Mayster Georgen Tannstetter.* No Place, No Date. Sig.: RII 8 11/4.

Gotha, Germany
Forschungsbibliothek

Tannstetter, Georg. *Ad laudem et gloriam Serenissimi et potentissimi principis Ferdinandi. Principis Hispaniarum. Archiducis Austrie. Dcis Burgunie etc. et ad usum commoditatemque provinciaum Austrie. Georgius Tannstetter Collimitius Lycoripensis Artium et Medicine Doctor. Vienne hoc prognosticum pro anno Christi 1522 edidit.* Landshut: Johannes Weißenburger, 1521. Sig.: Math.4° 145/3(3)R.

Halle, Germany
Marienbibliothek

Stabius, Johannes. *Pronosticon Ioann. Stanbii. Au. vatis. & Poc. Lau. ad annos domini. M.D.III.&IIII.* Nürnberg: Johannes Weissenberger, 1502. Sig.: Q3.105(19)

Heidelberg, Germany
Universitätsbibliothek

Cod. Pal. germ. 88

Innsbruck, Austria
Universitätsbibliothek

Codex 314

Krakow, Poland
Biblioteka Jagiellonska

BJ cod. 3225

Perlach, Andreas. *Usus almanach seu Ephemeridum: ex commentarijs Georgijs Tannstetter Colimitii / Preceptoris sui decerpti, & quinquaginta propositiones, per Magistrum Andream perlachium stirum / Redacti.* Vienna: Hieronimum Vietorem, 1518. Sig.: Cim 4735.

Tannstetter, Georg. *Judicium Astronomicum Viennense anni M.D.xix. ad Reverendissimi in choro patrem et illusttrissimi principem et d. d. Matteum S. Angeli sa. sa. Ro. ecclesie Cardinalem Burcennse coadiutorem Salisburgennse et in germania legatum. etc. do. suum gratiosissimi Geo. Tannstetter.* Vienna: Johannes Singrenium, 1518. Sig.: Cim 4730.

London, England
British Library

Proclus. *Procli Diadochi Sphaera. Astronomiam discere incipientibus utilissima. Thoma Linacro Britanno interprete.* Ed. Georg Tannstetter. Vienna: Hieronimum Vietorem, 1511. Shelfmark: 8561.b.6.

Stabius, Johannes. "Carmen saphicum ad Max. Ro. re. Trenice." In *Ludus Diane in modum Comedie coram Maximiliano Rhomanorum Rege. Kalendis Martijs et Ludis saturnalibus in arce Linsiana danubij acts: Clementissimo Ree et Regina ducibusque illustribus Mediolani totaque Regia curia spectatoribus: per petrum Bonumum Regi: Cancel. Joseph Grunpeckium Reg. Secre. Conradum Celten: Reg: Poe. Ulsenium Pfrisium: Vincentium Longinum in hoc Ludo Laurea donatum foeliciter e iucundissime representatus.* Ed. Conrad Celtis. Nuremberg: Hieronymo Hölcelio, 1501. Shelfmark: C.57.c.3.

Stabius, Johannes. *Horoscopion omni generaliter congruens climati.* 1512. Shelfmark: Printroom.

Tannstetter, Georg. *Als man zalt nach Christi gepurt M.CCCCC uñ XIIIIJ.* Augsburg: Erhardt Ölgin, 1513. Shelfmark: C.18.e.3(27)

Tannstetter, Georg. *Artificium de applicatione Astrologiae ad Medicinam, deque conuenientia earundem, Georgij Collimitij Tansteteri, Canones aliquot, et quaedam alia, quorum Catalogum reperies in proxima tabella.* Straßburg: G. Ulrich, 1531. Shelfmark: 1568/4204.

Warburg Institute

Carion, Johannes. *Bedeütnuss vnnd Offenbarung warer Hymlischer Influxion des Hocherfarnen Magistri Joannis Carionis Buetikaimensis: Churfürstlicher gnaden von Brandenburg rc. Mathematici von Jarn zu Jaren werende bis man schreybt, M.D. vnd xxxx. jar, Alle Landtschafft Stende vnd einflüss clarlich betreffend.* Berlin, 1526. Shelfmark: FAH 5540.

Tannstetter, Georg. *Wall-calendar.* 1523. Photo Collection Drawer Almanacs, Folder: Almanacs and Calendars—European.

Virdung, Johannes. *Wall-calendar.* 1495. Photo Collection Drawer: Prognostica; Folder: Dated Prognostica—1495.

Wellcome Institute

Peurbach, Georg. *Tabulae eclipsium.* Ed. Georg Tannstetter. Vienna: Johannes Winterberger, 1514. Shelfmark: Inc. 5.b.2.

Tannstetter, Georg. *Compendiariu[m] die[rum] obseruando[rum], pro venis incidendis, balneis, seminando plantandoq[ue] ad annum salutis nostre M.d. xvij.* 1516. Photo Collection Drawer: Almanacs; Folder: Almanacs and Calendars—European.

Munich, Germany
Bayerische Staatsbibliothek

Cgm 1502

Clm 19689

Clm 24103

Clm 24105

Grünpeck, Joseph. "Prognostikon für 1496–1499." Sig.: Cgm 3042

Perlach, Andreas. *Allmanach novum super anno Christi saluatoris nostri: M.D.XVIII. Ex tabulis doctiss. viri magistri Joannis de Gmunden studij quondam Wiennensis.* Vienna: Hieronimum Vietorem, 1518. Sig.: Res/4° Eph.Astr.38.

Perlach, Andreas. *Usus almanach seu Ephemeridum: ex commentarijs Georgijs Tannstetter Colimitii / Preceptoris sui decerpti, & quinquaginta propositiones, per Magistrum Andream perlachium stirum / Redacti.* Vienna: Hieronimum Vietorem, 1518. Sig.: Res/4° Eph.Astr.88d.

Perlach, Andreas. *Almanach novum super anno Christi saluatoris nostri. M.D.xviiij. Ex tabulis doctiss. viri magistri Joannis de Gmunden studij quondam Vienneň. alumni in officina Collimitiana per Magistrum Andream Perlachium Stirum ad Meridianum Vienneň. diligentissime supputatum.* Vienna: Ioannes Singrenium, 1519. Sig.: Res/4° Eph.Astr.39.

Perlach, Andreas. *Ephemerides Andree Perlachii pro anno M.D.XXIX. cum configuratiuonibus et habitudinibus planetarum inter se et cum stellis fixis utique insignioribus cum ex secundo tum ex prino mobili contingentibus, sub tali forma hactenus nemin visae.* Hieronimum Vietorem, [1528]. Sig.: Res/4°Eph.astr.88.

Stabius, Johannes. *Pronosticon Ioann. Stanbii. Au. vatis. & Poc. Lau. ad annos domini. M.D.III.&IIII.* Nürnberg: Johannes Weissenberger, 1502. Sig.: 2 Einbl. IV, 7a.

Stabius, Johannes. *Pronosticon Ioann. Stanbii. Au. vatis. & Poc. Lau. ad annos domini. M.D.III.&IIII.* Nürnberg: Johannes Weissenberger, 1502. Sig.: 2 Einbl. IV, 7b.

Stabius, Johannes. Astrolabium Imperatorium. 1515. Einbl. VIIII, 12.

Stöffler, Johannes, and Jacob Pflaum. *Almanach nova plurimis annis venturis inservientia.* Venice: Peter Lichtenstein, 1506. Sig.: Res/4° Eph.Astr.126.

Stöffler, Johannes, and Jacob Pflaum. *Almanach nova plurimis annis venturis inservientia.* Venice: Peter Lichtenstein, 1506. Sig.: Res/4° Eph.Astr.127.

Stöffler, Johannes, and Jacob Pflaum. *Almanach nova plurimis annis venturis inservientia.* Venice: Peter Lichtenstein, 1521. Sig.: Res/4° Eph.Astr.129.

Stöffler, Johannes, and Peter Pitatus. *Ephemeridum reliquiae Ioannis stoeffleri germani, superadditis novis usque ad annum Christi 1556. durantibus, Petri Pitatis Veronensis Mathematicis, una cum additionibus longe utilissimis eiusdem.* Tubingen: Ulricus Morhardus, 1548. Sig.: 4°Eph. astr. 132.

Tannstetter, Georg. *IVdicium Viennense Magistri Georgii Lanstetter Collimitii. Super anno M.cccc.v.* No Place, [1504]. Sig.: Res/4° Astr.p. 529, 2.

Tannstetter, Georg. *Judicium Viennense Anni Millesimiquingentesimi duodecimi. Ad illustrissimum principem Ludovicum bavarie Ducem Comitemque palatinum rheni.* Nürnberg: Wolfgang Huber, 1511. Sig.: Res/4° Astr.p. 510, 22.

Tannstetter, Georg. *Judcium Viennense Anni Millesimiquingentesimi duodecimi.*

Ad illustrissimum principem Ludovicum bavarie Ducem Comitemque palatinum rheni. Cologne: Henricus de Nussia, 1511. Sig.: Res/4° Astr.p. 510, 21.

Tannstetter, Georg. *Judicium Astronomicum pro anno Christi. M.cccc.xiij. per Magister Georgium Tannstetter collimitium astronomie professorem Viene Pannonie editum.* Nürnberg: Wolfgang Huber, 1512. Sig.: Res/4° Astr.p. 510, 25.

Tannstetter, Georg. *Ephemerides pro meridiano Viennensii supputate* [1513]. Sig.: Einbl. Kal. 1513a.

Tannstetter, Georg. *Ephemerides ad meridianum Olomuncesi. Anno a nativitate salvatoris et domini nostri iesu christi M.D.XXVII.* Vienna: Joh. Singriener, 1526. Sig.: Einbl. Kal. 1527.

Universitätsbibliothek
4 Cod. ms. 743

New York, New York
Columbia University Library
Brudzewo, Albert. Commentum in theoricas planetarum Georgii Purbachii. Milan: Ulrich Scinzenzeler, 1494. Call#: Goff B460.

Oxford, England
Bodleian Library
Proclus. *Sphaera. Astronomiam discere incipientibus utilissima. Thoma Linacro Britanno interprete.* Vienna: Johannes Singrenius, 1511. Sig.: Bod Vet D1 e.16.

Tannstetter, Georg. *Artificium de applicatione astrologiæ ad medicinam.* Strasbourg: Georgius Ulricherus, 1531. Sig.: Bod 8°E 105(4) Linc.

Regensburg, Germany
Stadtliche Bibliothek
Tannstetter, Georg. *Practica gemacht zu Wienn in Olsterreych aus der Kunst Astronomia: vnd ist wie ain außzug vnd ain sundere erklerung des grosseren Indici so vorhin auff das MDXXIIII. Jar von im ausgangen ist.* Vienna: Johannes Singrenium, 1523. Sig.: Philos. 2280 angeb. 1.

San Marino
The Huntington Library
Milich, Jakob. *Liber secundus C. Plinii de mundi historia cum commentariis Iacobi Milichii diligenter conscriptis & recognitis.* Schwäbisch Hall: Peter Brubach, 1538. Call#: 708383.

Salzburg, Austria
Stiftsbibliothek, St. Peter
Tannstetter, Georg. *Ephemerides ad meridianum Regie civitatis Budensis.* Vienna: Hieronymus Vietor & Johann Singriener, 1514. Sig.: Ink. Standort Nr. 1040/12.

South Bend, Indiana
University of Notre Dame

Ptolemy. *Quadriparti. Ptolo. Que hoc volumine continentur hec sunt. Liber Quadripartiti Ptolomei. Centiloquium eiusdem. Centiloquium Hermetis. Euisdem de stellis beibenijs. Centiloquium Bethem. et de horis planetarum. Eiusdem de significatione triplicitatum ortus. Centum quinquaginta propositiones Almansoris. Zahel de interrogationibus. Eiusdem de electionibus. Eiusdem de temporum significationibus in iudicijs. Messahallach de receptionibus planetarum. Eiusdem de interrogationibus. Epistola eiusdem cum duodecim capitulis. Eiusdem de revolutionibus annorum mundi.* Venice: Octaviani Scoti, 1519.

St. Gall
Vadianische Sammlung

Ms. 66

Vienna, Austria
Österreichische Nationalbibliothek

Cod. lat. 2476
Cod. lat. 3327
Cod. lat. 4224
Cod. lat. 5179
Cod. lat. 5275
Cod. lat. 5280
Cod. lat. 5318
Cod. lat. 5415
Cod. lat. 7419
Cod. lat. 8489
Cod. lat. 10358
Cod. lat. 12768
Cod. lat., S.n. 4265

Camillus, Egidius. *Practica Teutsch zu wienn gemacht auffs M.D.XXV. jar.* n.p.: 1524. Sig.: 19837-B Alt Mag.

Camillus, Egidius. *Prognosticon Viennense in Annum 1528.* Vienna: Johannes Singrenius, [1527]. Sig.: 19845-B Alt Mag.

Grünpeck, Joseph. Pronosticon sive (ut alij volunt) Judicium Ex coniuntione Saturni et Jovis Decennalique revolutione Saturni Ortu et fine antichristi ac alijs quibusdam interpositis prout exsequentibus claret praeambulis hic inseritur. Vienna: Johannes Winterburg, 1496. Sig.: Ink 17.H.13.

Muris, Johannes de. Arithmetica communis. Proportiones breves. De latitudinibus formarum. Algorithmus M. Georgii Peurbachii in integris. Algorithmus Magistri Joannis de Gmunden de minuciis phisicis. Johannes Singrenius, 1515. Sig.: 72.G.28 Alt Prunk.

Perlach, Andreas. *Usus almanach seu Ephemeridum: ex commentarijs Georgijs Tannstetter Colimitii / Preceptoris sui decerpti, & quinquaginta propositiones, per Magistrum Andream perlachium stirum / Redacti.* Vienna: Hieronimum Vietorem, 1518. Sig.: 72.V.77.

Perlach, Andreas. *Usus almanach seu Ephemeridum: ex commentarijs Georgijs Tannstetter Colimitii / Preceptoris sui decerpti, & quinquaginta propositiones, per Magistrum Andream perlachium stirum / Redacti.* Vienna: Hieronimum Vietorem, 1518. Sig.: *69.H.66.

Perlach, Andreas. *Ephemerides Andree Perlachii pro anno M.D.XXIX. cum configuratiuonibus et habitudinibus planetarum inter se et cum stellis fixis utique insignioribus cum ex secundo tum ex prino mobili contingentibus, sub tali forma hactenus nemin visae.* Vienna: Hieronimum Vietorem, [1528]. Sig.: 72.T.95.

Perlach, Andreas. *Ephemerides Andreae Perlachii Stiri ex Witschein, Artium & Philosophiae magistri, Mathematicarum disciplinarum studij Viennesis professoris ordinarij, pro Anno domini & Saluatoris nostri Iesu Christi M.D.XXXI. cum configurationibus & habitudinibus planetarum inter se, & cum stellis fixis, cum ex secuno, tum ex primo mobili contingentibus. Insuper adiunximus his nostris in hunc annum Ephemeridibus, Prognosticon, superioris anni eclipsium, quarum effectus hoc Anno apparebunt.* Vienna, [1530]. Sig.: 72.T.95*

Sacrobosco, Johannes de. Opusculum de sphaera clarissimi philosophi Ioannis de Sacro Busto. Theoricae planetarum Georgii Purbachii. Vienna: Johannes Singrenius, 1518. Sig.: 72.G.10(2) Alt. Prunk.

Stabius, Johannes. Practica Ingelstadiensis joannis Sabij phi ac mathematici benivolis et ingenuis lectroibus felicitatem. n.p.: [1499]. Sig.: Ink 8.H.85.

Tannstetter, Georg. *Wall-calendar.* Vienna: Johann Singriener, 1518. Sig.: Flugblättersammlung, 1519.

Tannstetter, Georg. Regiment für den Lauff der Pestilentz durch Georgen Tannstetter von Rain der siben freyen künst unnd Ertzney doctor: kurtzlich beschriben. Anno. 1521. n.p.: 1521. Sig.: 31.V.67.

Universitätsarchiv

UAW Cod. Ph. 9.

Universitätsbibliothek

Anonymous. *Canon Joannis de monte regio in Ephemerides.* Sig.: I.233.605 Adl.4.

Perlach, Andreas. *Usus almanach seu Ephemeridum: ex commentarijs Georgijs Tannstetter Colimitii / Preceptoris sui decerpti, & quinquaginta propositiones, per Magistrum Andream perlachium stirum / Redacti.* Vienna: Hieronimum Vietorem, 1518. Sig.: I.233.605 Adl. 5.

Tannstetter, Georg. *Judcium Astronomicum Viennense anni M.CCCCC.xvij. Ad nobiles et pro videntissimos dominos Magistrum civium et universum inclite urbis Vienne Senatum per Georgium Tannstetter Collimitium artium et medicine doctorem diligenter elaboratum.* Vienna: Johannes Singrenium, 1516. Sig.: I.545.631 ES.

Washington, DC
Smithsonian Institute Libraries
Tannstetter, Georg. Letter to Johannes Stabius, 18 April 1513. MSS 14444A, A.L.S.
(1513 April 16[?]).

Wroclaw, Poland
University Library
Tannstetter, Georg. "Artificium de applicatione Astrologiae ad Medicinam, deque
conuenientia earundem, Georgij Collimitij Tansteteri, Canones aliquot, et
quaedam alia, quorum Catalogum reperies in proxima tabella." Sig. R476 nr.4.

Würzburg, Germany
Universitätsbibliothek
Ms. I.t.q. 236

Zwickau, Germany
Ratsschulbibliothek
Stabius, Johannes. *Judicium Ingelstadiense Joannis Stabij philosophi ac Mathematici.
Sol dñs Anni Saturnus in regimine particeps.* 1498 Sig. Ni44.

Printed Sources

Abū Ma'šar. *De magnis coniunctionibus.* Venice: 1515.
Abū Ma'šar. *The Abbreviation of the Introduction to Astrology: Together with the Medieval
Latin Translation of Adelard of Bath.* Edited by Charles Burnett, Keiji Yamamoto, and
Michio Yano. Leiden: Brill, 1994.
Abū Ma'šar. *Liber introductorii maioris ad scientiam judiciorum astrorum.* Edited by
Richard Lemay. 9 vols. Naples: Instituto universitario orientale, 1995–96.
Abū Ma'šar. *On Historical Astrology. The Book of Religions and Dynasties (On the Great
Conjunctions).* Translated by Keiji Yamamoto, and Charles Burnett. 2 vols. Leiden:
Brill, 2000.
Albéri, Eugenio, ed. *Le Relazioni Degli Ambasciatori Veneti Al Senato Durante Il Secolo
Decimosesto* Florence: 1862.
Alcabitius. *Liber isagogicus Alchabitii de planetarum conjunctionibus.* Venice, 1485.
Alcabitius. *Al-Qabīsī (Alcabitius): The Introduction to Astrology.* Translated by Charles
Burnett, Keiji Yamamoto, and Michio Yano. Edited by Charles Burnett, Keiji
Yamamoto, and Michio Yano. Vol. 2, Warburg Institute Studies and Texts. London:
Warburg Institute, 2004.
Aretino, Pietro. *Un Pronostico satirico di Pietro Aretino (MDXXXIIII).* Edited by
Alessandro Luzio. Bergamo: Instituto Italiano d'Arti Graciche, 1900.
Brant, Sebastian. *Stultifera navis.* Basel, 1498.

Cardano, Giorlamo. *In Cl. Ptolemaei pelusiensis IIII de Astrorum Iudicijs, aut ut vulgo vocant, Quadripartitae.* Basel: Heinrich Petrus, 1554.

Carion, Johannes. *Beteutnuss und Offenbarung.* Berlin, 1526.

Celtis, Conrad. *Ludus Diane in modum Comedie coram Maximiliano Rhomanorum Rege Kalendis Martijs et Ludis saturnalibus in arce Linsiana danubij actus: Clementissimo Rege et Regina ducibusque illustribus Mediolani totaque Regia curia spectatoribus: per petrum Bonumum Regi: Cancel. Joseph Grunpeckium Reg. Secre. Conradum Celten: Reg: Poe. Ulsenium Pfrisium: Vincentium Longinum in hoc Ludo Laurea donatum foeliciter et iucundissime representatus.* Nuremberg: Hieronymus Hölecius, 1501.

Creutzer, Peter. *Außlegung Peter Creutzers / etwan des weytberümpten Astrologi M. Jo. Liechtenbergers discipels / uber den erschröcklichen Cometen / so im Westrich und umbligenden grenzen erschinen / am xj. tag Weinmonats / des M.D.xxvij. jars / zü eeren den wolgepornen Herrn / herr Johan / und Philips Granzen beyde / Will und Reungrauen etc.* Nuremberg: Georg Wachter, 1528.

Cuspianus, Johannes. *Tagebuch, 1502–1527.* Vol. 1: *Fontes Rerum Austriacarum. Österreichische Geschichts-Quellen. Scriptores.* Vienna: Kaiserl. Königl. Hof- und Staatsdruckerei, 1855.

della Mirandola, G. Pico. *Disputationes adversus astrologiam divinatricem, Libri I-V.* Edited by Eugenia Garin. Florence: Vallecchi Editore, 1946.

Ferrier, Orger. *A Learned Astronomical Discourse, of the Judgment of Nativities.* Translated by Thomas Kelway. London: Charlewoodhouse, 1593.

Fuchs, C. H., ed. *Die ältesten Schriftsteller über die Lustseuche in Deutchland, von 1495 bis 1510, nebst mehreren Anecdotis späterer Zeit, gesammelt und mit literarhistorischen Notizen und einer kurzen Darstellung der epidemischen Syphilis in Deutschland.* Göttingen: Dieterischschen Buchhandlung, 1843.

Grünpeck, Joseph. *Ein hübscher Tractat von dem ursprung des bösen Franzos. das man nennet die Wylden wärzen. Auch ein Regiment wie man sich regiren soll in diser zeyt.* Augsburg: Hans Schauer, 1496.

Grünpeck, Joseph. *Tractatus de pestilentiali Scora sive mala de Franzos. Originem. Remediaque eiusdem continens. compilatus a venerabili viro Magistro Joseph Grunpeck de Burckhausenn. super Carmina quedam Sebastiani Brant vtriusque iuris professoris.* Augsburg: Hans Schauer, 1496.

Grünpeck, Joseph. *Ein newe Außlegung der seltzamen Wunderzaichen und Wunderpürden.* Augsburg: Erhard Oeglin, 1507.

Grünpeck, Joseph. *Speculum naturalis cœlestis & propheticæ visionis: omnium calamitatum tribulationum & anxietatum: quæ super omnes status: stirpes & nationes christianæ reipublice: presertim quæ cancro & septimo climati subiecte sunt: proximis temporibus venture sunt.* Nuremberg: Georg Stuchs, 1508.

Grünpeck, Joseph. *Ain nuzliche Betrachtund der natürlichen, hymlischen und prophetischen, ansehungen aller trübsalen, angst, und not, die über all stände geschlechte und gemainden der Christenhait in kurzen tagen geen werden.* Augsburg: Hans Schönsperger, 1522.

Grünpeck, Joseph. *Pronostication Doctor Joseph Grünpecks Vom zway und dreyssigsten Jar an biß auff das Vierzigst Jar des aller durchleüchtigsten groß mächtisten Kaiser Carols des fünfften rc. und Begreifft in jr vil zükünfftiger Historien.* n.p., 1532.

Grünpeck, Joseph. *Practica der gegenwertigen grossen Trübsalen, und vilfaltiger Wunder.* Strasbourg: M. Jakob Cammerlander, 1533.

Grünpeck, Joseph. *Historia Friderici IV. et Maximiliani I. ab Jos. Grünbeck.* Edited by Joseph Chmel. n.p., 1838.

Gunther, R. T. *Chaucer and Messahalla on the Astrolabe.* Vol. 5: *Early Science at Oxford.* Oxford: Oxford University Press, 1929.

Hartmann, Georg. *Hartmann's Practika: A Manual for Making Sundials and Astrolabes with the Compass and Rule.* Edited by John Lamprey. Bellvue, CO: John Lamprey, 2002.

Heitz, Paul, and Konrad Haebler. *Hundert Kalender Inkunabeln.* Strasbourg: Heitz und Miendel, 1905.

Hochstetter, Christopher. *Practica Theutsch Christoffero Hochstetter Doctor der Ertzeney Beschriben auff das Tausend Fünffhundert und achtundzwantzigst jar.* Augsburg: Ulhart, 1527.

Lichtenberger, Johannes. *Prognosticatio latina anno LXXVIIJ ad magnam coniunctionem Saturni et Jovis qui fuit anno LXXXIIII ac eclipsim solis anni sequentis scilicet LXXXVconfecta ac nunc de novo emendata.* Mainz: Jakob Meydenbach, 1499.

Lilly, William. *Christian Astrology, Modestly Treated in Three Books.* London: John Partridge, 1647.

Locher, Jacob. *De cometa sub septentrionibus viso aquei coloris, carmen.* Augsburg: Hans Froschauer, 1506.

Luther, Martin. *D. Martin Luthers Briefwechsel.* Vol. 3: *1523–1525.* Weimar: Hermann Böhlaus, 1933.

Machiavelli, Niccolò. *The Prince.* Edited by Quentin Skinner and Russell Price. Cambridge: Cambridge University Press, 1988.

Maximilian. Undated Letter from Maximilian. In "Urkunden und Regesten." Edited by Heinrich Zimerman and Franz Kreyczi. *Jahrbuch der kunsthistorischen Sammlungen des Allerhochsten Kaiserhauses,* vol. 3 (1885), XVI #2419.

Maximilian. Letter from Maximilian, March 1506. In "Urkunden und Regesten." Edited by Heinrich Zimerman and Franz Kreyczi. *Jahrbuch der kunsthistorischen Sammlungen des Allerhochsten Kaiserhauses,* vol. 3 (1885), XXXI #2592.

Maximilian. Letter from Maximilian, September 1514. In "Urkunden und Regesten." Edited by Heinrich Zimerman. *Jahrbuch der kunsthistorischen Sammlungen des allerhöchsten Kaiserhauses,* vol. 1 (1883), LVII #332.

Maximilian. Letter from Maximilian to Jakob Villiner. In "Urkunden und Regesten." Edited by Heinrich Zimerman. *Jahrbuch der kunsthistorischen Sammlungen des allerhöchsten Kaiserhauses,* vol. 1 (1883), LX #352.

Maximilian. Letter from Maximilian, 1514. In *Jahrbuch der kunsthistorischen Sammlungen des allerhöchsten Kaiserhauses,* vol. 1 (1883), #446.

Maximilian. Letter from Maximilian. In "Urkunden und Regesten." Edited by Hans Petz. *Jahrbuch der kunsthistorischen Sammlungen des allerhöchsten Kaiserhauses,* vol. 10 (1889), XLIV #5827.

Maximilian. Letter from Maximilian. In "Urkunden und Regesten." Edited by Heinrich Zimerman and Franz Kreyczi. *Jahrbuch der kunsthistorischen Sammlungen des Allerhochsten Kaiserhauses,* vol. 3 (1885), XXVII #2551.

Maximilian. "Fragmente einer lateinischen Autobiographie Kaiser Maximilians I."
   Jahrbuch der kunsthistorischen Sammlungen des allerhöchsten Kaiserhauses 6 (1888):
   421–46.
Maximilian. *Kaiser Maximilians Weisskunig.* Edited by H. Th. Musper. 2 vols. Stuttgart:
   W. Kohlhammer Verlag, 1956.
Milich, Jakob. *Liber secundus C. Plinii de mundi historia cum commentariis Iacobi Milichii
   diligenter conscriptis & recognitis.* Frankfurt: Peter Brubach, 1543.
Molinet, Jean de. Doutrepont, Georges, and Omer Jodogne, eds. *Chroniques de Jean
   Molinet.* 3 vols. Brussels: Palais des Académies, 1935–37.
Muris, Johannes de. *Arithmetica communis. Proportiones breves. De latitudinibus
   formarum. Algorithmus M. Georgii Peurbachii in integris. Algorithmus Magistri Joannis
   de Gmunden de minuciis phisicis.* Vienna: Johannes Singrenius, 1515.
Opusculum divinarum revelationum et de vita Antichristi. Basel: Michael Furter, 1498.
Perlach, Andreas. *Usus almanach seu Ephemeridum: ex commentarijs Georgii Tannsteter
   Colimitii, Preceptoris sui decerpti, & in quinquaginta propositiones, per Magistrum
   Andream perlachium stirum redacti.* Vienna: Hieronimus Vietor, 1519.
Perlach, Andreas. *Commentaria ephemeridium clarissimi viri D. Andreæ Perlachii Stiri,
   Medicae artis doctoris, ac in academia Viennensi Ordinarij quondam Mathematici, ad
   usum stuiosorum ita fideliter conscripta, ut quisque absque Præceptore, ex sola lectione
   integram indo artem consequi possit.* Vienna: Egidius Aquila, 1551.
Peuerbach, Georg. *Tabulae eclypsium Magistri Georgii Peurbachii.* Vienna: Johannes
   Winterberger, 1514.
Peutinger, Konrad. *Konrad Peutingers Briefwechsel.* Ed. Erich König. Munich: C. H.
   Beck, 1923.
Pirckheimer, Willibald. *Willibald Pirckheimers Briefweschel.* Edited by Emil Reicke.
   Munich: C. H. Beck'sche Verlagsbuchhandlung, 1956.
Poulle, Emmanuel, ed. *Les tables alphonsines, avec Les canons de Jean de Saxe.* Paris:
   Éditions du Centre national de la recherche scientifique, 1984.
Proclus. *Sphaera. Astronomiam discere incipientibus utilissima. Thoma Linacro Britanno
   interprete.* Vienna: Johannes Singrenius, 1511.
Rantzau, Heinrich. *Catalogus, Imperatorum, regum, ac virorum illustrium, qui artem
   astrologicam amarunt, ornarunt & exercuerunt.* Leipzig: Georg Defner, 1584.
Regiomontanus, Johannes. *Der deutsche Kalender des Johannes Regiomontan.* Edited by
   Ernst Zinner. Leipzig: Otto Harrassowitz, 1937.
Sackur, Ernst, ed. *Sibyllinische Texte und Forschungen: Pseudomethodius, Adso und
   tiburtinische Sibylle.* Halle: Niemeyer, 1898.
Sacrobosco, Johannes de. *Opusculum de sphaera clarissimi philosophi Ioannis de Sacro
   Busto. Theoricae planetarum Georgii Purbachii.* Vienna: Johannes Singrenius, 1518.
Sanuto, Marino. *I Darii Di Marino Sanuto.* Venice, 1886.
Schedel, Hartmann. *The Nuremberg Chronicle. A Facsimile of Hartmann Schedel's Buch der
   Chroniken.* Nürmberg: Anton Koberger, 1493.
Schöner, Andreas. *Gnomonice Andreae Schoneri Noribergensis, Hoc est: De descriptionibus
   horologiorum sciotericorum Omnis Generis Proiectionibus circulorum Sphaericorum ad
   superficies.* Noribergae: Montanus & Neuberus, 1562.

Schöner, Johannes. *Coniectur odder abnemliche Außlegung Joannis Schöners uber den Cometen so im Augustmonat / des M.D.XXXj. jars erschinen ist / zu ehren einem erbern Rath / und gmainer burgerschafft der stat Nuremberg außgangen.* Nuremberg: Friderich Peypus, 1531.

Schöner, Johannes. *De iudiciis nativitatum. Libri tres.* Nuremberg: Ioannem Montanum & Ulricum Neuberum, 1545.

Schöner, Johannes. *Opera mathematica.* Nuremberg: Ioannem Montanum & Ulricum Neuberum, 1551.

Schonheinz, Jacobus. *Apologia astrologie.* Nuremberg: Georgius Schenk, 1502.

Scribanario, Marco. Iudicio de Marco Scribanario Bolognese supra la expositione de la apparente Cometa al Illustreier Excel. Meser Iohanne de Bentivolgi da Bologna. 1506.

Stabius, Johannes. *Practica Teutsch von Ingoldstat Joannis Stabij Philosophi und Mathematici auff Tausent funffhundert und ain Jar.* n.p., 1500.

Stabius, Johannes. *Horoscopion omni generaliter congruens climati (dedicated to Jakob Bannisius).* 1512.

Stabius, Johannes. *Horoscopion omni generaliter (dedicated to Matthäus Lang).* n.p., 1512

Stainpeis, Martin. *Liber de modo studendi seu legendi in medicina.* Vienna, 1520.

Stiborius, Andreas, and Georg Tannstetter. *Super requisitione Sanctissimi Leonis Papae.X. et Divi Maximiliani Imp. P. F. Aug. De Romani Calendarii correctione Consilium in Florentissimo studio Viennensi Austriae conscriptum & aeditum.* Vienna: Johannes Singrenius, 1514.

Stöffler, Johannes, and Jakob Pflaum. *Almanach nova plurimis annis venturis inservientia.* Venice: Peter Lichtenstein, 1506.

Stöffler, Johannes, and Jakob Pflaum. *Almanach nova plurimis annis venturis inservientia.* Venice: Peter Lichtenstein, 1507.

Stöffler, Johannes, and Peter Pitatus. *Ephemeridum reliquiae Ioannis stoeffleri germani, superadditis novis usque ad annum Christi 1556. durantibus, Petri Pitatis Veronensis Mathematicis, una cum additionibus longe utilissimis eiusdem.* Tubingen: Ulrich Morhard, 1548.

Tacitus. *Annals, Books IV–VI, XI–XII.* Edited by John Jackson. *Loeb Classical Library.* Cambridge, MA: Harvard University Press, 1937.

Tannstetter, Georg. *Iudicium Viennese Magistri Georgii Lanstetter Collimitii Super anno M.cccc.v.* n.p., 1504.

Tannstetter, Georg. *Practica Teutsch auff das M.cccc. und eylff Jar. durch Mayster Georgen Tannstetter.* Vienna, 1510.

Tannstetter, Georg. *Judcium Viennense Anni Millesimiquingentesimi duodecimi. Ad illustrissimum principem Ludovicum bavarie Ducem Comitemque palatinum rheni.* Nuremberg: Wolfgang Huber, 1511.

Tannstetter, Georg. *Judcium Viennense Anni Millesimiquingentesimi duodecimi. Ad illustrissimum principem Ludovicum bavarie Ducem Comitemque palatinum rheni.* Cologne: Henry de Nussia, 1511.

Tannstetter, Georg. *Judicium Astronomicum pro anno Christi. M.cccc.xiij. per Magister Georgium Tannstetter collimitium astronomie professorem Viene Pannonie editum.* Nuremberg: Wolfgang Huber, 1512.

Tannstetter, Georg. *Judicium Astronomicum Viennense. anni M.CCCC.xvij. Ad nobiles et providentissimos dños Magistrum civium et universum inclite urbis Vienne Senatum per Georgium Tannstetter Collimitium artium et medicine doctorem diligenter elaboratum.* Vienna: Johannes Singrenius, 1516.

Tannstetter, Georg. *Judicium astronomicum Viennense anni M.D.xix. ad Reverendissimi in choro patrem et illusttrissimi principem et d. d. Matteum S. Angeli sa. sa. Ro. ecclesie Cardinalem Burcennse coadiutorem Salisburgennse et in germania legatum. etc. do. suum gratiosissimi Geo. Tannstetter.* Vienna: Johannes Singrenius, 1518.

Tannstetter, Georg. *Regiment für den Lauff der Pestilentz durch Georgen Tannstetter von Rain der siben freyen künst unnd Ertzney doctor: kurtzlich beschriben. Anno. 1521.* n.p., 1521.

Tannstetter, Georg. *Ad laudem et gloriam am Serenissimi et Ptoentissimi principis Ferdinandi. Principis Hispaniarum. Archiducis Austrie. Ducis Burgundie etc. et ad usum commoditatemque provinciarum Austrie. Georgius Tannstetter Collimitius Lycoripensis Artium et Medicine Doctor. Vienne hoc prognosticum pro anno Christi 1522 edidit.* Landshut: Johann Weßemburger, 1521.

Tannstetter, Georg. *Libellus consolatorius, quo opinionem iam dudum annis hominum ex quorundam Astrologastrorum divinatione insidentem, de futuro diluuio & multis alijs horrendis periculis XXIIII. anni a fundamentis extirpare conatur.* Vienna: Johannes Singrenius, 1523.

Tannstetter, Georg. *Practica gemacht zu Wienn in Osterreych aus der Kunst Astronomia / durch Georgen annstetter von Rain am Lech / der freyen künsten und erzney Doctor / und ist wie ain außzugund ain sundere erklerung des grosseren Judicij so vorhin auff das M.D.XXIIII. Jar von im ausgangen ist.* Vienna: Johannes Singrenius, 1523.

Tannstetter, Georg. *Juditium Astronomicum Viennense Georgij Tannstetter Collimitij Lycoripensis artium et medicine doctoris in annum christi. 1525.* Vienna: Johannes Singrenius, 1524.

Tannstetter, Georg. *Artificium de applicatione astrologiæ ad medicinam.* Strasbourg: Georgius Ulricherus, 1531.

Temporal Des weitberümpten M. Johann Künigspergers / Natürlicher kunst der Astronomei / kurzer begriff / Von natürlichem einfluß der Gestirn / Planeten / und Zeichen / etc. Von den Complexionen / natur und Eigenschafft der menschen / Regiment durchs Jar über / mit Essen / Schlaffe / Baden / Purgieren / Aderlassen / etc. Auffs ordenlichst zugericht. Erfurt: Wolffgang Sthürmer, n.d.

Teutscher Kalender. Edited by Kurt Pfister. Munich: Verlag Dr. Albert Mundt, 1922.

Torrella, Gaspar. *Judicium Universale. De portentis: præsagiis: & ostentis: rerum admirabilium ac Solis et Lunee defectibus & Cometis.* Rome: Johannes Besicken, 1507.

Tractatus quidam de Turcis. Nuremberg: Conrad Zeninger, 1481.

Trithemius, Johannes. *Annalium Hirsavgiensium, Opus nunquam hactenus editum, & ab Eruditis semper desideratum. Complectens historiam Franciae et Germaniae, gesta imperatorum, regum, principum, episcoporum, abbatum, et illustrium virorum.* St. Gall: Joannes Georgius Schlegel, 1690.

Virdung, Johannes. *Practica deutsch uff das 1523. jare.* Nuremberg, 1522.

Virdung, Johannes. *Practica deutsch Meister Hansen Virdung von Haßfurt / uff das M.cccc. und xxiij. jare and Johann Virdung, Practica deütsch / des würdigen hochgelerten und*

weytbrümpted Doctor Johansen Virdungs von Haßfurt / auff das M.CCCCC. und.XXXI.
jare gemact. n.p., 1530.

Virdung, Johannes. *Explanatio maximarum et formidabilium rerum futurarum Anno Salutis*
M. D. VII. que per Cometem Anno 1506 in Climate nostro conspectum portenduntum.
Strasbourg, 1506.

Virdung, Johannes. *Der gesehen worden ist / im Augstmon im 1531. jare / durch Doctor*
Johansen Virdung von Haßfurt zü Eren Dem durchleuchtigisten hochgebornen
fürsten unnd herren Herrn Ludwigen Palzgrauenn bey Rheyn herzogen in Bayern des
heiligen Römischen Reichs Erz durchssessen und Churfürsten. Unnd zü eyner warnung
yedermenniglichen / wann es ist nye kein Comet gesehen worden / der nit etwas groß
erschröcklichs bracht / und bezeigt wie hernach volgt Capitels weyß. etc. Speier, 1531.

Werner, Johannes. *De triangulis sphericis libri quatuor. De meteoroscopiis libri sex.* Krakow:
Lazarus Andreae, 1557.

Secondary Literature

Abry, Josèphe-Henriette. "What Was Agrippina Waiting For? (Tacitus, Ann. XII,
68–69)." In *Horoscopes and Public Spheres. Essays on the History of Astrology,* Religion
and Society Series, vol. 42, edited by Günther Oestmann, H. Darrel Rutkin, and
Kocku von Stuckrad, 37–48. Berlin: Walter de Gruyter, 2005.

Alder, Ken. "A Revolution to Measure: The Political Economy of the Metric System in
France." In *The Values of Precision,* edited by M. Norton Wise, 39–71. Princeton, NJ:
Princeton University Press, 1995.

Altfahrt, Margit. "Die Politische Propaganda für Maximilians II. (Erster Teil)."
Mitteilungen des Instituts für österreichische Geschichtsforschung 88 (1980): 283–312.

Altfahrt, Margit. "Die Politische Propaganda für Maximilians II. (Zweiter Teil)."
Mitteilungen des Instituts für österreichische Geschichtsforschung 89 (1981): 53–92.

Anderson, Andrew Runni. *Alexander's Gate, Gog and Magog, and the Enclosed Nations.*
Cambridge, MA: Medieval Academy of America, 1932.

Angermeier, Heinz, ed. *Deutsche Reichstagsakten unter Maximilian I.* Vol. 5: *Reichstag von*
Worms 1495. Göttingen: Vandenhoek & Ruprecht, 1981.

Anglo, Sydney. *Images of Tudor Kingship.* Surrey: Seaby, 1992.

Antonicek, Theophil, Elisabeth T. Hilscher, and Harmut Krones, eds. *Die Wiener Hofmusik-*
kapelle. Georg von Slatkonia und die Wiener Hofmusikkapelle. Vienna: Böhlau, 1999.

Arrizabalaga, Jon, and Roger French. "Coping with the French Disease: University
Practitioners' Strategies and Tactics in the Transition from the Fifteenth to the
Sixteenth Century." In *Medicine from the Black Death to the French Disease,* edited
by Roger French, Jon Arrizabalaga, Andrew Cunningham, and Luis García-Ballester,
248–87. Brookfield, VT: Ashgate, 1998.

Arrizabalaga, Jon, John Henderson, and Roger French. *The Great Pox: The French Disease*
in Renaissance Europe. New Haven, CT: Yale University Press, 1997.

Ash, Eric H. *Power, Knowledge, and Experience in Elizabethan England.* Baltimore, MD:
Johns Hopkins University Press, 2004.

Ash, Eric H., ed. *Expertise. Practical Knowledge and the Early Modern State.* Chicago: University of Chicago Press, 2010.

Azzolini, Monica. "Reading Health in the Stars: Politics and Medical Astrology in Renaissance Milan." In *Horoscopes and Public Spheres. Essays on the History of Astrology,* Religion and Society Series, vol. 42, edited by Günther Oestmann, H. Darrel Rutkin, and Kocku von Stuckrad, 183–205. Berlin: Walter de Gruyter, 2005.

Azzolini, Monica. "The Politics of Prognostication: Astrology, Political Conspiracy and Murder in Fifteenth-Century Milan." *History of Universities* 22 (2008): 6–34.

Azzolini, Monica. "The Political Uses of Astrology: Predicting the Illness and Death of Princes, Kings and Popes in the Italian Renaissance." *Studies in History and Philosophy of Biological and Biomedical Sciences* 41, no. 2 (2010): 135–45.

Azzolini, Monica. *The Duke and the Stars. Astrology and Politics in Renaissance Milan.* Cambridge, MA: Harvard University Press, 2012.

Barnes, Robin Bruce. "Hope and Despair in Sixteenth-Century Almanacs." In *Die Reformation in Deutschland und Europa: Interpretationen und Debatten,* edited by Hans Guggisberg, Gottfried G. Krodel, and Hans Füglister, 440–61. Gütersloh: Gütersloher Verlagshaus, 1993.

Barton, Tamsyn. *Ancient Astrology.* London: Routledge, 1994.

Bauch, Gustav. *Die Anfänge des Humanismus in Ingolstadt: Eine litterarische Studie zur deutschen Universitätsgeschichte.* Munich: R. Oldenbourg, 1901.

Bauer, Barbara. "Die Rolle des Hofastrologen und Hofmathematicus als fürsterlicher Berater." In *Höfischer Humanismus,* edited by August Buck, 93–117. Weinheim: VCH Verlagsgesellschaft, 1989.

Bellot, Josef. "Konrad Peutinger und die literarisch-künsterischen Unternehmungen Kaiser Maximilians I." *Philobiblon* 11 (1967): 171–90.

Benesch, Otto, and Erwin M. Auer. *Die Historia Friderici et Maximiliani.* Berlin: Deutscher Verein für Kunstwissenschaft, 1957.

Benjamin, Francis S., Jr., and G. J. Toomer, eds. *Campanus of Novara and Medieval Planetary Theory. Theorica planetarum.* Madison: University of Wisconsin Press, 1971.

Bennett, Jim. "Presidential Address. Knowing and Doing in the Sixteenth Century: What Were Instruments For?" *British Journal for the History of Science* 36 (2003): 129–50.

Bernays, Edward. *Propaganda.* New York: H. Liveright, 1928.

Biagioli, Mario. *Galileo Courtier. The Practice of Science in an Age of Absolutism.* Chicago: University of Chicago Press, 1993.

Blair, Ann. *The Theater of Nature. Jean Bodin and Renaissance Science.* Princeton, NJ: Princeton University Press, 1997.

Blume, Dieter. *Regenten des Himmels. Astrologische Bilder in Mittelalter und Renaissance.* Vol. 3, *Studien aus dem Warburg-Haus.* Berlin: Akademie, 2000.

Borchardt, Frank L. *German Antiquity in Renaissance Myth.* Baltimore, MD: Johns Hopkins University Press, 1971.

Boudet, Jean-Patrice. "Simon de Phares et les rapports entre astrologie et prophétie à la fin du Moyen Âge." In *Les Textes prophétiques et la prophétie en occident (XIIe–XVIe siècle),* edited by André Vauchez. Rome: Ecole française de Rome, 1990.

Boudet, Jean-Patrice. *Entre Science et Nigromance: Astrologie, Divination et Magie Dans L'Occident Médiéval (Xiie–Xve Siècle)*. Paris: Publications de la Sorbonne, 2006.

Bourguet, Marie-Noëlle, Christian Licoppe, and H. Otto Sibum, eds. *Instruments, Travel and Science. Itineraries of Precision from the Seventeenth to the Twentieth Century*. London: Routledge, 2002.

Brann, Noel L. *Trithemius and Magical Theology. A Chapter in the Controversies over Occult Studies in Early Modern Europe*. Albany: State University of New York Press, 1999.

Brévart, Francis B. "The German *Volkskalender* of the Fifteenth Century." *Speculum* 63, no. 2 (1988): 312–42.

Brévart, Francis B. "Johann Blaubirers Kalender von 1481 und 1483: Traditionsgebundenheit und experimentelle Innovation." *Gutenberg Jahrbuch* 63 (1988): 74–83.

Brosseder, Claudia. *Im Bann der Sterne. Caspar Peucer, Philipp Melanchthon und andere Wittenberger Astrologen*. Berlin: Akademie Verlag, 2004.

Brosseder, Claudia. "The Writing in the Wittemberg Sky: Astrology in Sixteenth-Century Germany." *Journal of the History of Ideas* 66 (2005): 557–76.

Burger, Hans-Otto. "Der *Weisskunig* als Literaturdenkmal." In *Kaiser Maximilians Weisskunig*, edited by H. Th. Musper, 13–33. Stuttgart: W. Kohlhammer Verlag, 1956.

Burke, Peter. *The Fabrication of Louis XIV*. New Haven, CT: Yale University Press, 1992.

Burnett, Charles, and Gerrit Boss. *Scientific Weather Forecasting in the Middle Ages: The Writings of al-Kindi*. London and New York: Kegan Paul, 2000.

Byrne, James Steven. "A Humanist History of Mathematics? Regiomontanus's Padua Oration in Context." *Journal of the History of Ideas* 67, no. 1 (2006): 41–61.

Byrne, James Steven. "The Stars, the Moon, and the Shadowed Earth: Viennese Astronomy in the Fifteenth Century." PhD diss., Princeton University, 2007.

Campbell, Anna Montgomery. *The Black Death and Men of Learning*. New York: AMS, 1966.

Capp, Bernard. *English Almanacs, 1500–1800. Astrology and the Popular Press*. Ithaca, NY: Cornell University Press, 1979.

Carey, Hilary M. *Courting Disaster. Astrology at the English Court and University in the Later Middle Ages*. New York: St. Martin's Press, 1992.

Carey, Hilary M. "What Is the Folded Almanac? The Form and Function of a Key Manuscript Source for Astro-Medical Practice in Later Medieval England." *Social History of Medicine* 16 (2003): 481–509.

Carey, Hilary M. "Astrological Medicine and the Medieval English Folded Almanac." *Social History of Medicine* 17 (2004): 345–63.

Carr, Amelia J., and Richard L. Kremer. "Child of Saturn. The Renaissance Church Tower at Niederaltaich." *Sixteenth Century Journal* 17 (1986): 401–34.

Carrar, Daniela Mugnai. *La biblioteca di Nicolò Leoniceno. Tra Aristotele e Galeno: cultura e libri di un medico umanista*. Florence: L.S. Olschki, 1991.

Castigloni, Arturo. "The School of Ferrara and the Controversy on Pliny." In *Science, Medicine and History: Essays on the Evolution of Scientific Thought and Medical Practice Written in Honour of Charles Singer*, edited by E. Ashworth Underwood, 269–79. London: Oxford University Press, 1953.

Christianson, J. R. "Tycho Brahe's German Treatise on the Comet of 1577: A Study in Science and Politics." *Isis* 70 (1979): 110–40.

Cole, Mary Hill. *The Portable Queen. Elizabeth I and the Politics of Ceremony.* Amherst: University of Massachusetts Press, 1999.

Cole, R. G. "The Reformation Pamphlet and Communication Processes," In *Flugschriften Als Massenmedium Der Reformationszeit,* edited by Hans-Joachim Köhler. Stuttgart: Klett-Cotta, 1981.

Cooper, J. P. D. *Propaganda and the Tudor State. Political Culture in the Westcountry.* Oxford: Oxford University Press, 2003.

Coutu, Joan. *Persuasion and Propaganda. Monuments and the Eighteenth-Century British Empire.* Montreal: McGill-Queen's University Press, 2006.

Crystal, Malcolm Lee. "Medicine in Vienna in the Sixteenth and Seventeenth Centuries." PhD diss., University of Virginia, 1994.

Csapodi, Csaba. *The Corvinian Library: History and Stock.* Budapest: Akademai Kiado, 1973.

Csapodi, Csaba, and Klára Csapodi-Gárdonyi. *Bibliotheca Corviniana. The Library of King Matthias Corvinus of Hungary.* New York: Frederick A. Praeger, 1969.

Csendes, Peter, and Ferdinand Opll, eds. *Wien. Geschichte einer Stadt.* Vol. 1: *Von den Anfängen bis zur Ersten Türkenbelagerung.* Vienna: Böhlau, 2001.

Cull, Nicholas J. *The Cold War and the United States Information Agency. American Propaganda and Public Diplomacy, 1945–1989.* Cambridge: Cambridge University Press, 2008.

Cuneo, Pia F. "Images of Warfare as Political Legitimization: Jörg Breu the Elder's Rondels for Maximlian I's Hunting Lodge at Lermos (ca. 1516)." In *Artful Armies, Beautiful Battles. Art and Warfare in Early Modern Europe,* History of Warfare, edited by Pia F. Cuneo, 87–105. Leiden: Brill, 2002.

Curry, Patrick. *Prophecy and Power. Astrology in Early Modern England.* Princeton, NJ: Princeton University Press, 1989.

Curry, Patrick. "Astrology in Early Modern England: The Making of a Vulgar Knowledge." In *Astrology, Religion and Politics in Counter-Reformation Rome,* edited by Stephen Pumphrey, Paolo L. Rossi, and Maurice Slawinski, 274–91. Manchester: Manchester University Press, 1991.

Curtius, Ernst Roberts. *European Literature and the Latin Middle Ages.* Princeton, NJ: Princeton University Press, 1988.

Czerny, Albin. "Der Humanist und Historiograph Kaiser Maximilians I. Joseph Grünpeck." *Archiv für österreichische Geschichte* 73 (1888): 315–64.

Daston, Lorraine, and Katherine Park. *Wonders and the Order of Nature, 1150–1750.* New York: Zone Books, 1998.

Davies, Martin. "Making Sense of Pliny in the Quatrocento." *Renaissance Studies* 9, no. 2 (1995): 240–57.

Dean, Paul. "Tudor Humanism and the Roman Past: A Background to Shakespeare." *Renaissance Quarterly* 41, no. 1 (1988): 84–111.

Demaitre, Luke. "The Art and Science of Prognostication in Early University Medicine." *Bulletin of the History of Medicine* 77 (2003): 765–88.

Denis, Michael. *Wiens Buchdruckergeschichte bis zum Jahre 1560.* Vienna, 1782.

Diederichs, Peter. "Kaiser Maximilian I als politischer Publizist." PhD diss., University of Heidelberg, 1931.

Dietrichs, Peter. "Kaiser Maximilian I. (1459–1519)." In *Deutsche Publizisten das 15. bis 20. Jahrhunderts,* edited by Heinz-Dietrich Fischer, 35–42. Munich: Verlag Dokumentation, 1971.

Dixon, C. Scott. "Popular Astrology and Lutheran Propaganda in Reformation Germany." *History* 84 (1999): 403–18.

Dodgson, Campbell. "Drei Studien." *Jahrbuch der kunsthistorischen Sammlungen des Allerhochsten Kaiserhauses* 29 (1910/11): 1–20.

Dodgson, Campbell. "An Unknown MS. of Freydal." *The Burlington Magazine* 48, no. 278 (1926): 235–42.

Dodgson, Campbell. "More Freydal Drawings." *Burlington Magazine* 53, no. 307 (1928): 170–73.

Dohrn-van Rossum, Gerhard. *History of the Hour. Clocks and Modern Temporal Orders.* Translated by Thomas Dunlap. Chicago: University of Chicago Press, 1996.

Dooley, Brendan. *Morandi's Last Prophecy and the End of Renaissance Politics.* Princeton, NJ: Princeton University Press, 2002.

Ducos, Joëlle. *La méteorologie en Français au Moyen Âge (XIII–XIVe siècles).* Paris: Champion, 1998.

Dupède, Jean. "L'Ars notoria et la polémique sur la divination et la magie." In *Divination et controverse religieuse en Frnace au XVIe siècle, Collection de l'École Normale Supérieur de Jeunes Filles,* 123–34. Paris: École Normale Supérieur de Jeunes Filles, 1987.

Dupède, Jean. "L'écriture chez l'eremite Pelagius." In *Le texte et son inscription,* edited by Roger Lauter, 113–54. Paris: CNRS, 1989.

Durling, Richard J. "An Early Manual for the Medical Student and the Newly Fledged Practitioner: Martin Steinpeis, 'Liber de modo studendi seu legendi in medicina' (Vienna, 1520)." *Clio Medica* 5 (1970): 7–33.

Eagleton, Catherine. *Monks, Manuscripts and Sundials. The Navicula in Medieval England.* Vol. 11, Medieval and Early Modern Science. Leiden: Brill, 2010.

Eastwood, B. S. "Plinian Astronomy in the Middle Ages and Renaissance." In *Science in the Early Roman Empire: Pliny the Elder, His Sources and Influence,* edited by Roger French and Frank Greenaway, 197–251. Totowa, NJ: Barnes & Noble, 1986.

Edwards, Mark. *Printing, Propaganda, and Martin Luther.* Berkeley: University of California Press, 1994.

Eisenstein, Elizabeth L. *The Printing Revolution in Early Modern Europe.* Cambridge: Cambridge University Press, 1983.

Eisenstein, Elizabeth L. "Reply." *American Historical Review* 107 (2002): 126–28.

Eisenstein, Elizabeth L. "An Unacknowledged Revolution Revisited." *American Historical Review* 107 (2002): 87–105.

Eisermann, Falk, and Volker Honemann. "Die ersten typographischen Einblattdrucke." *Gutenberg Jahrbuch* 75 (2000): 88–131.

Eisler, Robert. "The Frontispiece to Sigismondo Fanti's Triompho di Fortuna." *Journal of the Warburg and Courtauld Institutes* 10 (1947): 155–59.

Ellul, Jacques. *Propaganda. The Formation of Men's Attitudes.* New York: Knopf, 1969.

Emerson, Catherine. *Olivier de La Marche and the Rhetoric of 15th-Century Historiography.* Woodbridge: Boydell Press, 2004.

Ernst, Germanna. "Astrology, Religion and Politics in Counter-Reformation Rome." In *Science, Culture and Popular Belief in Renaissance Europe,* edited by Stephen Pumphrey, Paolo L. Rossi, and Maurice Slawinski, 249–73. Manchester: Manchester University Press, 1991.

Esser, Thilo. "Die Pest—Strafe Gottes oder Naturphänomen? Eine frömmigkeitsge-schichtliche Untersuchung zu Pesttraktaten des 15. Jahrhunderts." *Zeitschrift für Kirchengeschichte* 108 (1997): 32–57.

Evans, R. J. W. *Rudolf II and His World. A Study in Intellectual History, 1576–1612.* Oxford: Oxford University Press, 1973.

Feldman, Walter. "The Celestial Sphere, the Wheel of Fortune, and Fate in the Gazels of Naili and Baki." *International Journal of Middle East Studies* 28, no. 2 (1996): 193–215.

Fichtner, Paula Sutter. *Ferdinand I of Austria: The Politics of Dynasticism in the Age of the Reformation. East European Monographs.* New York: Columbia University Press, 1982.

Fichtner, Paula Sutter. *Emperor Maximilian II.* New Haven, CT: Yale University Press, 2001.

Findlen, Paula. *Possessing Nature. Museums, Collecting, and Scientific Culture in Early Modern Italy.* Berkeley: University of California Press, 1994.

Flint, Valerie. *The Rise of Magic in Early Medieval Europe.* Princeton, NJ: Princeton University Press, 1991.

Frasca-Spada, Marina, and Nicholas Jardine, eds. *Books and the Sciences in History.* Cambridge: Cambridge University Press, 2000.

French, Roger. "Pliny and Renaissance Medicine." In *Science in the Early Roman Empire: Pliny the Elder, His Sources and Influence,* edited by Roger French and Frank Greenaway, 252–81. Totowa, NJ: Barnes & Noble, 1986.

French, Roger. "Astrology in Medical Practice." In *Practical Medicine from Salerno to the Black Death,* edited by Luis García-Ballester, Roger French, and Jon Arrizabalaga, 30–59. Cambridge: Cambridge University Press, 1994.

French, Roger. "Foretelling the Future: Arabic Astrology and English Medicine in the Late Twelfth Century." *Isis* 87, no. 3 (1996): 453–80.

French, Roger. *Canonical Medicine. Gentile da Foglio and Scholasticism.* Leiden: Brill, 2001.

French, Roger. *Medicine before Science. The Business of Medicine from the Middle Ages to the Enlightenment.* Cambridge: Cambridge University Press, 2003.

Fučíková, Eliška. "The Collection of Rudolf II at Prague: Cabinet of Curiosity or Scientific Museum?" In *The Origins of Museums. The Cabinet of Curiosities in Sixteenth- and Seventeenth-Century Europe,* edited by Oliver Impey and Arthur MacGregor, 47–53. Oxford: Clarendon Press, 1985.

Fučíková, Eliška, ed. *Rudolf II and Prague. The Court and the City.* London: Thames and Hudson, 1997.

Fürst, Dietmar, and Jürgen Hamel. *Johann Carion (1499–1537): Der erste Berliner Astronom.* Berlin: Archenhold-Sternwarte, 1988.

Füssel, Stephan. *Gutenberg und seine Wirkung.* Frankfurt: Insel Verlag, 1999.

Füssel, Stephan. *Kaiser Maximilian und die Medien seiner Zeit. Der Theuerdank von 1517. Ein kulturhistorische Einführung. Die Abenteuer des Ritters Theuerdank.* Cologne: Taschen, 2003.

Füssel, Stephan. *Gutenberg and the Impact of Printing.* Translated by Douglas Martin. Aldershot: Ashgate, 2005.

Gack-Scheiding, Christine. *Johannes de Muris Epistola super reformatione antiqui kalendarii. Ein Beitrag zur Kalendarreform im 14. Jahrhundert.* Hannover: Hahnische Buchhandlung, 1995.

Garin, Eugenio. *Astrology in the Renaissance. The Zodiac of Life.* Translated by Carolyn Jackson. London: Routledge & Kegan Paul, 1976.

Gaunt, Sarah. "Visual Propaganda in England in the Later Middle Ages." In *Propaganda. Political Rhetoric and Identity 1300–2000,* edited by Bertrand Taithe and Tim Thornton, 27–39. Gloucestershire: Sutton, 1999.

Geneva, Ann. *Astrology and the Seventeenth-Century Mind: William Lilly and the Language of the Stars.* Manchester: Manchester University Press, 1995.

Genuth, Sara Schechner. *Comets, Popular Culture, and the Birth of Modern Cosmology.* Princeton, NJ: Princeton University Press, 1997.

Gingerich, Owen. "Astronomical Paper Instruments with Moving Parts." In *Making Instruments Count. Essays on Historical Scientific Instruments Presented to Gerard L'Estrange,* edited by R. J. W. Anderson, J. A. Bennett, and W. F. Ryan, 63–74. Aldershot: Variorum, 1993.

Gingerick, Virginia. "The *Ludus Diane* of Conrad Celtis." *Germanic Review* 15 (1940): 159–80.

Glowatzki, E., and Helmut Göttsche. *Die Tafeln des Regiomontanus: Ein Jahrhundertwerk.* Munich: Institut für Geschichte der Naturwissenschaften, 1990.

Goldman, Arthur. *Die Wiener Universität, 1519–1740.* Vienna: Adolf Holzhausen, 1916.

Gottlieb, Theodor. *Büchersammlung Kaiser Maximilians.* Leipzig, 1900.

Gottschalk, H. B. "The Conclusion of Brant's 'De corrupto ordine vivendi pereuntibus.'" *Modern Language Review* 77 (1982): 348–50.

Gow, Andrew. "Gog and Magog on Mappaemundi and Early Printed World Maps: Orientalizing Ethnography in the Apocalyptic Tradition." *Journal of Early Modern History* 2 (1998): 61–88.

Graf-Stuhlhofer, Franz. "Georg Tannstetter (Collimitius). Astronom, Astrologe und Leibarzt bei Maximilian I. und Ferdinand I." *Jahrbuch des Vereins der Stadt Wien* 37 (1981): 7–49.

Graf-Stuhlhofer, Franz. "Andrej Perlach kot ucenec jurija Tannstetterja na dunaju." *Casopis za Zgodovino in Narodopisje St.* 2 (1991): 280–83.

Graf-Stuhlhofer, Franz. *Humanismus zwischen Hof und Universität. Georg Tannstetter (Collimitius) und sein wissenschaftliches Umfeld im Wien des frühen 16. Jahrhunderts.* Vienna: WUV-Universitätsverlag, 1996.

Graf-Stuhlhofer, Franz. "Zu Den Hofastronomen Kaiser Maximilians. Über Das Jahrzehntelange Fortwirken Historischer Irrtümer." *Bibliothèque d'Humanisme et Renaissance* 60, no. 2 (1998): 413–19.

Graf-Stuhlhofer, Franz. "Das Weiterbestehen des Wiener Poetenkollegs nach dem Tod Konrad Cetis' (1508)." *Zeitschrift für historische Forschung* 26, no. 2 (1999): 393–407.

Grafton, Anthony. "The Importance of Being Printed." *Journal of Interdisciplinary History* 11 (1980): 265–86.

Grafton, Anthony. *Cardano's Cosmos. The Worlds and Works of a Renaissance Astrologer.* Cambridge, MA: Harvard University Press, 1999.

Grafton, Anthony. "Geniture Collections, Origins and Uses of a Genre." In *Books and the Sciences in History,* edited by Marina Frasca-Spada and Nick Jardine, 49–68. Cambridge: Cambridge University Press, 2000.

Grafton, Anthony. "Starry Messengers: Recent Work in the History of Western Astrology." *Perspectives on Science* 8, no. 1 (2000): 70–83.

Grafton, Anthony. *Bring Out Your Dead. The Past as Revelation.* Cambridge, MA: Harvard University Press, 2001.

Grafton, Anthony. "Introduction." *American Historical Review* 107 (2002): 84–86.

Green, Jonathan. *Printing and Prophecy. Prognostication and Media Change 1450–1550.* Ann Arbor: University of Michigan Press, 2012.

Green, Louis. "Historical Interpretations in Fourteenth-Century Florentine Chronicles." *Journal of the History of Ideas* 28, no. 2 (1967): 161–78.

Grendler, Paul. *The Universities of the Italian Renaissance.* Baltimore, MD: Johns Hopkins University Press, 2002.

Grössing, Helmuth. "Johannes Stabius. Ein Beitrag zur Kulturgeschicht der Zeit Kaiser Maximilians I." PhD diss., University of Vienna, 1964.

Grössing, Helmuth. "Johannes Stabius. Ein Oberösterreicher im Kreis der Humanisten um Kaiser Maximilian I." *Mitteilungen des Oberösterreichischen Landesarchiv* 9 (1968): 239–64.

Grössing, Helmuth. "Astronomus poeta. Georg von Peuerbach als Dichter." *Jahrbuch des Vereins für Geschichte der Stadt Wien* 34 (1978): 54–66.

Grössing, Helmuth. "Der Humanist Regiomontanus und sein Verhältnis zu Georg von Peuerbach." In *Humanismus und Naturwissenschaften,* edited by Rudolf Schmitz and Fritz Krafft, 69–82. Boppard: Harald Boldt Verlag, 1980.

Grössing, Helmuth. *Humanistische Naturwissenschaft: Zur Geschichte der Wiener mathematischen Schulen des 15. und 16. Jahrhunderts.* Baden-Baden: Valentin Koerner, 1983.

Grössing, Helmuth. "Die Wiener Universität im Zeitalter des Humanismus von der Mitte des 15. bis zur Mitte des 16. Jahrhunderts." In *Das alte Universitätsviertel in Wien, 1385–1985,* Schriftenreihe de Universitätsarchivs. edited by Günther Hamann, Kurt Mühlberger, and Franz Skacel, 37–45. Vienna: Universitätsverlag für Wissenschaft und Forschung, 1985.

Grössing, Helmuth. "Naturwisseschaften in Österreich im Zeitalter des Humanismus." In *Verdrängter Humanismus. Verzögerte Aufklärung,* Band 1: Philosophie in Österreich (1400–1650), edited by Michael Benedikt, Reinhold Knoll, and Josef Rupitz, 249–62. Klausen-Leopoldsdorft: Editura Triade, 1996.

Grössing, Helmuth, and Franz Stuhlhofer. "Versuch einer Deutung der Rolle der Astrologie in den persönlichen Entscheidungen einiger Habsburger des Spätmittelalters." *Anzeiger der phil.-hist. Klasse der Österreichischen Akademie der Wissenschaften* 117 (1980): 267–83.

Großmann, Karl. "Die Frühzeit des Humanismus in Wien bis zu Celtis Berufung 1497." *Jahrbuch für Landeskunde von Niederösterreich* 22 (1929): 152–325.

Grotefend, Hermann. *Taschenbuch der Zeitrechnung des deutschen Mittelalters und der Neuzeit.* 2 vols. Hannover: Hahnische Buchhandlung, 1891.

Gruber, Gernot. "Beginn der Neuzeit." In *Musikgeschichte Österreichs. Von den Anfängen zum Barock,* edited by Rudolf Flotzinger, 169–213. Vienna: Böhlau Verlag, 1995.

Gunther, R. T. *Chaucer and Messahalla on the Astrolabe.* Vol. 5: *Early Science at Oxford.* Oxford: Oxford University Press, 1929.

Günther, Siegmund. *Geschichte des mathematischen Unterrichts.* Vol. 3: *Monumenta Germaniae Paedagogica.* Berlin: A. Hofmann & Comp., 1887.

Gutas, Dimitri. *Greek Thought, Arabic Culture. The Graeco-Arabic Translation Movement in Baghdad and Early 'Abbāsid Society (2nd–4th/8th–10th centuries).* London: Routledge, 1998.

Hammerstein, Notker. "Relations with Authority." In *Universities in Early Modern Europe (1500–1800),* A History of the University in Europe, vol. 2, edited by Hilde de Ridder-Symoens, 114–53. Cambridge: Cambridge University Press, 1996.

Harrison, E. L. "Virgil, Sebastian Brant, and Maximilian I." *Modern Language Review* 76 (1981): 99–115.

Harwell, Gregory Todd. "Aurea Condet Saecula (Per Arva Saturno Quondam). Imperial Habsburg Medals from the Coronation of Frederick III (1452) until the Succession of Maximilian I (1494). Art and Legitimacy between Feudalism and Absolutism." PhD diss., Princeton University, 2005.

Hauber, A. *Planetenkinderbilder und Sternbilder zur Geschichte des menschlichen Glaubens und Irrens.* Vol. 194: *Studien zur deutchen Kunstgeschichte.* Strasbourg: Heitz & Mündel, 1916.

Hayton, Darin. "Joseph Grünpeck's Astrological Explanation of the French Disease." In *Responding to Sexual Disease in Early-Modern Europe,* edited by Kevin Siena, 241–74. Toronto: Centre for Renaissance and Reformation Studies, 2005.

Hayton, Darin. "Astrology as Political Propaganda: Humanist Responses to the Turkish Threat in Early-Sixteenth-Century Vienna." *Austrian History Yearbook* 38 (2007): 61–91.

Hayton, Darin. "Martin Bylica at the Court of Matthias Corvinus: Astrology and Politics in Renaissance Hungary." *Centaurus* 49 (2007): 185–98.

Hayton, Darin. "Expertise *Ex Stellis*: Comets, Horoscopes, and Politics in Renaissance Hungary." *Osiris* 25 (2010): 27–46.

Hellman, C. Doris. *The Comet of 1577: Its Place in the History of Astronomy.* New York: Columbia University Press, 1944.

Hellmann, Gustav. *Meteorologische Volksbücher : Ein Beitrag zur Geschichte der Meteorologie und zur Kulturgeschichte.* Berlin: H. Paetel, 1895.

Hellmann, Gustav. *Beiträge zur Geschichte der Meteorologie.* Berlin: Behrend & Co., 1917.

Hellmann, Gustav. *Die Meteorologie in den deutschen Flugschriften und Flugblättern des XVI. Jahrhunderts. Ein Beitrag zur Geschichte der Meteorologie.* Berlin: Akademie der Wissenschaften, 1921.

Hellmann, Gustav. *Versuch einer Geschichte der Wettervorhersage im xvi. Jahrhundert.* Berlin: Akademie der Wissenschaften, 1924.

Hoffmann, Leonhard. "Almanache des 15. und 16. Jahrhunderts und ihre Käufer." In *Beiträge zur Inkunabelkunde,* edited by Hans Lülfing, Ursula Altmann, and Heinrich Roloff, 130–43. Berlin: Akademie Verlag, 1983.

Holleger, Manfred. *"Erwachen vnd Aufsten als ein starcker Stryter.* Zu Formen und Inhalt der Propaganda Maximilians I." *Propaganda, Kommunikation und Öffentlichkeit 6* (2002): 223–34.

Hönig, Edeltraud. "Kaiser Maximilian I als politischer Publizist." PhD diss., University of Graz, 1970.

Ineichen-Eder, Christine E. "A Computus Notebook by Sebastian Brant (CLM 26618)." *Scriptorium* 35, no. 1 (1981): 91–95.

Jansen, Sharon L. *Political Protest and Prophecy under Henry Viii.* Woodbridge: Boydell Press, 1991.

Jardine, Lisa, and Anthony Grafton. "'Studied for Action': How Gabriel Harvey Read His Livy." *Past and Present* 129 (1990): 30–78.

Jenks, Stuart. "Astrometeorology in the Middle Ages." *Isis* 74, no. 2 (1983): 185–210.

Jervis, Jane L. *Cometary Theory in Fifteenth-Century Europe.* Boston: Kluwer Academic, 1985.

Johns, Adrian. *The Nature of the Book. Print and Knowledge in the Making.* Chicago: University of Chicago Press, 1998.

Johns, Adrian. "How to Acknowledge a Revolution." *American Historical Review* 107 (2002): 106–25.

Jowett, Garth S., and Victoria O'Donnell. *Propaganda & Persuasion.* London: Sage, 2011.

Kagan, Richard L. *Lucretia's Dreams. Politics and Prophecy in Sixteenth-Century Spain.* Berkeley: University of California Press, 1990.

Kassell, Lauren. *Medicine & Magic in Elizabethan London. Simon Forman: Astrologer, Alchemist, & Physician.* Oxford: Clarendon Press, 2005.

Kaufmann, Thomas Da Costa. "Astronomy, Technology, Humanism, and Art at the Entry of Rudolf II into Vienna, 1577: The Role of Paulus Fabritius." In *The Mastery of Nature. Aspects of Art, Science, and Humanism in the Renaissance,* edited by Thomas Da Costa Kaufmann, 136–50. Princeton, NJ: Princeton University Press, 1993.

Kaufmann, Thomas Da Costa. "From Mastery of the World to Mastery of Nature: The *Kunstkammer*, Politics, and Science." In *The Mastery of Nature. Aspects of Art, Science, and Humanism in the Renaissance,* edited by Thomas Da Costa Kaufmann, 174–94. Princeton, NJ: Princeton University Press, 1993.

Kink, Rudolf. *Geschichte der kaiserlichen Universität zu Wien.* Vol. 1. Geschichliche Darstellung, Vienna: Carl Gerold & Sohn, 1854.

Kink, Rudolf. *Geschichte der kaiserlichen Universität zu Wien.* Vol. 2. Statutenbuch der Universität, Vienna: Carl Gerold & Sohn, 1854.

Kleehoven, Hans Ankwicz. *Der Wiener Humanist Johannes Cuspinian. Gelehrter und Diplomat zur Zeit Kaiser Maximilian I.* Graz: Hermann Böhlaus, 1959.

Klemm, Hans Gunther. *Georg Hartmann aus Eggolsheim (1489–1564). Leben und Werk eines Fränkischen Mathematikers und Ingenieurs.* Forchheim: Gürtler-Druck, 1990.

Klug, R. *Johannes von Gmunden. Der Begrunder der Himmelskunde auf deutschem Bodem.* Vienna: Hölder-Pichler-Tempsky, 1943.

Kohler, Hans-Joachim. "Die Flugschriften. Versuch der Präzisierung eines geläufigen Begriffs." In *Festgabe für Ernst Walter Zeeden zum 60. Geburtstag,* edited by Horst Rabe, Hansgeorg Molitor, and Hans-Christoph Rublack, 36–61. Munster: Aschendorff, 1976.

Kohler, Hans-Joachim. *Flugschriften Als Massenmedium der Reformationszeit: Beiträge Zum Tübinger Symposion 1980.* Stuttgart: Klett-Cotta, 1981.

Kohler, Hans-Joachim. "The Flugschriften and Their Importance in Religious Debate: A Quantitative Approach." In *"Astrologi hallucinati": Stars and the End of the World in Luther's Time,* edited by Paolo Zambelli, 153–75. Berlin: Walter de Gruyter, 1986.

Kokott, Wolfgang. "The Comet of 1533." *Journal for the History of Astronomy* 12 (1981): 95–111.

Kokott, Wolfgang. *Die Kometen der Jahre 1531 bis 1539 und ihre Bedeutung für die spätere Entwicklung der Kometenforschung.* Stuttgart: Verlag für Geschichte der Naturwissenschaften und der Technik, 1994.

Kokott, Wolfgang. "Peter Creutzers 'Komet' vom 11. Oktober 1527: Zur Langlebigkeit von Fehldatierungen der Sekundärliteratur." In *Cosmographica et gegraphica: Festschrift für Heribert M. Nobis,* edited by Bernhard Fritscher and Gerhard Brey, 249–54. Munich: Institute für Geschichte der Naturwissenschaft, 1994.

Kraus, H. P. *Choice Books and Manuscripts from Distinguished Library, Catalog 126.* New York, 1971.

Kraus, H. P. *Important Works in the Field of Science, Catalog 137.* New York, 1973.

Kren, Claudia. "Astronomical Teaching at the Late Medieval University of Vienna." *History of Universities* 3 (1983): 15–30.

Kren, Claudia. "Patterns in Arts Teaching at the Medieval University of Vienna." *Viator* 18 (1987): 321–27.

Kühnel, Harry. *Mittelalterliche Heilkunde in Wien.* Vol. 5: *Studien zur Geschichte der Universität Wien.* Graz: Verlag Hermann Böhlaus Nachf., 1965.

Kuper, Michael. *Johannes Trithemius: Der schwarze Abt.* Berlin: Clemens Zerling, 1998.

Kurze, Dietrich. "Prophecy and History: Lichtenberger's Forecasts of Events to Come (From the Fifteenth to the Twentieth Century); Their Reception and Diffusion." *Journal of the Warburg and Courtauld Institutes* 21 (1958): 63–85.

Kurze, Dietrich. "Johannes Lichtenberger (†1503): Eine Studie zur Geschichte der Prophetie und Astrologie." *Historische Studien* 379 (1960).

Kurze, Dietrich. "Popular Astrology and Prophecy in the Fifteenth and Sixteenth Centuries: Johannes Lichtenberger." In *"Astrologi hallucinati." Stars and the End of the World in Luther's Time,* edited by Paolo Zambelli, 177–93. Berlin: Walter de Gruyter, 1986.

Kusukawa, Sachiko. *The Transformation of Natural Philosophy: The Case of Philip Melanchthon.* New York: Cambridge University Press, 1995.

Lammer, Günther. "Literaten und Beamte im publizistischen Dienst Kaiser Maximilians I. 1477–1519." PhD diss., Karl-Franzens Universität, 1983.

Lang, Helmut. "Wiener Wandkalender des 15. und 16. Jahrhunderts." *Biblos* 17 (1968): 40–50.

Langer, Eduard. *Bibliographie der österreichischen Drucke des 15. und 16. Jahrhunderts.* Vienna: 1913.

Laschitzer, Simon. "Die Geneologie des Kaisers Maximilian I." *Jahrbuch der kunsthistorischen Sammlungen des allerhöchsten Kaiserhauses* 7 (1888): 1–200.

Lerner, Robert E. "Antichrists and Antichrist in Joachim of Fiore." *Speculum* 60 (1985): 553–70.

Lewis, Suzanne. *The Rhetoric of Power in the Bayeux Tapestry.* Cambridge: Cambridge University Press, 1999.

Lhotsky, Alphons. "Kaiser Friedrich III.: Sein Leben und seine Persönlichkeit." In *Aufsätze und Vorträge,* vol. 2: *Das Haus Habsburg, edited by Hans Wagner, and Heinrich Koller,* 119–63. Vienna: Verlag für Geschichte und Politik, 1971.

Lippencott, Kristen. "Two Astrological Ceilings Rconsidered: The Sala di Galatea in the Villa Farnesina and the Sala del Mappamundo at Caprarola." *Journal of the Warburg and Courtauld Institutes* 53 (1994): 185–207.

Louthan, Howard. *The Quest for Compromise. Peace Makers in Counter Reformation Vienna.* Cambridge: Cambridge University Press, 1997.

Luh, Peter. *Der Allegroische Reichsadler von Conrad Celtis und Hans Burgkmair.* Frankfurt am Main: Peter Lang, 2002.

Madar, Heather Kathryn Suzanne. "History Made Visible: Visual Strategies in the Memorial Project of Maximilian I." PhD diss., University of California, 2003.

Maddison, Francis, and Anthony Turner. "The Names and Faces of the Hours." In *Between Demonstration and Imagination. Essay in the History of Science and Philosophy Presented to John D. North,* edited by Lodi Nauta and Arjo Vanderjagt, 124–56. Leiden: Brill, 1999.

Manzalaoui, M. A., ed. *Secretum secretorum: Nine English Versions.* Early English Text Society no. 276. Oxford: Oxford University Press, 1977.

Markowski, Mieczyslaw. "Die Astrologie an der Krakauer Universität in den Jahren 1450–1550." In *Magia, Astrologia e Religione nel Rinascimento,* 83–89. Wrocław, 1974.

Mathäus, K. "Zur Geschichte des Nürnberger Kalenderwesens. Die Entwicklung der in Nürnberg gedruckten Jahreskalender in Buchform." *Archiv für Geschichte des Buchwesens* 9 (1969): 965–1396.

Mayer, Anton. *Wiens Buchdruckergeschichte, 1482–1882.* Vol. 1. Vienna: F. Jasper, 1882–87.

McCluskey, Stephen C. *Astronomies and Cultures in Early Medieval Europe.* Cambridge: Cambridge University Press, 1998.

McDonald, William C. "Maximilian I of Habsburg and the Veneration of Hercules: On the Revival of Myth and the German Renaissance." *Journal of Medieval and Renaissance Studies* 6 (1976): 139–54.

Mertens, Dieter. "Maximilian I. und das Elsass." In *Die Humanisten in ihrer politischen und sozialen Umwelt,* edited by Otto Herding and Robert Stupperich, 177–201. Boppard-am-Rhein: Harald Boldt Verlag, 1976.

Mielke, Hans. *Albrecht Altdorfer. Zeichnungen. Deckfarbenmalerei. Druckgraphik.* Berlin: Reimer Verlag, 1988.

Misch, Georg. "Die Stilisierung des eigenen Lebes in dem Ruhmeswerk Kaiser Maximilians, des letzten Ritters." *Nachrichten von der Gesellschaft der Wissenschaften zu Göttingen* (1930): 435–59.

Monfasani, John. *Fernando of Cordova. A Biographical and Intellectual Profile.* Vol. 28: *Transactions of the American Philosophical Society.* Philadelphia, PA: American Philosophical Society, 1992.

Moran, Bruce T. "Princes, Machines and the Valuation of Precision in the 16th Century." *Sudhoffs Archiv* 61 (1977): 209–28.

Moran, Bruce T. "German Prince-Practitioners: Aspects in the Development of Courtly Science, Technology, and Procedures in the Renaissance." *Technology and Culture* 22 (1981): 253–74.

Moran, Bruce T. *The Alchemical World of the German Court. Occult Philosophy and Chemical Medicine in the Circle of Moritz of Hessen (1572–1632)*. Stuttgart: Franz Steiner Verlag, 1991.

Moran, Bruce T. "Patronage and Institutions: Courts, Universities, and Academies in Germany. An Overview 1550–1750." In *Patronage and Institutions. Science, Technology, and Medicine at the European Court, 1500–1750*, edited by Bruce T. Moran, 169–83. Woodbridge: Boydell Press, 1991.

Mosley, Adam. *Bearing the Heavens. Tycho Brahe and the Astronomical Community of the Late Sixteenth Century*. Cambridge: Cambridge University Press, 2007.

Mühlberger, Kurt. "Die Gemeinde der Lehrer und Schüler—Alma Mater Rudolphina." In *Wien. Geschichte einer Stadt. Band I: Von den Anfängen bis zur ersten Wiener Türkenbelagerung (1529)*, edited by Peter Csendes and Ferdinand Opll, 319–410. Vienna: Böhlau, 2000.

Müller, Jan-Dirk. "Poet, Prophet, Politiker: Sebastian Brant als Publizist und die Rolle der laikalen Intelligenz um 1500." *Zeitschrift für Literaturwissenschaft und Linguistik* 10 (1980): 102–27.

Müller, Jan-Dirk. *Gedechtnus. Literatur und Hofgesellschaft um Maximilian I*. Munich: Wilhelm Fink Verlag, 1982.

Nauert, Charles G. "Caius Plinius Secundus." *Catalogus Translationum et Commentariorum* 4 (1980): 297–422.

Nelson, Alan H. "Mechanical Wheels of Fortune, 1100–1547." *Journal of the Warburg and Courtauld Institutes* 43 (1980): 227–33.

Niccoli, Ottavia. *Prophecy and People in Renaissance Italy*. Translated by Lydia G. Cochrane. Princeton, NJ: Princeton University Press, 1990.

North, J. D. "Kalenderes Enlumyned Ben They. Part III." *The Review of English Studies* 20, no. 80 (1969): 418–44.

North, J. D. "Astrolabes and the Hour-Line Ritual." *Journal for the History of Arabic Sciences* 5 (1981): 113–14.

North, J. D. "The Western Calendar—'Intolerabilis, horribilis, et derisiblis': Four Centuries of Discontent." In *Gregorian Reform of the Calendar. Proceedings of the Vatican Conference to Commemorate Its 400th Anniversary, 1582–1982*, edited by G. V. Coyne, M. A. Hoskin, and O. Pedersen, 75–113. Vatican: Pontifical Academy of Sciences, 1983.

North, J. D. *Horoscopes and History*. London: Warburg Institute, 1986.

North, J. D. "Astrology and the Fortunes of Churches." In *Stars, Minds and Fate: Essays in Ancient and Medieval Cosmology*, edited by J. D. North, 59–89. London: Hambledon, 1989.

Nummedal, Tara. *Alchemy and Authority in the Holy Roman Empire*. Chicago: University of Chicago Press, 2007.

Nutton, Vivian. "Medicine at the German Universities, 1348–1500: A Preliminary Sketch." In *Medicine from the Black Death to the French Disease*, edited by Roger French, Jon Arrizabalaga, Andrew Cunningham, and Luis García-Ballester, 85–109. Aldershot: Ashgate, 1988.

Nutton, Vivian. "Hellenism Postponed: Some Aspects of Renaissance Medicine, 1490–1530." *Sudhoffs Archiv* 81, no. 2 (1997): 158–70.

O'Boyle, Cornelius. *Medieval Prognosis and Astrology. A Working Edition of the Aggregationes de crisi et creticis diebus, with Introduction and English Summary.* Cambridge: Wellcome Unit for the History of Medicine, 1991.

O'Boyle, Cornelius. "Astrology and Medicine in Later Medieval England. The Calendars of John Somer and Nicholas of Lynn." *Sudhoffs Archiv* 89 (2005): 1–22.

Oberman, Heiko A. "Zwischen Agitation und Reformation: Die Flugschriften als 'Judenspiegel.'" In *Flugschriften als Massenmedium der Reformationszeit. Beitäge zum Tübinger Symposium 1980,* Spätmittelalter und Frühe Neuzeit, vol. 13, edited by Hans-Joachim Kohler, 269–89. Stuttgart: Klett-Cotta, 1981.

Oestmann, Günther. *Heinrich Rantzau und die Astrologie.* Vol. 2, *Disquisitiones Historiae Scientiarum. Braunschweiger Beiträge zur Wissenschaftsgeschichte.* Braunschweig: Braunschweigisches Landesmuseum, 2004.

Oestmann, Günther, H. Darrel Rutkin, and Kocku von Stuckrad, eds. *Horoscopes and Public Spheres. Essays on the History of Astrology.* Berlin: Walter de Gruyter, 2005.

O'Shaughnessy, Nicholas Jackson. *Politics and Propaganda. Weapons of Mass Seduction.* Ann Arbor: University of Michigan Press, 2004.

Park, Katherine, and Lorraine J. Daston. "Unnatural Conceptions: The Study of Monsters in Sixteenth- and Seventeenth-Century France and England." *Past and Present* 92 (1981): 20–54.

Parker, Geoffrey. "The Political World of Charles V." In *Charles V and His Time. 1500–1558,* edited by Hugo Soly. Antwerp: Mercatorfonds, 1999.

Parshall, Peter. "Prints as Objects of Consumption in Early Modern Europe." *Journal of Medieval and Early Modern Studies* 28 (1998): 19–36.

Pawlik, Christian. "Martin Stainpeis: Liber de modo studendi seu legendi in medicina. Bearbeitung und Erläuterung einer Studienanleitung für Mediziner im ausgehenden Mittelalter." PhD diss., Technische Universität Munich, 1980.

Peacey, Jason. *Politicians and Pamphleteers. Propaganda during the English Civil Wars and Interregnum.* Aldershot: Ashgate, 2004.

Peacey, Jason. *Print and Politics in the English Revolution.* Cambridge: Cambridge University Press, 2013.

Pedersen, Olaf. "Some Astronomical Topics in Pliny." In *Science in the Early Roman Empire: Pliny the Elder, His Sources and Influence,* edited by Roger French and Frank Greenaway, 162–96. Totowa, NJ: Barnes & Noble, 1986.

Pérez-Higuera, Teresa. *Medieval Calendars.* London: Weidenfeld and Nicolson, 1998.

Perger, Richard. "Der organisatorische und wirtschaftliche Rahmen." In *Wien. Geschichte einer Stadt,* Vol: I: *Von Anfängen bis zur ersten Wiener Türkenbelagerung (1529).* Vienna: Böhlau, 2001.

Pettegree, Andrew. *The Book in the Renaissance.* New Haven, CT: Yale University Press, 2010.

Pettegree, Andrew, and Matthew Hall. "The Reformation and the Book: A Reconsideration." *Historical Journal* 47 (2004): 785–808.

Posset, Franz. *Renaissance Monks. Monastic Humanism in Six Biographical Sketches.* Leiden: Brill, 2005.

Potter, David. *Prophets and Emperors. Human and Divine Authority from Augustus to Theodosius*. Cambridge, MA: Harvard University Press, 1994.

Puddington, Arch. *Broadcasting Freedom: The Cold War Triumph of Radio Free Europe and Radio Liberty*. Lexington: University Press of Kentucky, 2000.

Quinlan-McGrath, Mary. "The Astrological Vault of the Villa Farnesina Agostino Chigi's Rising Sign." *Journal of the Warburg and Courtauld Institutes* 47 (1984): 91–105.

Quinlan-McGrath, Mary. "The Villa Farnesina, Time-Telling Conventions and Renaissance Astrological Practice." *Journal of the Warburg and Courtauld Institutes* 58 (1995): 52–71.

Rabelais, François. *Pantagrueline prognostication pour l'an 1533. Avec Les almanachs pour les ans 1533, 1535 et 1541. La grande et vraye Pronostications nouvelle de 1544*. Edited by M.-A. Screech. Geneva: Librairie Droz, 1974.

Rady, Martyn. "The Corvina Library and the Lost Royal Hungarian Archive," In *Lost Libraries. The Destruction of Great Book Collections since Antiquity*, edited by James Raven. New York: Palgrave, 2004.

Ramminger, Johann. "Humanist Poetry and Its Classical Models: A Collection from the Court of Emperor Maximilian I." In *Acta Conventus Neo-Latini Torontonensis. Proceedings of the Seventh International Congress of Neo-Latin Studies, Toronto 8 August to 13 August 1988*, Medieval & Renaissance Texts & Studies, edited by Alexander Dalzell, Charles Fantazzi, and Richard J. Schoeck, 581–93. Binghamton, NY: Arizona Center for Medieval and Renaissance Studies, 1991.

Rashdall, Hastings. *The Universities of Europe in the Middle Ages*. Vol. 1: *Salerno, Bologna, Paris*. Oxford: Clarendon Press, 1936.

Rashdall, Hastings. *The Universities of Europe in the Middle Ages*. Vol. 2: *Italy, Spain, France, Germany, Scotland*. Oxford: Clarendon Press, 1936.

Reeves, Marjorie. *The Influence of Prophecy in the Later Middle Ages: A Study in Joachimism*. South Bend, IN: University of Notre Dame, 1993.

Reisinger, Reiner. *Historische Horoskopie. Das iudicium magnum des Johannes Carion für Albrecht Dürers Patenkind*. Wiesbaden: Harrassowitz, 1997.

Richardson, Brian. *Printing, Writers and Readers in Renaissance Italy*. Cambridge: Cambridge University Press, 1999.

Riegle, Alois. "Die Holzkalender des Mittelalters und der Renaissance." *Mittheilungen des Instituts für österreichische Geschichtsforschung* 9 (1888): 82–103.

Riha, Otrun. *Meister Alexanders Monatsregeln. Untersuchung zu einem spätmittelalterlichen Regimen duodecim mensium mit kritischer Textausgabe*. Würzburg: Böhler Verlag, 1985.

Robinson, David M. "The Wheel of Fortune." *Classical Philology* 41, no. 4 (1946): 207–16.

Roland, Ingrid D. *The Culture of the High Renaissance. Ancients and Moderns in Sixteenth-Century Rome*. Cambridge: Cambridge University Press, 1998.

Rosenfeld, Hellmut. "Kalender, Einblattkalender, Bauernkalender und Bauernpraktik." *Bayerisches Jahrbuch für Volkskunde* (1962): 7–24.

Rosenfeld, Hellmut. "Bauernkalender und Mandlkalender als literarisches Phänomen des 16. Jahrhunderts und ihr Verhältnis zur Bauernpraktik." *Gutenberg Jahrbuch* 38 (1963): 88–96.

Rosenfeld, Hellmut. "Die Titelholzschnitte der Bauernpraktik von 1508–1600 als soziologische Selbsinterpretation." In *Festschrift für Joseph Benzing zum sechzigsten Geburtstag. 4. Februar 1964,* edited by Elizabeth Geck and Guido Pressler, 373–89. Wiesbaden: Guido Pressler Verlag, 1964.

Röttel, Karl, ed. *Peter Apian. Astronomie, Kosmographie und Mathematik am Beginn der Neuzeit.* Eichstätt: Polygon-Verlag, 1995.

Rupprich, Hans, ed. *Der Briefwechsel des Konrad Celtis.* Munich: C. H. Beck, 1934.

Russel, Paul Albert. "Syphilis, God's Scourge or Nature's Vengeance? The German Printed Response to a Public Problem in the Early Sixteenth Century." *Archiv für Reformationsgeschichte* 80 (1989): 286–306.

Russel, Paul Albert. "Astrology as Popular Propaganda: Expectations of the End in the German Pamphlets of Joseph Grünpeck (†1533?)." In *Forme e destinazione del messaggio religioso. Aspetti della propaganda religiosa nel cinquecento,* edited by Antonio Rotondò, 165–95. Florence: Leo S. Olschki, 1991.

Rutkin, H. Darrel. "Astrology." In *Early Modern Science,* edited by Katherine Park and Lorraine Daston, 541–61. Cambridge: Cambridge University Press, 2006.

Ryan, Michael. *A Kingdom of Stargazers: Astrology and Authority in the Late Medieval Crown of Aragon.* Ithaca, NY: Cornell University Press, 2011.

Ryan, W. F., and Charles Schmitt, eds. *Pseudo-Aristotle, The Secret of Secrets: Sources and Influences.* Vol. 9: *Warburg Institute Surveys and Texts.* London: Warburg Institute, 1982.

San Martin, A. Bonilla y. *Fernando de Córdoba (1425–1486?) y los orígenes del Renacimiento filosófico en España.* Madrid, 1911.

Saxl, Fritz. "The Villa Farnesina." In *Lectures I,* 189–99. London: Warburg Institute, 1957.

Scheicher, Elisabeth. "The Collection of Archduke Ferdinand II at Schloss Ambrass: Its Purpose, Composition and Evolution." In *The Origins of Museums. The Cabinet of Curiosities in Sixteenth- and Seventeent-Century Europe,* edited by Oliver Impey and Arthur MacGregor, 29–38. Oxford: Clarendon Press, 1985.

Schiller, Peter. "Die himmelskundliche Ikonographie der Decke der Sala di Galatea in der Villa Farnesina in Rom." In *Die okkulten Wissenschaften in der Renaissance,* edited by August Buck, 255–88. Wolfenbüttel: Herzog August Bibliothek, 1992.

Schmid, Alfred. *Augustus und die Macht der Sterne. Antike Astrologie und die Etablierung der Monarchie in Rom.* Cologne: Böhlau, 2005.

Schmid, Alois. "'*Poeta et orator a Caesare laureatus.*' Die Dichterkrönungen Kaiser Maximilians I." *Historisches Jahrbuch* 109 (1989): 56–108.

Schmidt, Suzanne Karr. "Art. A User's Guide: Interactive and Sculptural Printmaking in the Renaissance." PhD diss., Yale University, 2006.

Schmidt, Suzanne Karr. *Altered and Adorned. Using Renaissance Prints in Daily Life.* New Haven, CT: Yale University Press, 2011.

Schmidt, Suzanne Karr. "Printed Instruments." In *Prints and the Pursuit of Knowledge in Early Modern Europe,* edited by Susan Dackerman, 267–315. New Haven, CT: Yale University Press, 2011.

Schöner, Christoph. *Mathematik und Astronomie an der Universität Ingolstadt im 15. und 16. Jahrhundert.* Vol. 13: *Ludovico Maximlianea.* Berlin: Duncker & Humblot, 1994.

Schrauf, Karl, ed. *Acta facultatis medicae universitatis Vindobonensis.* Vol. 3: *1490–1558.* Vienna: Verlag des Medicinischen Doktorenkollegiums, 1904.

Schwitalla, Johannes. "Deutsche Flugschriften im ersten Viertel des 16. Jahrhunderts." *Freiburger Universiätsblätter* 76 (1982): 37–58.

Scribner, Robert W. "Oral Culture and the Diffusion of Reformation Ideas." In *Popular Culture and Popular Movements in Reformation Germany*, edited by Robert W. Scribner, 49–70. London: Hambledon Press, 1988.

Scribner, Robert W. *Popular Culture and Popular Movements in Reformation Germany*. London: Hambledon Press, 1988.

Scribner, Robert W. *For the Sake of Simple Folk. Popular Propaganda for the German Reformation*. Oxford: Oxford University Press, 1994.

Seethaler, Josef. "Das Wiener Kalenderwesen von seinen Anfängen bis zum Ende des 17. Jahrhunderts. Ein Beitrag zur Geschichte des Buchdrucks." PhD diss., University of Vienna, 1982.

Shank, Michael. "How Shall We Practice History? The Case of Mario Biagioli's *Galileo Courtier*." *Early Science and Medicine* 1, no. 1 (1996): 106–50.

Shank, Michael. "Academic Consulting in 15th-Century Vienna: The Case of Astrology." In *Texts and Contexts in Ancient and Medieval Science. Studies on the Occasion of John E. Murdoch's Seventieth Birthday*, edited by Edith Sylla and Michael McVaugh, 245–70. Leiden: Brill, 1997.

Sharpe, Kevin. *Reading Revolutions. The Politics of Reading in Early Modern England*. New Haven, CT: Yale University Press, 2000.

Sharpe, Kevin. *Selling the Tudor Monarchy. Authority and Image in Sixteenth-Century England*. New Haven, CT: Yale University Press, 2009.

Sharpe, Kevin. *Image Wars. Promoting Kings and Commonwealths in England 1603–1660*. New Haven, CT: Yale University Press, 2010.

Sherman, William H. *John Dee. The Politics of Reading and Writing in the English Renaissance*. Amherst: University of Massachusetts Press, 1995.

Silver, Larry. "Forest Primeval: Albrech Altdorfer and the German Wilderness Landscape." *Simiolus* 13 (1983): 4–43.

Silver, Larry. "Prints for a Prince: Maximilian, Nuremberg, and the Woodcut." In *New Perspectives on the Art of Renaissance Nuremberg*, edited by Jeffrey Chips Smith, 7–21. Austin: University of Texas at Austin, 1985.

Silver, Larry. "Paper Pageants: The Triumphs of Emperor Maximilian I." In *"All the World's a Stage": Art and Pageantry in the Renaissance and Baroque*, vol. 1, edited by Barbara Wollesen-Wisch and Susan Munshower, 292–331. University Park: Pennsylvania State University Press, 1990.

Silver, Larry. "Power of the Press: Maximilian's Arch of Honor." In *Albrecht Dürer in the Collection of the National Gallery of Victoria*, edited by Irena Zdanowicz, 45–64. Melbourne: National Gallery of Victoria, 1994.

Silver, Larry. "Germanic Patriotism in the Age of Dürer." In *Dürer and His Culture*, edited by Dagmar Eichberger and Charles Zika, 38–68. Cambridge: Cambridge University Press, 1998.

Silver, Larry. "Nature and Nature's God: Landscape and Cosmos of Albrecht Altdorfer." *Art Bulletin* 81 (1999): 194–214.

Silver, Larry. "Shining Armor: Emperor Maximilian, Chivalry, and Warfare." In *Artful*

Armies, Beautiful Battles. Art and Warfare in Early Modern Europe, History of Warfare, edited by Pia F. Cuneo, 61–85. Leiden: Brill, 2002.

Silver, Larry. *Marketing Maximilian. The Visual Ideology of a Holy Roman Emperor.* Princeton, NJ: Princeton University Press, 2008.

Simon, Eckhard. *The Türkenkalender (1454) Attributed to Gutenberg and the Strasbourg Lunation Tracts.* Cambridge, MA: Medieval Academy of America, 1988.

Siraisi, Nancy. *Medieval and Early Renaissance Medicine: An Introduction to Knowledge and Practice.* Chicago: University of Chicago Press, 1990.

Slattery, Sarah. "Astrologie, Wunderzeichen und Propaganda. Die Flugschriften des Humanisten Joseph Grünpeck." In *Zukunftsvoraussagen in der Renaissance,* edited by Klaus Bergdolt and Walther Ludwig, 329–47. Wiesbaden: Harrassowitz, 2005.

Smith, Pamela H. *The Business of Alchemy: Science and Culture in the Holy Roman Empire.* Princeton, NJ: Princeton University Press, 1994.

Smith, Pamela H. "Science on the Move: Recent Trends in the History of Early Modern Science." *Renaissance Quarterly* 62 (2009): 345–75.

Smoller, Laura Ackerman. *History, Prophecy, and the Stars. The Christian Astrology of Pierre d'Ailly, 1350–1420.* Princeton, NJ: Princeton University Press, 1994.

Smoller, Laura Ackerman. "*Teste Albumasare cum Sibylla*: Astrology and the Sibyls in Medieval Europe." *Studies in History and Philosophy of Biological and Biomedical Sciences* 41, no. 2 (2010): 76–89.

Spitz, Lewis W. *Conrad Celtis. The German Arch-Humanist.* Cambridge, MA: Harvard University Press, 1957.

Strong, Roy. *Splendor at Court. Renaissance Spectacle and the Theater of Power.* Boston: Houghton Mifflin, 1973.

Strong, Roy. *The Cult of Elizabeth: Elizabethan Portraiture and Pageantry.* London: Pimlico, 1999.

Szaivert, Willy, and Franz Gall, eds. *Die Matrikel der Universität Wien.* Vol. 3. Graz: Hermann Böhlaus, 1971.

Taithe, Bertrand, and Tim Thornton. "Propaganda: A Misnomer of Rhetoric and Persuasion?" In *Propaganda. Political Rhetoric and Identity 1300–2000,* edited by Bertrand Taithe and Tim Thornton, 1–24. Gloucestershire: Sutton, 1999.

Talkenberger, Heike. *Sintflut. Prophetie und Zeitgeschehen in Texten und Holzschnitten astrologicher Flugschriften, 1488–1528.* Tübingen: Max Niemeyer Verlag, 1990.

Tanner, Marie. *The Last Descendant of Aeneas. The Habsburgs and the Mythic Image of the Emperor.* New Haven, CT: Yale University Press, 1993.

Thorndike, Lynn. *A History of Magic and Experimental Science.* 8 vols. Vol. I: *During the First Thirteen Centuries of Our Era.* New York: Columbia University Press, 1923.

Thorndike, Lynn. *A History of Magic and Experimental Science.* 8 vols. Vol. IV: *Fourteenth and Fifteenth Centuries.* New York: Columbia University Press, 1958.

Thorndike, Lynn. *A History of Magic and Experimental Science.* 8 vols. Vol. V: *Sixteenth Century.* New York: Columbia University Press, 1958.

Tihon, Anne, Régine Leurquin, and Claudy Scheuren. *Une version Byzantine du traité sur l'astrolabe du Pseudo-Messahalla.* Louvain-la-Neuve: Bruylant-Academia, 2001.

Trevor-Roper, Hugh. *Princes and Artists: Patronage and Ideology at Four Habsburg Courts, 1517–1633.* New York: Harper and Row, 1976.

Ulmann, Heinrich. *Kaiser Maximilian I. Auf urkundlicher Grundlage dargestellt.* Stuttgart: Cotta, 1884.

van Nouhuys, Tabitta. *The Age of Two-Faced Janus. The Comets of 1577 and 1618 and the Decline of the Aristotelian World View in the Netherlands.* Leiden: Brill, 1998.

Vanden Broecke, Steven. *The Limits of Influence: Pico, Louvain, and the Crisis of Renaissance Astrology.* Leiden: Brill, 2003.

Veenstra, Jan R. *Magic and Divination at the Courts of Burgundy and France. Text and Context of Laurens Pignon's Contre les devineurs (1411).* Leiden: Brill, 1998.

Verger, Jacques. *Men of Learning in Europe at the End of the Middle Ages.* South Bend, IN: University of Notre Dame, 2000.

Veronese, Julien. "L'Ars notoria: une tradition théurgico-magico au Moyen Age (XII–XVIe siècle)." PhD diss., Paris-X Nanterre, 1999.

Vocelka, Karl. *Die Politische Propaganda Kaiser Rudolfs II. (1576–1612).* Vienna: Verlag der Österreichischen Akademie der Wissenschaften, 1981.

von Aschbach, Joseph Ritter. *Geschichte der Wiener Universität im ersten Jahrhunderte ihres Bestehens.* Vol. 1. Vienna: 1865.

von Aschbach, Joseph Ritter. *Die Wiener Universität und ihre Humanisten im Zeitalter Kaiser Maximilians I.* Vienna: Wilhelm Braumüller, 1877.

Voss, W. "Eine Himmelskarte vom Jahre 1503 mit den Wahrzeichen des Wiener Poettenkollegiums als Vorlage Albrecht Dürers." *Jahrbuch der Preußischen Kunstsammlungen* 64 (1943): 89–150.

Waas, Glenn Elwood. *The Legendary Character of Kaiser Maximilian I.* New York: Columbia University Press, 1941.

Wade, Marjorie Dale. "The Education of the Prince: A Mirror of Reality and Romance in Maximilian's *Weisskunig.*" PhD diss., University of Michigan, 1974.

Warner, Deborah Jean. "What Is a Scientific Instrument, When Did It Become One, and Why?" *British Journal for the History of Science* 23 (1990): 83–93.

Watanabe-O'Kelly, Helen. *Court Culture in Dresden from Renaissance to Baroque.* Houndmills: Palgrave Macmillan, 2002.

Webster, Charles. *From Paracelsus to Newton: Magic and the Making of Modern Science.* Cambridge: Cambridge University Press, 1982.

Weiss, Edmund. "Albrecht Dürers Geographische, Astronomische und Astrologische Tafeln." *Jahrbuch der kunsthistorischen Sammlungen des allerhöchsten Kaiserhauses* 7 (1888): 207–20.

Weller, Emil. *Repertorium typographicum. Die deutsche Literatur im ersten Viertel des sechzehnten Jahrhunderts.* 2 vols. Nordlingen: Beck'schen Buchhandlung, 1864–85.

Wenskus, Otta. "Columellas Bauernkalender zwischen Mündlichkeit und Schriftlichkeit." In *Gattungen wissenschaftlicher Literatur in der Antike,* edited by Wolfgang Kullmann, Jochen Althoff, and Markus Asper, 253–62. Tübingen: Gunter Narr Verlag, 1996.

Westfall, Richard. "Science and Patronage. Galileo and the Telescope." *Isis* 76 (1985): 11–30.

Westman, Robert S. "The Astronomer's Role in the Sixteenth Century: A Preliminary Study." *History of Science* 18 (1980): 105–47.

Westman, Robert S. *The Copernican Question. Progonostication, Skepticism, and Celestial Order*. Berkeley: University of California Press, 2011.

Wiesflecker, Hermann. *Joseph Grünpecks Commentaria und Gesta Maximiliani Romanorum Regis. Die Entdeckung eines verlornen Geschichtswerkes*. Graz: Jos. A. Kieneich, 1965.

Wiesflecker, Hermann. "Joseph Grünpecks Redaktion der lateinischen Autobiographie Maximilians I." *Mitteilungen des Instituts für österreichische Geschichtsforschung* 78 (1970): 416–31.

Wiesflecker, Hermann. *Kaiser Maximilian I. Das Reich, Österreich und Europa an der Wende zur Neuzeit*. 5 vols. Munich: R. Oldenbourg Verlag, 1975–86.

Williams, Steven J. *The Secret of Secrets: The Scholarly Career of a Pseudo-Aristotelian Text in the Latin Middle Ages*. Ann Arbor: University of Michigan Press, 2003.

Wilson, F. A. "Some English Mock-Prognostications." *Library* 19 (1938): 6–43.

Wilson, F. A. "Introduction." In *Pantagruel's Prognostication*. Oxford: Luttrell Society, 1947.

Wise, M. Norton, ed. *The Values of Precision*. Princeton, NJ: Princeton University Press, 1995.

Wollesen, Jens T. "Sub specie ludi.: Text and Images in Alfonso El Sabio's Libro de Acedrex, Dados e Tablas." *Zeitschrift für Kunstgeschichte* 53, no. 3 (1990): 277–308.

Wood, Christopher S. "Maximilian I as Archeologist." *Renaissance Quarterly* 58 (2005): 1128–74.

Wood, Christopher S. *Forgery, Replica, Fiction. Temporalities of German Renaissance Art*. Chicago: University of Chicago Press, 2008.

Wünsch, Josef. "Wiener Kalender-Einblattdrucke des XV., XVI. und XVII. Jahrhunderts." *Berichte und Mittheilungen des Alterthums-Vereines zu Wien* 44 (1911): 67–81.

Wussin, Joh. "Alte Wiener Drucke." *Berichte und Mittheilungen des Alterthums-Vereines zu Wien* 26 (1890): 75–82.

Wuttke, Dieter. *Die Histori Herculis des Nürnberger Humanisten und Freundes der Gebrüder Vischer, Pangratz Bernhaubt gen Schwenter*. Cologne: Böhlau Verlag, 1964.

Wuttke, Dieter. "Sebastian Brants Verhältnis zu Wunderdeutung und Astrologie." In *Studien zur deutschen Literatur und Sprache des Mittelalters. Festschrift für Hugo Moser zum 65. Geburtstag*, edited by Werner Besch, Günther Jungbluth, Gerhard Meissburger, and Eberhard Nellmann, 272–86. Berlin: Erich Schmidt Verlag, 1974.

Wuttke, Dieter. "Sebastian Brant und Maximilian I. Eine Studie zu Brants Donnerstein-Flugblatt des Jahres 1492." In *Die Humanisten in ihrer politischen und sozialen Umwelt*, edited by Otto Herding and Robert Stupperich, 141–76. Soppard: Harald Boldt Verlag, 1976.

Wuttke, Dieter. "Wunderdeutung und Politik. Zu den Auslegungen der sogenannten Wormser Zwillinge des Jarhes 1495." In *Landesgeschichte und Geistesgeschichte*, edited by Kaspar Elm, Eberhard Gönner, and Eugen Hillenbrand, 217–44. Stuttgart: W. Kohlhammer Verlag, 1977.

Wuttke, Dieter. "Erzaugur des heiligen römischen Reiches deutscher Nation: Sebastian Brant deutet siamesische Tiergeburten." *Humanistica Lovaniensia* 43 (1994): 106–31.

Zafran, Eric. "Saturn and the Jews." *Journal of the Warburg and Courtauld Institutes* 42 (1979): 16–27.

Zambelli, Paolo. "Fine del mondon o inizio della propaganda? Astrologia, filosofia della storria e propaganda politico-religiosa ne dibattito sulla congiunzione del 1524." In *Scienze, credenze occulte, livelli di cultura,* 291–368. Florence: L. S. Olschki, 1982.

Zambelli, Paolo. "Many Ends for the World. Luca Gaurico Instigator of the Debate in Italy and in Germany." In *"Astrologi hallucinati." Stars and the End of the World in Luther's Time,* edited by Paolo Zambelli, 239–63. Berlin: Walter de Gruyter, 1986.

Zappert, Georg. "Über das Badewesen mittelaltlicher und späterer Zeit." *Archiv für Kunde österreichischer Geschichtsforschung* 21 (1859): 3–166.

Zinner, Ernst. *Astronomische Instrumente des 11. bis 18. Jahrhunderts.* Munich: C. H. Beck, 1956.

Zinner, Ernst. *Geschichte und Bibliographie der Astronomischen Literatur in Deutschland zur Zeit der Renaissance.* Stuttgart: Anton Hiersemann, 1964.

Zinner, Ernst. *Leben und Werken des Joh. Müller von Königsberg gennant Regiomontanus.* Osnabrück: Otto Zeller, 1968.

INDEX

Note: Page numbers in *italic* refer to figures.

Abenragel, Haly, 44, 88, 132, 135, 157, 218n26, 255n44
Abubacher, 44, 218n26
Aegidius of Tebaldis, 255n44
Aeneas, 31, 38
Aeneid (Virgil), 55, 100
Agrippina, 17
Albert, Duke, 134
Albert of Brudwezo, 80
Albrecht, Duke, 36
Albumasar, 17, 42–43, 44, 53, 132, 135, 164, 183–84, 217n21, 218n26; conjunctions and, 46; Latin translations of, 218n32; translation of, 45
Alcabitius, 43–44, 45, 88, 218n26, 218n28
alchemy, 200, 268n11
Alexander VI, Pope, 49, 50, 62
Alexander (Hartlieb), 25, 213n54
Alexander the Great, 25, 28, 41, 60, 176–77; birth of, 27
Alexander's Gates, *176*, 177
Alfonso X, King, 74
Alfraganus, 88
Ali ibn Khalaf, 74
Allegory of the Imperial Eagle (Burgkmair), woodcut from, *69*
Almanach nova (Stöffler and Pflaum), 122, 180–81, 191

Almanach novum (Perlach), 146, 148, 153–54, 159, 246n33; table from, *147*
almanacs, 10, 12, 93, 96, 126–27, 151
Almansor, 44, 218n26
Alphonsine tables, 84
Altdorfer, Albrecht, 19, 22–23
Altfahrt, Margit: pamphlets and, 6
Angelus, Johannes, 79, 131, 244n13
Anglo, Sydney: propaganda and, 205n23
Antichrist, 50, 174–77, 192, 260n18
Apelles, 55
Apian, Peter, 96
Apologie astrologie (Schonheintz), 250n85
Aquinas, 77
"Arabic 'Almanac,' Latin 'Diale' or 'Diurnale,' Greek 'Ephemerides'" (Perlach), 145
Aretino, Pietro, 131, 248n51
Aries, 52–53, 123, 133–34, 145, 155
Aristotle, 28, 49, 55, 77, 188
Arnold of Friburg, 44, 218n28
Artificium de applicatione astrologiae ad medicinam (Tannstetter), 89
astrolabes, 36, 72, 73, 90, 100; brass, 226n14; described, 234n116; paper-wood, 71, 226n16; types of, 76, 101; universal, 71, 237n15; using, 74, 92
Astrolabium Imperatorium (Stabius), 113, *114*, 115

303

Libellus consolatorius (Tannstetter), 12, 141–43, 180–81, 185, 194; title page from, *182*
Liber de modo studendi (Stainpeis), 89, 90
"Liber horologium" (Stiborius), 76
Liber introductorii maioris ad scientiam judiciorum astrorum (Albumasar), 44, 135
"Liber umbrarum" (Stiborius), 71, 72–73, 76
Libri umbrarum, 101
Libros del Saber de Astronomia (Alfonso X), 74
Lichtenberger, Johannes, 49, 51, 260n16; astrology and, 260n17; Joachimist position of, 260n18; prophetic interpretations and, 51, 211n29; Stabius and, 174; theory of conjunctions of, 45
literacy rates, 4, 247n40
Lord of the Ascendant, 33
Lord of the Year, 123, 134, 136, *138*, 140–41, 191; Jupiter as, 137; Mercury as, 135; misunderstanding of, 165
Louis XIV, 5, 204n11
Ludus Diane (Celtis), 172
lunar cycle, 124, 125
Luther, Martin, 170, 258n1

Machiavelli, 6–7, 85
Mainz, Archbishop of, 49
marginal glosses, 158, 159
Marketing Maximilian (Silver), 8
Mars, *29*, 41–42, 66, 85, 123, 131, 134, 136, 141–42, 162, 168, 173, 177, 184, 192; characteristics of, 193; Furies and, 174; malefic effects of, 163; motions of, 254n34; personification of, 28; prominence for, 16; Saturn and, 44–46, 53–54
Mary of Burgundy, 25, 30, 35
Maternus, Julius Firmicus, 52–53
mathematics, 70, 77–79, 93, 98
mathematicus, 3
Maximilian I, Emperor: agenda of, 10, 63; astrologers and, 12, 14, 16, 81; astrological instruments and, 36, *105*, 116; astrology and, 3, 9, 10, 18, 20, 21, 24, 31, 33–36, 47, 92, 152, 171, 196, 197, 199–201; autobiography of, 18–19, 20–21, 22, 25–26, 35, 58, 62, 74, 197; birth of, 14, 16; character of, 16, 20, 21, 26; coinage practices of, 5, 205n17; communications by, 1, 8; death of, 96, 97, 147, 154, 196, 241n54, 241n58; drawing of, *61*; effectiveness of, 8, 213n65; family of, 37, 38, 101, 168, 259n5; financial situation of, 47, 135, 168; genealogy of, 38, 78; geniture of, *15*, 16, 29; as German Hercules, *59*; horoscope for, 14–15, 17, 33; image of, 11, 35; knowledge/skills of, 23–24; legacy of, 12–13, 104, 199–200; letters by, 1–2,

207n40; liberal arts and, 21; military situation of, 1, 47, 48, 96, 100, 135, 141, 168; natural world and, 3–4; papal recognition for, 63–65; patronage by, 13, 78–79, 137, 154; politics and, 3, 7, 8, 11–12, 35–36, 39, 40, 49, 68, 85, 97, 107, 116, 119–20, 141, 169, 172–73, 198–99; propaganda by, 2, 31, 57, 178, 198; reforms by, 39, 46, 47, 57, 66, 93, 198; science and, 13, 200, 207n42, 236n1; self-portrayal of, 19, 197–98; woodcut of, *59*
Maximilian II, Emperor, 6, 200, 235n1
medical issues, 74, 89–92, 151, 152, 233n103
medicine, 120, 127, 152, 231n78; astrological, 96; taking, 128, 157
medium coeli, 53, 73–75, 90, 102, 142, 158
Melanchthon, Philipp, 88, 118–19, 127, 209n7, 232n91
Mennel, Jakob, 19, 24, 37, 38
Mercury, 17, *29*, *32*, 134, 136–37, 142; astrology and, 31; as Lord of the Year, 135; Maximilian I and, 29; personification of, 28; role of, 33
Messahalah, 44, 71, 218n26, 225n10
meteorites, 39, 40, 170, 171
Meteorologia (Aristotle), 80
Methodian prophecies, 45, 51, 174, 175–76
Michelangelo, 65
Milich, Jakob, 88, 232n91, 232n93
Minerva, 55
Molinet, Jean, 21
Monatsregeln, 121, 244n12
Moran, Bruce, 236n8, 238n26, 241n55, 267n4
Moritz of Hessen-Kassel, Landgrave, 201, 268n11
Mosley, Adam, 268n11
Muntz, Johannes, 121

natural knowledge, 2, 9, 98, 200
natural order, 50, 61
natural world, 3–4, 17, 18–19, 200
Neidelhart, 30
Nero, 17
Neuman, Johannes: calendars by, 244n13
New Testament, 181, *182*
Niccoli, Ottavia, 263n59, 263n62
Nihili, Johannes, 17
Nissus, King, 46
Noah, 38, 181
Nummedal, Tara: alchemy and, 268n11
Nuremberg, 104, 115, 135, 155, 241n58
Nussia, Henricus de, 139–40, 250n81
Nutton, Vivian: medicine and, 231n80

O'Donnell, Victoria, 203n10
Old Testament, 64, 181, *182*
On Great Conjunctions (Albumasar), 17